D1570940

Multiobjective
Programming and Planning

This is Volume 140 in
MATHEMATICS IN SCIENCE AND ENGINEERING
A Series of Monographs and Textbooks
Edited by RICHARD BELLMAN, *University of Southern California*

The complete listing of books in this series is available from the Publisher upon request.

Multiobjective Programming and Planning

Jared L. Cohon

Department of Geography
and Environmental Engineering
The Johns Hopkins University
Baltimore, Maryland

ACADEMIC PRESS New York San Francisco London 1978
A Subsidiary of Harcourt Brace Jovanovich, Publishers

ACADEMIC PRESS, INC.
111 Fifth Avenue, New York, New York 10003

United Kingdom Edition published by
ACADEMIC PRESS, INC. (LONDON) LTD.
24/28 Oval Road, London NW1 7DX

Library of Congress Cataloging in Publication Data

Cohon, Jared L
Multiobjective programming and planning.

(Mathematics in science and engineering;)
Includes bibliographical references.
1. Planning——Data processing. 2. Programming
(Mathematics) 3. Decision-making——Data processing.
I. Title. II. Series
HD30.28.C64 309.2'12'0184 78-12985
ISBN 0-12-178350-2

PRINTED IN THE UNITED STATES OF AMERICA

To my wife Bunny and my daughter Hallie
and
To my parents, Delbert and Ruth Cohon,
who have always made decisions as if their
children were their only objective

Contents

CHAPTER 9 Multiobjective Analysis of Water Resource Problems

CHAPTER 10 Multiobjective Analysis of Facility Location Problems

CHAPTER 11 Summary and Prospects for Future Development

Preface

Multiobjective programming has evolved in the past two decades into a recognized specialty of operations research. The attainment of that status is evidenced by the specialty conferences and symposia that have been held for discussions on multiobjective programming, vector optimization, multiple-criteria decision making, and multiattribute problems, all of which are different names for the same problem area. As the field has left its infancy and started to mature, the need for a book that brings together a large number of the several multiobjective methods has become apparent. This book is an attempt to satisfy that need.

While the book takes a broad view of multiobjective programming, not all methods are covered. The book emphasizes those classes of methods that are useful for continuous problems, by which I mean problems that are characterized by an infinite number of solutions or planning alternatives. Multiobjective methods that deal with problems with a finite number of predefined alternatives are not covered. Such methods are discussed in other books, which are cited in Chapter 1.

My objective in this book is to review multiobjective programming methods in the context of public decision-making problems to which the methods may be applicable. This objective provides the "planning" aspect of the book, in that public decision-making problems will be discussed to develop the applicable problem contexts. The discussion of planning and public decision-making problems will suggest a basis for categorizing and evaluating the methods. In this manner, the review is a structured one that provides the reader with an appreciation of how the methods relate to the real world and to each other, as well as an understanding of the mathematical details of the methods. The practical aspects of using the methods are stressed with the discussions of planning issues and some actual applications of multiobjective programming.

The planner who has no training in mathematics beyond high-school algebra can use this book by concentrating on Chapters 1 and 2, in which multiobjective planning is discussed; Chapter 3, in which linear programming (a special case of mathematical programming) is reviewed; Chapters 4 and 5, where the multiobjective programming problem is established and the methods are categorized; a selective reading of Chapters 6–8, in which several methods are discussed with varying degrees of mathematical detail; and Chapters 9 and 10, which present a detailed discussion of applications of multiobjective programming to water resource and facility location (fire stations and power plants) problems. Chapter 11 presents conclusions and an enthusiastic review of the prospects for future theoretical development and application.

The reader with a mathematical background that includes a course in linear programming can skim Chapter 3 and skip some of the mathematically advanced sections, indicated by an asterisk (*) on the section numbers. The reader who has also had a course in linear algebra should concentrate on Chapters 4–8.

The book is intended as both a reference and a textbook. Much of it has been written from my class notes for a graduate level course from which this book takes its name. Thus, the book can be used as a text in a one-semester course on multiobjective programming. My own experience has been that the material in Chapters 1, 2, and 4–7 can fill a 14-week semester. More selective reading of these chapters and a greater emphasis on applications are another possibility for a one-semester course. The book can also be used as a reference for a portion of a course in public policy analysis or mathematical programming.

Acknowledgments

This book has evolved over a period of seven years, a time during which the development of the ideas was impacted by several people. I was introduced into the area of multiobjective analysis during my graduate study at MIT. Professors David Marks and David Major provided formal instruction, guidance, encouragement, and the opportunity to learn in the context of the Argentina project (discussed in Chapter 9). Tomas Facet and Anders Haan were my fellow conspirators during the Argentina project; they taught me a great deal with their friendly advice and clever ideas.

There are several people in the Department of Geography and Environmental Engineering at Johns Hopkins who deserve acknowledgment for their help in the preparation of this book. Professor Charles ReVelle, my good friend and colleague, was involved in the development of many of the ideas in this book through his active participation in the applications discussed in Chapter 10 and his being a willing listener whenever I needed one. Professor M. Gordon Wolman, Chairman of the Department, has made it so easy to do good things and to have fun at Hopkins that I am still suspicious. I also wish to thank him for his advice and guidance during the initial stages of the book. Thelma Barrett and Julie Cole were remarkably patient and efficient in their typing of the manuscript through its several drafts and midcourse corrections. I must also thank all of those students from the Departments of Geography and Environmental Engineering and Mathematical Sciences at Johns Hopkins for teaching me so much during the four years that I have taught my course on multiobjective programming and planning.

My wife Bunny and I are the only ones who can know how important she has been during the preparation of this book. Whatever was wrong, there was always something right.

CHAPTER 1

Introduction

The focus of this book is the theory of multiobjective programming, a new specialization of mathematical programming, and its application to real-world public decision-making problems. We can trace the seeds of what is now a strong branch of operations research to the early work of Kuhn and Tucker (1951) and Koopmans (1951). Multiobjective programming was not really considered a separate speciality, however, until a 1972 conference in South Carolina (Cochrane and Zeleny, 1973). Multiobjective programming is indeed new, but it is a very rich field. It is an area that has attracted an enormous amount of attention [see the imposing reference lists of Cochrane and Zeleny (1973) and Zeleny (1976)] because it is so useful for real-world decision making.

Multiobjective programming and planning is concerned with decision-making problems in which there are several conflicting objectives. Such problems are ubiquitous, particularly in the public sector, which must be concerned with society's objectives. A few examples of multiobjective public decision-making problems are offered in Section 1.2. Detailed discussions of multiobjective water resource and facility location problems are presented in Chapters 9 and 10. For now it is sufficient to observe that such phrases as "tradeoff" and "conflict resolution" are now part of every public decision maker's vocabulary.

In the remainder of this chapter the value of multiobjective analysis is considered before the discussion of example problems. A brief history of multiobjective analysis is then provided and the scope of the book is defined. The chapter concludes with a plan for the remainder of the book.

1.1 VALUE OF MULTIOBJECTIVE APPROACHES

Multiobjective programming and planning represents a very useful generalization of more traditional, single-objective approaches to planning problems. The consideration of many objectives in the planning process accomplishes three major improvements in problem solving. First, multiobjective programming and planning promotes more appropriate roles for the participants in the planning and decision-making processes. Second, a wider range of alternatives is usually identified when a multiobjective methodology is employed. Third, models (if they are used) or the analyst's perception of a problem will be more realistic if many objectives are considered. Each of these aspects is discussed in more detail below.

1.1.1 Multiobjective Approaches and the Public Decision-Making Process

The public decision-making process is complex and, apparently, poorly understood. There is little agreement as to how it works: Is the process guided by a knowledge of the public interest, by selfish pursuit of personal interests, or by a haphazard intermingling of special interests? It is not the role of this book to discuss alternative models of the public decision-making process. Rather, our interest in the process is motivated by our concern with the impact of analytical techniques on that process.

For our purposes, it suffices to hypothesize two types of actor in the decision-making process: analysts (or planners) and decision makers. Analysts are technicians who provide information about a problem to decision makers who decide which course of action to take. This simple model can take on many forms, depending on the context. For example, within a federal bureau the analysts are the staff economists, engineers, systems analysts, and sociologists who develop alternatives and investigate their relative impacts on measures of effectiveness. The decision maker, in this case, may be the office director, an agency administrator, or a member of the secretary's office. It is conceivable that Congress may be the ultimate decision maker, although it is usually called on to accept or reject a previous decision.

Programming and planning techniques are tools which analysts may use to develop useful information for the decision makers. It is the contention here that traditional single-objective approaches often expand the analyst's role, resulting in a decrease in the decision maker's control of decision situations. Single-objective models require that all project effects be measurable in terms of a single unit. Project evaluation can be accomplished only by subsuming all impacts such as environmental degradation or income redistribution under a single measure of effectiveness such as net economic efficiency benefits. Thus, one-dimensional approaches place the burden of decision making squarely on the shoulders of the analyst. It is the analyst who must decide the monetary equivalent of a specific environmental impact. (If the impact is ignored, this is equivalent to placing a value of zero on environmental quality.) It may appear that this decision regarding the monetary value of a nonmonetary project impact is of minor significance since it concentrates on a so-called "design parameter." This would be a serious mistake, however, since project selection is critically sensitive to the choice of such parameters.

Multiobjective approaches pursue an important decision-making result: an explicit consideration of the relative value of project impacts. By systematically investigating project alternatives, the range of choice and the relationship between alternatives and the relative values of the objectives are identified. In this manner the responsibility of assigning relative values remains where it belongs: with the decision makers.

The multiobjective result is more desirable because analysts are not required to make important value judgments that they are not in a good position to articulate. There is a tendency for bureaucratic staff to confuse agency objectives with special objectives. One would expect, for example, that an engineer employed by a federal design agency would be under some pressure to evaluate social objectives in a manner that would promote the construction of large projects. This is easy to understand, particularly when the evaluation of objectives is done implicitly, as with single-objective approaches.

An agency administrator or departmental secretary would be subject to the same criticism and skepticism. If the decision-making process were open to public scrutiny, however, one would expect more responsiveness to social objectives on the part of these decision makers. They are in a better (more lofty) position from which socially consistent value judgments can be made. The pressures to which these decision makers may be subjected are from more scattered origins.

Regardless of the actual nature of the public decision-making process, multiobjective approaches can be useful in promoting the explicit consideration of the value judgments which are implicitly made in the application of single-objectives approaches. In addition, multiobjective methods allow decision makers and analysts to maintain appropriate roles in the process. The analyst is in the position of generating alternatives and tradeoffs among objectives. Important value judgments regarding the relative significance of the objectives are made by the decision makers.

1.1.2 Multiobjective Approaches and the Generation of Alternatives

Most project and program design and planning problems are characterized by a large (frequently infinite) number of alternatives. Single-objective methods are predicated on a unique measure of effectiveness, so that they lead to the unambiguous identification of an optimal alternative. Decision makers will therefore be in the position of accepting or rejecting this single alternative identified as the best.

Multiobjective methods are used to generate and evaluate more than a single alternative. The number of alternatives that are ultimately presented to decision makers varies from one multiobjective programming technique to another. It is generally true, however, that multiobjective approaches will indicate to decision makers a range of choice larger than the one "optimal" project identified by single-objective methods. This larger range of choice is due to the articulation of value judgments regarding the objectives by decision makers in the project selection phase rather than during analysis.

This aspect of multiobjective programming and planning is perceived as an advantage. A general rule for decision making which is assumed here is that more information (carefully presented) is better than less information. The decision to accept or reject a single optimal alternative is an uninformed decision. Informed, rational decision making requires a knowledge of the full range of possibilities. This can be provided by multiobjective analysis.

1.1.3 Multiobjective Approaches and Analytical Reality

Perhaps the strongest support for the use of multiobjective methods comes from reality: Real-world problems *are* multiobjective. Public action generally impacts many different groups and social concerns. The imposition of a single-objective approach on such problems is overly restrictive and unrealistic. Multiobjective analysis allows several noncommensurable effects to be treated without artificially combining them. This is clearly a significant improvement in analytical capability.

1.2 EXAMPLES OF MULTIOBJECTIVE PROBLEMS

Multiobjective programming and planning is applicable to a wide range of problems in both the private and public sectors. Each of our lives is filled with daily multiobjective problems, decision situations with noncommensurable objectives. Should I take the car or the bus? Well, the bus is cheaper (when the cost of gasoline, maintenance, and insurance are computed for the car), but the car is more convenient, particularly since I should stop at the store on my way home from work. The bus will save energy, but I can listen to the radio in the car. There are probably other *attributes* or objectives in addition to cost, convenience, energy consumption, and comfort that might be considered in choosing between the car and the bus. The point is that there is no single measure of what is best, like dollars. Instead, there are several measures or objectives of importance and making a decision requires the decision maker (the author, in this case) to articulate value judgments, at least implicitly, on the relative importance of the objectives. Problems like this abound.

Our interest is in the public sector where multiobjective problems are the rule rather than the exception, due primarily to the multiplicity of interests that are embodied by social welfare or the so-called public interest. In the remainder of this section a few public sector problems are mentioned. The goal here is to tempt you, to excite your imagination, to give you a sense of the richness of multiobjective analyses in the public sector so that you may have the will to wade through some of the heavier going in later chapters.

In an attempt to span the range of public sector problems, we will talk about three types: public investment, regulation and control of economic activity, and programs and policy. These arbitrary categories were invented for this section only; they were chosen for convenience.

1.2.1 Public Investment Problems

Public investment is the problem area which has attracted the attention of most multiobjective analysts. It was in this area, in the context of water resource planning, that Marglin (1962, 1967) and Major (1969) pioneered the development of multiobjective analysis. Most of the practical applications have also been accomplished for public investment decision making—primarily for water resource systems (Miller and Byers, 1973; Cohon and Marks, 1973; Major, 1974; Haimes, 1977) and in transportation (Hill, 1973; Keeney, 1973b). Multiobjective analyses of water resource problems are considered in detail in Chapter 9.

In his evaluation of urban transportation plans, Hill (1973, pp. 69–77) presents a list of 14 objectives. Some of these are the reduction of air pollution, noise, and accidents; the increase of accessibility; fiscal efficiency; and the attainment of a more equitable income distribution. These six objectives confiict and there is no obvious method for collapsing all of them into a single monetary measure. For example, the most direct route for an urban highway will usually maximize accessibility (measured as time of travel) and fiscal efficiency, but give high levels of air pollution and noise impacts. A circuitous route, on the other hand, will lessen air pollution and noise impacts but cost more and require longer travel times.

Which route is better when there are conflicting objectives? One cannot say which route is better without making value judgments about the relative importance of the objectives. Given that the various participants in the decision-making process will usually evaluate the relative importance of the objectives differently, the resolution of the conflicts among objectives will usually require a political process. It would be a mistake for the planner to select only one of these objectives, such as efficiency, or to assume relative values for the objectives. This would preempt the political process and most certainly lead to later impasses in the planning process. Manheim (1974) makes a very strong case for participatory transportation planning for just these reasons.

The location of public facilities represents another area of public investment to which multiobjective planning is applicable. In addition to an investment decision, a governmental entity must also decide on the location for the activity. Such problems present spatial conflicts among the areas served or impacted by the facility. For example, in locating a fire station, should the station be sited near the central business district, which is characterized by high property value and low (nighttime) population, or should it be located near a low-value densely populated area? Again, the resolution of this conflict is the focus of multiobjective planning. Examples of multiobjective facility location problems are presented in Chapter 10.

1.2.2 Regulation and Control of Economic Activity

Recent work has applied multiobjective analysis to the public regulation and control of private economic activities. Perhaps the classic case of government intervention into the private sector is the abatement of water pollution. It is the perceived role of government in water quality control to develop regulations and programs that will induce private interests to treat, or otherwise withhold, their wastes. There is, of course, a public investment aspect to this problem as well, as evidenced by the federal contribution to municipal sewage treatment plants (see Water Pollution Control Act Amendments of 1972, Public Law 92-500). Public investment, however, represents only one of the several tools available to the government in the control of water pollution.

Brill *et al.* (1976) considered the development of an effluent charge program for the Delaware River Basin. An effluent charge is a tax levied on a polluter's wastes that are discharged into a stream. The tax is intended to induce the polluter to withhold or treat the wastes, thereby avoiding the charge. Brill and his colleagues considered the objectives of economic efficiency and equity. That is, one can develop a charge program that will minimize total costs but the differences in charges from one polluter to the next may be very different; alternatively, a charge scheme which promotes equality of charges results in higher total costs. Once again, the resolution of the conflicting objectives is a political problem that does not permit the planner to identify a single optimal solution.

The current United States law covering water pollution (Public Law 92-500 cited above) exhibits a multiobjective nature. The part of the law dealing with attainment of the 1983 goal of "fishable and swimmable" waters requires dischargers to implement the best available treatment technology (BAT) that is "economically achievable." It is the responsibility of the U.S. Environmental Protection Agency (EPA) to interpret what BAT means in each case. To do this, the EPA must weigh the environmental benefits of a given technology, i.e., the improvement in several measures of water quality, against the cost of implementing the technology. This is a multiobjective problem since the environmental benefits defy reliable quantification in monetary units.

Many other regulatory problems are inherently multiobjective. Other areas of pollution control such as air quality and solid-waste management present conflicts similar to those just discussed: environmental gains versus cost; efficiency versus equity. The National Environmental Policy Act (U.S. Code, Volume 42, Section 4321, 1970) requires the preparation of an environmental impact statement (EIS) for virtually every construction project that uses federal funds, regardless of the project's purposes. The implication of an EIS is that environmental concerns are ultimately brought into play in the final decisions on the project.

Land-use planning is another rich regulatory context for multiobjective analyses and there have been many examples presented in the literature. Vedder (1970) considered approaches to multiobjective planning problems generally. Stuart (1970) used multiobjective analysis to evaluate the effectiveness of the U.S. Department of Housing and Urban Development's Model Cities program. Orne and Wallace (1974) analyzed the conflicts among objectives in new-town developments. Schinnar (1976) used multiobjective analysis for the evaluation of development in an economic demographic framework. Werczberger (1976) analyzed industrial locations when air pollution and economic considerations conflict. Bammi *et al.* (1976) and Barber (1976) analyzed community development in the face of conflicts among environmental impacts, land-use incompatibilities, accessibility of facilities, and energy consumption.

The analysis of energy problems is a new area to which multiobjective analysis is applicable. The familiar efficiency–equity–environment conflicts are present both in investment programs for new technologies and in regulatory approaches such as tax incentives for conservation.

1.2.3 Programs and Policy

This last area is the most general and the least explored among our hypothesized three classes of problems. One could legitimately claim that all public actions fit into this category, so that the first two problem areas are merely subsets of the present one. This is probably true, but the programs and policy area has been included to capture all of those public activities that do not fit conveniently into either the public investment or regulation categories.

In the program area we can include any of the social programs that the government pursues: welfare, aid to education, social security, and health care. In the development of each of these social programs the government is faced with a set of conflicting objectives. In the case of welfare, for example, one can see conflicts between an attempt to reduce administrative costs and the desire to minimize the occurrence of fraudulent claims. Further, the goal to include work incentives for welfare recipients may conflict with the objective of providing support to those who really cannot work—the efficiency–equity conflict.

A controversial social program in the United States is school busing to achieve racial desegregation of public schools. There have been many computer-based analyses of this problem, but the application of multiobjective analysis to busing is a recent event. Lee and Moore (1977) and Silverstein (1977) have developed multiobjective models for school busing. Among the objectives are the minimization of student-miles traveled and the achievement, as closely as possible, of racial balances preset by government policies.

In the general area of policy formulation, not including regulatory policy, perhaps one of the most intriguing arenas for multiobjective analysis is international relationships, especially in the context of negotiations on treaties, nonaggression pacts, and arms limitations. The conflicts in these areas arise from the multiple participants in the negotiations, each with their own objective or objectives. Howard (1971), using the "theory of metagames," takes a hypothetical look at the Cuban Missile Crisis and at a nuclear test ban among the United States, the Soviet Union, and the People's Republic of China. Such multiple-decision-maker problems are discussed further in Chapter 8.

1.3 HISTORICAL PERSPECTIVES AND SCOPE OF THE BOOK

The analysis of multiobjective problems has evolved rapidly over the last two decades. Its progress has occurred primarily in three disciplines: operations research, economics, and psychology. Within operations research, the early theoretical work of Kuhn and Tucker (1951) provided the basis for later algorithmic developments of mathematical programming. Gass and Saaty (1955) provided the first approach applicable to multiobjective programming problems.

The economists' first concern that could be characterized as multiobjective was with the efficient allocation of resources (Koopmans, 1951). Later, Marglin (1962, 1967) introduced multiobjective analysis as an alternative to traditional benefit/cost analysis for public investment decision making.

The psychologists' contribution to multiobjective approaches stems from their concern with how an individual chooses from among a set of multidimensional alternatives. Torgerson's (1958) landmark work has spawned many so-called "scaling" methods for aiding the individual decision maker.

As we look at recent developments in multiobjective methods we see continuing advances in each of these disciplines and extensions of previous work into new areas. The interest in multiobjective programming among operations researchers continues; indeed, it has grown. The first conference devoted exclusively to multiobjective programming was held at the University of South Carolina in 1972. The proceedings from that conference were published in Cochrane and Zeleny (1973). More recently, the proceedings of the 1975 multiobjective sessions at The Institute of Management Sciences international meeting in Kyoto, Japan, have been published in Zeleny (1976). In addition, Ijiri (1965), Lee (1972), and Ignizio (1976) have prepared books on goal programming (see Chapter 7); Zeleny (1974a) has published a book on the multiobjective simplex method (see Chapter 6); and Haimes *et al.* (1975) presented a book devoted to the surrogate worth tradeoff method (see Chapter 7). A special issue of *Management Science* on multiobjective programming has also been published (Starr and Zeleny, 1977).

In economics the use of multiobjective analysis for public investment decision making has been formalized by the United Nations Industrial Development Organization (UNIDO) (1972) and Major (1977). These two books present complete and detailed accounts of multiobjective planning processes and the development of criteria for the evaluation of public investments.

The individual decision-making concepts, originally developed by psychologists, have found their way, in new and extended forms, into both private and public sector multiobjective decision making. The books by Johnsen (1968) and Easton (1973) present many of these scaling methods for the decision making of private organizations, while Hill (1973) considers them in the context of transportation planning.

Keeney and Raiffa (1976) present a complete review of the development of multiattribute utility theory. This represents a hybrid area of analysis, drawing from all three of the disciplines mentioned above.

Where do all of these developments and recent books leave the material presented here? To explain the scope of this book it is necessary to go a little further into the differences among the various directions pursued by mutli-objective researchers.

It is convenient to think of three problem areas: continuous multiobjective problems, discrete multiobjective problems, and multiobjective planning. The first two problem areas differ only by the nature of the alternatives under consideration. In discrete problems, a predefined set of alternatives is available before the multiobjective part of the analysis begins. Transportation plans, for example, may exhibit this characteristic to the extent that a finite set of, say, alternative routes will be considered in any given problem. MacKrimmon (1973) and the works cited above by Easton, Hill, and Johnsen are directed at these kinds of problem. In addition, the ELECTRE method, proposed by Benayoun *et al.* (1966) [see also Roy (1971)] and applied by David and Duckstein (1976) and by Nijkamp and Vos (1977) to water resource development, falls into this category.

Continuous multiobjective problems are not characterized by a predefined set of alternatives. Instead, a model that includes decision variables, constraints, and multiple objective functions must be formulated in order to generate the alternatives. The problem is continuous in the sense that decision variables, which are representative of the alternatives, can take on any values that still meet the constraints of the system under study. It is this problem to which multiobjective programming and this book are directed.

Finally, the third problem area is multiobjective planning, the primary concern of the economists who have developed multiobjective approaches. Multiobjective planning is not problem specific, i.e., it is concerned with both continuous and discrete problems. It is concerned, however, with the process by which public decisions are made, and that is also the concern of this book.

The scope of this book, then, is continuous multiobjective problems set in a real-world public sector planning environment. We shall concentrate on multiobjective programming methods—their theory and algorithms—and place them in the context of multiobjective planning processes. Methods applicable to discrete multiobjective problems are not discussed here. The interested reader is referred to the previously cited references.

None of the books cited above matches the scope of this book. This is the first book to attempt a detailed review of multiobjective programming and the first to consider the practical implications of the range of methods that have been proposed.

1.4 PLAN FOR THE REMAINDER OF THE BOOK

The primary topic of this book is multiobjective programming in theory and practice. Before we enter the mathematics of the methods, however, we will spend some time discussing multiobjective planning in Chapter 2. A six-step planning procedure is then offered, after which a detailed discussion of the identification and quantification of objectives—the crucial first step in the planning procedure—is presented.

Chapters 3 and 4 are preparation for the remainder of the book. Linear programming, a special case of mathematical programming, is reviewed in Chapter 3. The presentation is complete but rather elementary so that linear algebra is not a prerequisite for getting through most of the chapter. Those elements of linear programming that are essential for the discussion in subsequent chapters are emphasized. Chapter 4 presents the terminology and notation of multiobjective programming as it is used throughout the remainder of the book.

The presentation of multiobjective methods begins in Chapter 5 with a categorization of the methods discussed in Chapters 6–8. The criteria used for the categories are derived from planning considerations: the roles for analysts and decision makers dictated by a method, the information requirements, and the directions in which information flows. An evaluation of the methods is offered, but it is intentionally cursory so as not to interject too much of this author's bias into the reading of Chapters 6–8.

The first category of methods is presented in Chapter 6. These methods are called generating techniques because they emphasize the generation of the range of choice which is presented to decision makers for their consideration. Preferences are not explicitly articulated by decision makers; rather, preferences are expressed implicitly by virtue of the decision makers' choice of a preferred alternative. Four generating techniques are presented theoretically and then applied to a small sample problem.

In Chapter 7 preference-based methods are reviewed. These techniques require an explicit statement of preferences by decision makers either prior to the solution process or in an iterative fashion. The preferences are expressed in many different forms for the various techniques: weights, a multiattribute utility function, goals and priorities, and others. The theory that underlies several methods is presented and each technique is applied to our sample problem.

Chapter 8 will take us into the mysterious and complex world of multiple-decision-maker problems. The emphasis here is on the prediction of political outcomes. We will consider: (1) the interactions among many decision makers with game theory; (2) the aggregation of individual preferences into a social welfare function which will take us into welfare economics; and (3) the applicability of multiobjective programming methods to such problems. This area of analysis is the least well developed and, therefore, the most speculative of all of the methods we will consider.

In Chapters 9 and 10 real-world experiences with the application of multiobjective programming are discussed. Water resource problems and the analysis of the development of the Rio Colorado in Argentina are the topics of Chapter 9. Chapter 10 considers multiobjective facility location problems. Two case studies are discussed: fire station location in Baltimore, Maryland, and the analysis of energy facility siting problems.

Chapter 11 concludes the book with a summary of the key elements of the discussion. It presents an optimistic view of the promise which multiobjective programming holds for future problem solving.

One note on references is in order. All of the references cited in this book are, of course, included in the list at the end of the book. The reference list is not, however, exhaustive; only those works relevant to the book, and of which this author is aware, are included. Lengthy bibliographies may be found in the conference proceedings edited by Cochrane and Zeleny (1973) and Zeleny (1976).

CHAPTER 2

The Multiobjective Planning Problem

Planning in the public sector is a complex procedure that is pursued at all levels of government and in many different problem areas. All of the various planning contexts will typically exhibit multiple conflicting objectives. In an attempt to make some general observations about multiobjective planning, a step-by-step planning procedure is presented. A detailed discussion is devoted to the first step in that procedure—the identification and quantification of objectives. In pursuing this issue of identification and quantification the reader will also see some of the objectives typically considered in public decision-making problems and some additional specific examples of multiobjective public sector problems.

We begin with a discussion aimed at drawing a distinction between programming and planning.

2.1 MULTIOBJECTIVE PROGRAMMING AND PLANNING

It is useful for explanatory reasons to make a distinction between multiobjective programming and multiobjective planning. Indeed, such a distinction exists in theory: Economists and operations researchers pursue their own disciplines and theories for solving the same problems, frequently without heed of the other. Yet, in practice, i.e., in the implementation of these theories, planning and programming inevitably become interwoven, as indeed they must if effective analysis is to be performed.

Planning is the process by which analysts perceive a problem, define it, collect data about it, formulate it (perhaps mathematically as a model), and generate and evaluate alternatives for solving it, leading to the end of the process when decision makers choose an alternative for implementation. Thus, planning is defined as the sum total of the activities of analysts and decision makers from problem perception to project implementation.

The above definition is intended to be general, but planning is a loosely structured exercise that need not conform to the above model. For example, there may be loops in the process, e.g., the problem may be redefined after data collection or after the generation of alternatives. Multiobjective planning is problem specific and it is this characteristic that distinguishes it most from multiobjective programming.

Multiobjective programming models may be a piece of the planning process; they address the generation and evaluation phases. Multiobjective programming is a specific form of mathematical programming, a highly structural and formal mathematical procedure for finding the optimal solution to a decision problem. Mathematical programs are models which have the form of a constrained optimization (see Chapter 3), i.e., some objective or criterion of system performance is maximized or minimized subject to constraints on system behavior.

Programming is obviously in marked contrast with planning in one important respect. The former is highly structural in that a problem must be pushed into a special form before it can be treated as a mathematical programming problem. On the other hand, planning is loosely structured and not nearly as formal. The two, however, are definitely related. Multiobjective programming models may serve as the focal point for multiobjective planning activities. Planning can be much more effective when models are part of the methodology. This dependence goes in both directions, however, since a model can be correctly formulated and implemented only when the other pieces of the planning process are successfully performed. Thus, programming and planning are quite different but must be closely related for effective analysis.

The claim that multiobjective models should be a focal point for planning activities requires some justification. The assumption which underlies this contention and which, in fact, supports all of systems analysis, is that an orderly, systematic, structured approach to problem solving is likely to be more efficient and effective than an informal approach. Further, models allow many more considerations to be weighed simultaneously than would otherwise be possible. Models and systems analysis cannot, however, perform all of the activities that are critical for effective planning. Models have a role in the planning process; they cannot replace it.

Models can have a beneficial influence on planning activities such as problem formulation, data collection, and the generation and evaluation of alternatives. If analysts attempt to devise a model for the planning problem under consideration, then more time and effort will be demanded for formulation since mathematical formalism demands that system relationships and interactions be carefully stated. Data collection is usually more efficient when models are employed because the opportunity exists to test the sensitivity of the solutions to changes in data. Generation and evaluation of alternatives are significantly more efficient when they are accomplished with the aid of a model. Mathematical programs, in particular, allow the analysts to consider implicitly an infinite number of possible solutions.

Recent history requires that the virtue claimed for modeling in the above paragraphs be qualified. The rather dismal success rate for the use of systems analysis in solving real problems is a sobering fact. The literature of systems analysis is filled with new mathematical techniques, theoretical discussions and extensions of existing methods, and the application of these approaches to neat hypothetical problems. There are strikingly few reports of the successful application of systems analysis to real-world public planning problems. (A critical distinction exists between real-world and hypothetical problems. Somebody in a decision-making capacity can use the results from the analysis of a real-world problem.)

There are probably several reasons why systems analysis has not been as successful as analysts had hoped. The success that this author and his colleagues have achieved, demonstrated in part in Chapters 9 and 10, has led to the conclusion that the single-objective character of most previous attempts is a major cause for their failure. Since models play a central role in the planning process, the type of model used has a profound effect on the nature of planning activities. Single-objective programming models dictate the use of a single measure of effectiveness, which entails data- and parameter-estimation difficulties and a very limited range of choice for decision makers. It has been found that multiobjective methodologies can go a long way in restoring credibility to systems analysis and in promoting the successful application of systems-analytic procedures. Decision makers like to decide; multiobjective approaches allow them to do so. Thus, multiobjective programming has a significant role to play in the planning process.

2.2 A PLANNING METHODOLOGY

The range of planning problems is clearly large (see Section 1.2), but a general approach to problem solving does seem possible. Multiobjective

TABLE 2-1

STEPS FOR A PLANNING METHODOLOGY

(1) Identification and quantification of objectives
(2) Definition of decision variables and constraints
(3) Data collection
(4) Generation and evaluation of alternatives
(5) Selection of a preferred alternative
(6) Implementation of the selected alternative

planning processes in the context of national public investment problems have been discussed in UNIDO (1972) and Major (1977). A general methodology which consists of six steps toward the resolution of a multiobjective planning problem was presented by deNeufville and Stafford (1971). A similar methodology is displayed in Table 2-1.

The methodology begins with the identification and quantification of objectives and the definition of decision variables and constraints. That is, the determinants of what is important (the objectives), the controls which decision makers have available to them (the decision variables), and the limits on the range of the controls (the constraints) must be identified first in any planning exercise. If a mathematical model is to be used, then steps (1) and (2) correspond to model formulation.

After data are collected in step (3), alternatives that are feasible in terms of the constraints are generated and evaluated for their impact on the objectives. If a mathematical model is used, then generation and evaluation can be done in one step, as shown in Table 2-1. Without a model this phase would require two steps.

In step (5), a preferred alternative is selected by decision makers through a political selection process. This can happen in several different ways, depending on the decision-making context: for example, a single decision maker, such as the administrator of the U.S. Environmental Protection Agency, may select an alternative; a group of decision makers, such as the members of a river basin commission, may select an alternative by consensus; or a group of decision makers, such as a legislative body, may select through a voting mechanism.

It is interesting to consider the implications for the proposed methodology of a single-objective approach. If a single-objective model were used, then steps (4) and (5) would be combined into one. That is, the existence of a single measure of what is best makes the decision makers redundant, since the model will tell us what is optimal. Thus, the differences in methodology provide a clear distinction between multiobjective and single-objective models.

In the final step of the methodology in Table 2-1, the chosen alternative is implemented.

Each of these methodological steps is worthy of detailed consideration. Indeed, thousands of pages have been devoted to selection by economists and political scientists, while management scientists have spent a great deal of conceptual effort on implementation. The formulation of models, their data requirements, and their use in generating and evaluating alternatives have also been considered at length elsewhere. The one major issue that has not been fully considered, with the exception of Major's (1977, Chapter 4) work, is the first step in the methodology: the identification and quantification of objectives. This step is truly critical since the choice of objectives has profound implications for the evaluation and selection steps. For this reason the types of objective that may be important for each of the three types of planning problem mentioned above, as well as those approaches the analyst can take to identify and quantify them, are discussed further in the next section.

Before turning to a detailed discussion of the first step of the methodology, a very important point must be made. The methodology discussed above is intended as a framework, not as a schedule for planning. Thus, many of the steps may happen simultaneously or at least they may overlap. Furthermore, planning is frequently an iterative process: After initial formulations are established, results should be generated to get decision makers' reactions to the approach; these reactions should then be incorporated into a new formulation and the process repeated. Iteration is essential for effective planning because planning is an educational process for both planners and decision makers. The knowledge generated by the process must be fed back into the process to achieve optimal yield from planning.

2.3 THE IDENTIFICATION AND MEASUREMENT OF OBJECTIVES

The place to begin this discussion is with a definition of what constitutes an objective. We will make a distinction between what Hill (1973) calls an "ideal"—a general evaluative statement with which almost all people will agree—and an objective (or criterion)—an operationally useful statement that is consistent with an underlying ideal, but with which all may not agree. The crucial point is that the distinction between an ideal and an objective is based on operational usefulness. An objective may lead to a mathematical statement in terms of the decision variables of a problem, i.e., an objective function, while an ideal will never lead to such quantification. Thus, for example, the maximization of social welfare is an ideal while the maximization of net economic efficiency benefits is an objective.

The major theme of multiobjective programming and planning is that many objectives must be considered in the decision-making process. All of the objectives—even those that conflict—may derive from the same underlying ideal. Economic efficiency, environmental quality, and equity are three conflicting objectives that spring from the same ideal—the maximization of social welfare. While virtually all of society would agree on the ideal, one would expect little agreement as to which of the three objectives is most important in a particular decision context. [Many neoclassical economists still cling to the notion that economic efficiency alone is an adequate surrogate for social welfare. This is a theoretically orthodox view that has been challenged by other economists—see Marglin (1967) and Major (1969)—and by practitioners—see Cohon and Marks (1975, p. 209)].

Objectives, not ideals as they are defined here, are the items of interest for the analyst and planner. They are the things that get included in models and that allow the evaluation of alternative solutions. In addition, the selection of a preferred alternative is directly related to the set of objectives. Therefore the operational usefulness of an objective as an evaluative criterion is a key point.

The practical implication of operationally useful objectives is that the analyst may have to expend a great deal of time and effort on the identification and measurement of objectives. Since decision makers may initially perceive evaluation in terms of ideals, the planner must persist until a representative set of objectives has been developed. The planner must continually ask whether a particular statement is specific enough to allow measurements in terms of the decision variables of the problem. Thus, environmental quality is not sufficiently specific to be operationally useful. The analyst must consider what environmental quality means in the problem context. It may mean maximizing dissolved oxygen at several different points in a water body or it may mean minimizing the concentration of sulfur dioxide at several monitoring points in an airshed. It may frequently occur that operation usefulness will result in the decomposition of an ideal into several objectives, e.g., environmental quality may require an objective for each of several air pollutants.

Operational usefulness is not defined only in terms of quantitative convenience, i.e., the ease with which a particular objective may be represented in a model. Another very important determinant is the degree to which a particular statement of an objective is meaningful to decision makers. Thus, equity may mean different things to different people and, furthermore, there may be alternative mathematical formulations of the same specific statement of the objective. Equity may be defined spatially, as for the distribution of benefits and/or costs among regions or political jurisdictions. Alternatively, equity may relate to the distribution of public impacts among income classes. Even with a given meaning for the equity objective, the analyst is still faced with the problem of defining a metric (a method of measurement) for the objective.

2.3.1 Identifying Objectives

The process by which objectives can be identified by an analyst deserves attention. Ideally, the analyst would ask the decision maker for a statement of the appropriate objectives, but this can be done only in situations in which there is a single decision maker, i.e., one person who will make a final choice from among the set of alternatives. However, such situations frequently do not and should not exist in the public sector. Society (and government in its image) is made up of several different interest groups, each with its own objectives.

There are various sources and mechanisms for the definition of objectives that the analyst has at his disposal. The analyst's knowledge of the decision problem is an obvious first source. An awareness of the history behind a proposed public sector project can be invaluable, as it was in the analysis of the Rio Colorado in Argentina discussed in Chapter 9. The several Argentine participants in the study were frequently consulted because of their knowledge of the historical interprovincial controversy over the waters of the Rio Colorado and of the then-current stances of the concerned provincial governments.

Another possible source of information is the decision makers themselves. A serious pitfall, however, may be the identification of the decision makers in that who they are or where they may be is frequently not apparent in a public decision-making context. In spite of this complexity, the analyst should emphasize contact with decision makers or people who can speak for or correctly assess the decision makers' reactions. The Rio Colorado case study taught the analysts involved this lesson. One formulation of the planning objectives that apparently captured an equity issue was discarded due to the reactions of people close to the decision makers. Somehow, the mathematical elegance of the formulation (see Chapter 9) was not fully appreciated. But, of course, elegance is not important in a real application: The most crucial task of the analysis is to achieve a representation of the problem that is meaningful to the decision makers.

The analysis of fire station locations in Baltimore City, presented in Chapter 10, was characterized by a very close working relationship with the Fire Chief. The many discussions with the Chief and several deputy chiefs led to the identification of the important objectives. In this case, there was truly one major decision maker who was easily identified and relatively accessible.

This fortuitous situation should not be expected to exist in most public decision-making problems.

A third origin of information that may be useful for identifying objectives is published material relative to the decision problem. The approach that uses this source exclusively, however, is unlikely to succeed since published official documents frequently may not present the whole range of issues necessary for an understanding of the decision process. Nevertheless, when the personal approaches mentioned above cannot be pursued, or after they have been exhausted, government documents may be the only alternative. The analysts in the Argentina study used a published report of an interprovincial committee that was established to control the development of the Rio Colorado. The committee report presented areas of agreement and disagreement among the provinces in the document and was quite useful in steering the analysts toward an appropriate statement of the objectives. The success in using the document was due as much to the *interpretation* of the report in light of the analysts' knowledge of the problem and prior contact with people close to the decision makers as it was to the report itself.

It is interesting to note that in the fire station problem of Chapter 10, published documents *alone* would have been misleading. The accepted institutionalized objective for fire station location, as published by the Insurance Services Office, was based entirely on property value. This is in sharp contrast to the objectives which were finally used: A strong weight was put on the protection of life hazard, as articulated by the Fire Chief. Thus, a reliance on documents would have been terribly misleading in this case.

2.3.2 Quantifying Objectives

Identification is only part of the analyst's responsibility. The quantification of objectives, i.e., the statement of an objective as a mathematical function of decision variables, is necessary for the analysis to proceed. (Of course, this can be done only after decision variables are defined in Step 2 of the methodology.) It is impossible to generalize about objectives since there are several problem types and because objectives are problem specific. A few of the most common objectives are discussed below. Our primary concern is with the mathematical form of the objectives since this is most relevant for the later discussion of programming methods. Marglin (1967), UNIDO (1972), and Major (1977) presented detailed discussions of the theoretical and practical aspects of measuring the impact of decision variables on the most common public objectives.

Economic Efficiency

The most common objective for public sector problems is economic efficiency. It is the only objective which is pursued by traditional public planning efforts. In an engineering sense, economic efficiency means the minimization of costs to achieve a particular design. For economists, economic efficiency is the maximization of net national income benefits. It is claimed, from neoclassical economic theory, that the attainment of economic efficiency assures a state of maximum social welfare when the assumptions of pure competition are not violated [see Samuelson (1965, Chapter 8)]. Since this is never the case in the real world, i.e., purely competitive markets do not exist, the maximization of net national income benefits, although an important indicator of the social value of an alternative, does not guarantee the maximization of social welfare. This observation was one of the original motivations for multiobjective planning (Marglin, 1962, 1967).

The measurement of economic efficiency benefits is usually straightforward in a theoretical sense although many difficulties may arise in application. Prest and Turvey (1965) and Howe (1971) discuss many of these issues. It is interesting to note that many of the measurement difficulties result when project effects that are not appropriate for economic efficiency accounting are forced into that objective. A classic example discussed by Prest and Turvey is the economic value of a human life, required by an economic efficiency benefit/cost accounting whenever the saving or taking of lives is a project impact. It is much easier and more appropriate simply to define a "life" objective, i.e., maximize lives saved or minimize lives lost, than to go through the awkward and embarassing manipulations required to place a value on a life (which is still incorrect). Thus, the use of multiple planning objectives can reduce measurement difficulties. Further, regardless of the power analysts may take onto themselves, they are not the decision makers who must ultimately accept or reject a project and all of the value judgments made by analysts. Thus, placing a monetary value on life, or similar controversial value judgments made by analysts, may jeopardize project acceptance.

Equity Objectives

Equity implications may arise from many governmental actions, including all of the problem types mentioned in Section 1.2. One should expect, therefore, that distributional objectives will be required in many planning problems. Unlike the economic efficiency objective, measured as net national income benefits, which is virtually universal, the measurement of equity objectives is a case-by-case problem since distributional questions tend to be project specific.

There are two general types of distributional objectives: geographic distributional objectives and distribution based on a demographic or social characteristic such as income class, race, or sex. The former has been considered frequently in water resource investment problems [see Major (1969), Cohon and Marks (1973), and U.S. Water Resources Council (1973)], apparently due to the inherent nature of regional distribution in river basin planning. Class, race, or sex distributional issues may be of even greater importance, but they may also be more politically sensitive in their consideration: A government which pursues income redistribution among classes is a government which is admitting the existence of a maldistribution of income. Major *et al.* (1974) has considered distributional objectives for income classes in a river basin planning context.

When geographic distributional objectives have been defined, their measurement is typically in terms of discounted net regional income benefits when this is appropriate. The measurement of these benefits proceeds by tracing the flow of project benefits and costs to each region affected by the project. Metrics for class distributional objectives are similar: Major *et al.* (1974) measured net project benefits accruing to each income class.

Quantifying distributional objectives appears straightforward, but there are two issues that remain: the definition of regions or classes and the measurement of equity. The first issue is somewhat easier to address. It is an obviously important issue since different definitions of regions or classes will, in general, yield different estimates of benefits and costs. Indeed, "maximize net benefits accruing to families with $10,000 annual income" is a different objective from "maximize net benefits accruing to families with $5000 annual income."

One cannot and should not generalize as to how classes or regions should be defined for analysis. It is an issue which must be settled on a problem-by-problem basis. It can be concluded, however, that the definition should be performed so as to coincide with the characteristics of the decision problem. Thus, in one instance it may be appropriate to consider adjacent provinces as a single region when there is no differential impact on the two areas, while in another case adjacent towns or even two manufacturing plants may require separate treatment due to differing project impacts.

The second issue—the measurement of equity—is a critical one. The identification of several distributional objects, i.e., several regional or class income accounts, provides decision makers with the information necessary for the assessment of the distributional impacts of a project, but there are two pressing reasons to go further in the analysis. As discussed in Chapters 5 and 6, the computational requirements of most multiobjective programming methods increase exponentially with the number of objectives. Thus, there is a computational motivation for decreasing the number of objectives.

Second, there may be an analytical benefit to be gained by getting more specific in the statement of the equity objective. Perhaps the decision-making process can be made more efficient by focusing on a single equity objective. In order to present methods for quantifying equity we will begin to introduce the notation of mathematical programming.

There are many ways to reduce several distributional objectives to a single equity objective. Brill (1972) and Brill *et al.* (1976) have discussed ways in which this could be done for the case of water quality control. Suppose there is a single objective for every "region" (this may be a manufacturer or any other appropriate unit) so that there are as many distributional objectives as there are regions, say N of them. One approach, then, is simply to maximize each of these objectives individually,

$$\text{maximize} \quad Z_k, \qquad k = 1, 2, \ldots, N \tag{2-1}$$

where Z_k is the benefit accruing to the kth region. (The difficulties encountered in maximizing N objectives simultaneously will be discussed throughout most of the remainder of this book.) It is the reduction of these N regional objectives to a single equity objective which is sought.

One possible approach was offered by Cohon and Marks (1973). In this case, an interprovincial equity question was interpreted to imply that the closer to equality of benefits[†] among the regions the better. To achieve a mathematical statement of such an equity objective, first define the average regional income benefit as

$$\bar{Z} = \frac{1}{N} \sum_{k=1}^{N} Z_k \tag{2-2}$$

The objective, then, is to minimize absolute deviations from a distribution of benefits which gives all regions the mean regional benefit. This can be written mathematically as

$$\text{minimize} \quad D = \sum_{k=1}^{N} |Z_k - \bar{Z}| \tag{2-3}$$

Keep in mind that the purpose of this reduction is *not* to do away with the multiobjective nature of the problem. After the development of this single equity objective other objectives such as economic efficiency remain to be traded off against equity.

The computational benefits to be gained from the new formulation result from the reduction of the objective function's dimensionality from $N + L$ to $L + 1$, where L is the number of other objectives under consideration. There are analytical benefits besides, in that the tradeoff between, say, efficiency and equity as measured by D in (2-3) can now be presented in a concise, graphical manner. Of course, the new formulation is of value only if the new equity objective and its metric, the "sum of deviations from the average benefit" in this case, are meaningful to the decision makers. It is important to keep in mind that elegance does not ensure usefulness. The

[†] Cohon and Marks (1973) considered Z_k as the amount of water, not net benefits, accruing to region k. The formulation, however, is general.

particular formulation presented above was used in the case study of Chapter 9, but was found not to be very meaningful to the Argentine decision makers, in part because of the arbitrary choice of an equal distribution as the base distribution. Positions over the formulation had solidified before it could be pointed out that the deviations in D can be measured relative to any base distribution.

Other equity objectives may also be formulated. It may be desirable to maximize the minimum regional net benefit. This can be done mathematically by first defining the N constraints,

$$Z_k \geq M, \qquad k = 1, 2, \ldots, N \tag{2-4}$$

where M is allowed to vary and will therefore be no larger than the smallest Z_k. M is the quantity to be maximized. The computational and decision-making benefits noted for (2-3) accrue to this formulation as well. The caveat as to relevance, however, also applies.

Another formulation that corresponds to the first equity objective is the minimization of the range of regional benefits. While the same results would be expected, this formulation may be more appealing to the decision makers' intuitions. To formulate the objective, define the minimum regional net benefits M, as in (2-4), and the maximum regional benefits,

$$Z_k \leq B, \qquad k = 1, 2, \ldots, N \tag{2-5}$$

The quantity $(B - M)$ then defines the range of benefits among the regions and is the quantity to be minimized to achieve equity; i.e.,

$$\text{minimize} \quad (B - M) \tag{2-6}$$

Equity questions are obviously very difficult to answer since equity issues are politically sensitive for any government to address. The analyst should not be surprised, therefore, when attempts at explicitly quantifying equity–efficiency tradeoffs are shunned. It is imperative, however, that the issue not be buried by the decision-making process, for whether it is explicitly recognized or not, the ultimate decision to pursue or to reject any public project or policy frequently has an impact on distribution. It is the government's responsibility to weigh that impact before taking action.

Environmental Quality Objectives

Environmental quality objectives tend to be difficult to quantify because of the large number of parameters that may define environmental quality. The statements "maximize environmental quality" and "minimize environmental impacts" may mean very different things, depending on the problem context, since such statements tend to be ideals rather than objectives. For Major (1974), it meant minimizing the number of acres of ecologically valuable land inundated by a proposed reservoir impoundment. For Miller and Byers (1973), it meant minimizing sediment phosphorous, pool sediment, and sediment nitrogen, simultaneously, due to a proposed reservoir. In effect, Miller and Byers were faced with three environmental quality objectives.

The measurement of environmental quality presents a problem similar to the equity objective, in that the multiplicity of environmental quality indicators presents computational and analytical complexity. It was already noted that the computational requirements of many multiobjective programming methods increase exponentially with the number of objectives. Furthermore, the ability of decision makers to digest the tradeoffs among more than three or four objectives is questionable. For environmental quality, as for the equity objectives, there is a motivation for attempting to collapse several environmental quality objectives into a single objective.

Miller and Byers (1973) performed this reduction by measuring a gain in environmental quality as the percent reduction in all quality parameters simultaneously. Thus, the metric for the composite environment quality objective was the percent reduction in each parameter, i.e., phosphorus, pool sediment, and nitrogen. The Miller and Byers problem then became a two-objective problem, maximizing economic efficiency benefits and environmental quality rather than a four-objective problem. This ingenious approach to reducing the number of objectives captures the computational and decision-making benefits to be gained from a two-objective representation. As the authors pointed out, however, the approach assumes that the quality parameters are of equal importance. In general, this will not be the case. Of course, one could apply their approach to those quality indicators which can be taken as approximately equal in importance, treating any remaining parameters separately. A reduction in dimensionality which is less than the maximum possible would then be realized.

The issue of developing a composite environmental quality indicator has not been resolved; nor should we expect it to be in the near future. The problem has motivated research into so-called "quality-of-life" indicators [see Dalkey *et al.* (1972); U.S. Environmental Protection Agency (1973); Andrews *et al.* (1973)].

Other Planning Objectives

Another problem which may confront the analyst is to find any metric at all for some objectives. Some environmental quality objectives and the so-called "social well-being" objective may present this problem due to the vagueness or generality of the goal.

Aesthetic value is an example of an environmental quality objective which is extremely difficult to measure. How does one quantify aesthetic value? Even if a metric can be devised there is no guarantee that it is a meaningful measure for all or even most individuals. One person may see the river as an aesthetically valuable scene while another may value the trees which stand at the river's banks. Still a third person may value both, and he may not be willing to sacrifice either.

Aesthetic quality is also an example of an objective which is "all or nothing"; i.e., in some cases, either there is aesthetic value or there is not. Objectives such as these can be quantified, but not in the continuous manner that allows the graphical displays with which analysts like to work.

Another example of an "all-or-nothing" objective is the complementary energy objective of the Rio Colorado study in Chapter 9. The analysts identified an objective which stated that although hydroelectric energy was a feasible water use, its development should not preclude any beneficial irrigation development. This complementary energy objective was quantified by comparing model solutions in which energy development was allowed to compete with irrigation with model results for which only noncompetitive energy production was allowed. This representation of the objective was considered quite satisfactory by the decision makers.

Another objective which is difficult to quantify is the so-called "social well-being" objective [see Andrews *et al.* (1973)]. Like environmental quality, one would expect social well-being to mean different things in different problem contexts. The measurement of such an objective, regardless of its setting, will remain difficult.

An interesting new development is the quantification of risk as an objective. Risk is a particularly important consideration for large public investments that are directed at problems in an uncertain environment. Water resource problems are typical in this regard, since streamflow—its quantity and quality—is a probabilistic process. Haimes *et al.* (1975) identified a separate risk objective, allowing decision makers to consider risk and its tradeoff with other objectives explicitly.

CHAPTER **3**

Review of Linear Programming

In this chapter the notation, formulation, and solution of linear programming problems are reviewed. The treatment to be found here is concise with an emphasis on an intuitive appreciation of linear programming. Readers who have studied the topic will find this to be a review while those with no linear programming background can consider it an introduction to the topic. Excellent presentations of linear programming theory are in Hillier and Lieberman (1967, Chapter 5) and Wagner (1969, Chapters 2–4). For a complete development of the theory, for which a background in linear algebra is recommended, the reader is referred to Hadley (1962).

The chapter is organized for selective reading. Beginners should read through the sections on: definitions for mathematical programming (Section 3.1); linear programming assumptions (Section 3.2); the sample problem (Section 3.3); the activity analysis problem (Section 3.4); graphical solution techniques (Section 3.5); the characterization of extreme points (Section 3.6); and computational procedures briefly discussed in Section 3.7. The ambitious novice can continue on through the presentation of the simplex method (Section 3.8) and duality (Section 3.9). Those with prior training in linear programming may wish to skim the earlier sections. Confident math programmers with a background in linear algebra may wish to skip to the last section on the Kuhn–Tucker conditions (Section 3.10).

Much of the material in this chapter is presented in a manner that is consistent with later discussions of multiobjective programming methods. All of the techniques require, at a minimum, a familiarity with the structure and formulation of mathematical programming problems (Sections 3.1 and 3.3). An understanding of many of the techniques, or portions of their presentations, will require a familiarity with specific sections in this chapter. In those cases adequate section citations are provided.

3.1 MATHEMATICAL PROGRAMMING DEFINITIONS

Mathematical programming addresses optimization problems which possess a specific structure: Maximize (or minimize) an objective function[†] subject to a set of constraints which define feasibility. The objective function and the constraints are mathematical functions of decision variables and parameters. Decision variables are those aspects of a system which are controllable, while parameters are *givens*—quantities that are not controllable. In structuring and solving a mathematical program, the analyst is attempting to discover decisions about the system under study that are in some sense best.

A collection of values for each of the decision variables is called a solution to the mathematical program. A solution which satisfies all of the constraints is called a feasible solution. In general, there is an enormous number of feasible solutions. In linear programming problems there is an infinite number of feasible solutions. The role of the objective function is to provide a basis for the evaluation of the feasible solutions. That feasible solution which gives the best (lowest or highest) value of the objective function is called the optimal solution.

It is useful to state the general mathematical programming problem with mathematical symbolism. Let us say that there are n decision variables for which values must be selected, i.e., there are n controls such as reservoir sizes, number of highway lanes, and amounts of commodities to produce. The decision variables will be denoted with the variable x and subscripted so as to distinguish one decision variable from another. Thus, x_1 is the first decision variable, x_2 the second, and so on up to x_n. Alternatively, each decision variable could be assigned a different letter, but this is cumbersome when n is a large number. The general form of the mathematical programming

[†] For now, we shall assume that there is a single objective function. In all other chapters the case of multiple objective functions will be considered.

problem is

$$\begin{aligned} \text{maximize (or minimize)} \quad & Z(x_1, x_2, \ldots, x_n) \\ \text{subject to} \quad & g_1(x_1, x_2, \ldots, x_n) = 0 \\ & g_2(x_1, x_2, \ldots, x_n) = 0 \\ & \vdots \\ & g_m(x_1, x_2, \ldots, x_n) = 0 \end{aligned} \tag{3-1}$$

where $Z(\quad)$ is the objective function and $g_1(\quad), g_2(\quad), \ldots, g_m(\quad)$ are the constraints, i.e., there are m constraints. Note that although the constraints are written as equalities, mathematical programming does admit inequality constraints as well. The parameters of the problem are implicit in the symbols for a function.

All mathematical programs include decision variables, constraints, and at least one objective function. It is useful to consider a few problem settings for which mathematical programming problems may be structured. Suppose the problem at hand is the determination of the capacity of a reservoir which is to be built to satisfy a known monthly demand for water. The decision variables in this problem are reservoir release and storage in each month and reservoir capacity. The parameters include a characterization of the streamflows into the reservoir in each month. If the model is deterministic, then historical mean monthly streamflow or streamflows which occurred during a critical (dry) period may be used. If a probabilistic analysis is pursued, then statistical parameters are required. Other parameters such as evaporation rate from the reservoir surface and the monthly demand for water are required. The constraints are included to insure physical feasibility: Monthly release cannot exceed the inflow plus storage; storage cannot exceed capacity. Another set of constraints, which require a certain level of performance, state that release must be at least as large as demand. Finally, the objective function may be the minimization of reservoir capacity, which is a surrogate for minimizing costs. The optimal solution to this mathematical program will yield the smallest reservoir and monthly reservoir releases and storages required to meet water demands.

Another problem setting that is ubiquitous in the public sector is the capital budgeting problem. In this situation a decision maker, e.g., a budget director, is attempting to choose from among a set of projects those projects that are to be implemented. There is a limited budget, so all of the projects may not be selected. Those that are selected should provide the most benefits possible. This problem is often referred to as the "knapsack" problem because it is analogous to the situation in which a hiker must choose from a set of items those to be packed. All of the items cannot be taken since that would exceed the capacity of the knapsack.

The decision variables for each element in the capital budgeting and knapsack problems are of the "yes–no" or "go-no-go" variety; the budget director must decide whether or not to accept each project. The parameters include cost per project, the available budget, and the benefit to be gained from each project if it is implemented. There is only one constraint in this case: The sum of the costs of the projects selected for implementation may not exceed the available budget. The objective function is the maximization of the benefits from the selected projects.

There are many more settings for which mathematical programs can be formulated. Books such as those by deNeufville and Marks (1974) and Salkin and Saha (1975) show the impressive range of problems to which mathematical programming has been applied. While all mathematical programs have the form of (3-1), there are several special cases defined by the assumptions one makes about the decision variables, objective function, and constraints. The most widely used specialization of mathematical programming is linear programming, which is the topic of the remainder of this chapter.

3.2 THE ASSUMPTIONS OF LINEAR PROGRAMMING

Linear programming is a special case of mathematical programming in which all of the functions in (3-1) are linear. This is a rather restrictive assumption since the analyst is limited to relatively special functions that are frequently difficult to use because the world is, unfortunately, nonlinear. There is a collection of techniques subsumed under the heading of nonlinear programming that are used to treat cases when some or all of the functions are nonlinear. Nonlinear programming techniques are relatively (sometimes prohibitively) expensive to implement. Linear programming problems, on the other hand, are inexpensive to solve.

Linearity implies that all of the constraints and the objective function look like sums of decision variables, each of which is multiplied by a coefficient. A constraint in a problem with three decision variables may look like

$$2x_1 + 9x_2 - 7.5x_3 \leq 18 \tag{3-2}$$

where this is an inequality ("less-than-or-equal-to") constraint in which the "right-hand side" is 18. The objective function in the same problem could be

$$\text{maximize} \quad Z = 1.5x_1 + 3.67x_2 + 6x_3 \tag{3-3}$$

Note that decision variables may not be raised to a power other than one (or trivially, zero). Furthermore, decision variables may not be multiplied together.

It is also assumed in linear programming that all decision variables are continuous. This means that the decision variables are not discrete; they can take on any value between some lower bound and some upper bound. The lower bound must always be greater than or equal to zero, i.e., the decision variables must be nonnegative. Any nonnegative number may be used as an upper bound: Positive infinity is assumed by most solution methods unless otherwise indicated.

The assumed continuous nature of the decision variables may be unrealistic in some cases; the real world is discrete. For example, a third of a chair and a half of a dam, although feasible, are not serious options. Integer programming is a specialization of mathematical programming which addresses the discreteness of decision variables.

3.3 A SAMPLE PROBLEM

For the purposes of explanation, a simple example of a linear programming problem will be formulated in this section. It will be used again later to discuss the solution of linear programs.

The Airtight Plastics Corporation manufactures plastic containers for home use and plastic tubing for commercial use. It can earn $6 for every 100-ft length of tubing and $1 for every box of containers it makes. (We are assuming it can sell all that it makes.) Everyday the company receives 12 cartons of polyethylene which it uses in its production process. Each box of containers produced requires two cartons of polyethylene; each 100-ft length of tubing requires three cartons of the material. There is one machine that produces both products. This common machine can run 24 hr/day. It takes 6 hr of machine time to produce a full box of containers and 3 hr to produce a 100-ft length of tubing. We shall structure a linear program to allocate Airtight's resources (polyethylene and machine time) so as to maximize the company's profits.

The first step is to define the decision variables and parameters. The decisions are the amounts of the products (boxes of containers and 100-ft lengths of tubing) that the company should produce each day. We shall make the following definitions:

$x_1 \equiv$ the number of boxes of containers to produce each day
$x_2 \equiv$ the number of 100-ft lengths of tubing to produce each day

The parameters of the problem include

(1) the polyethylene required per unit of product,
(2) the machine time required per unit of product,
(3) the amount of each resource available each day, and
(4) the profit earned per unit of product.

Values for all of these parameters have been given in this example. In practice a great deal of effort may be required to obtain the necessary information.

The next step is to state the objective function and constraints. Each of these mathematical statements are functions of the decision variables x_1 and x_2. The objective function is

maximize (daily profits)

or

maximize (daily profits from containers + daily profits from tubing)

or

maximize [(profit/box of containers)(boxes of containers produced per day) + (profit/length of tubing) (lengths of tubing produced per day)]

which is a statement in terms of the parameters and decision variables. Mathematically the objective function is

$$\begin{array}{lccccc} \text{maximize} & Z & = & 1x_1 & + & 6x_2 \\ & [\$/\text{day}] & = & [\$/\text{day}] & + & [\$/\text{day}] \end{array} \tag{3-4}$$

where Z is just a symbol that has been assigned to the value of the objective function. Note that the units of Z and the two terms on the right of the equality are the same, as they must always be for consistency.

The constraints in this problem are on the available resources: In producing containers and tubing we cannot use more of each resource than is available. The constraints take the form

$$[\text{resources used}] \leq [\text{resources available}]$$

or

$$\begin{aligned} &[(\text{resources/box of containers})(\text{boxes of containers produced per day})] \\ &\quad + [(\text{resources/length of tubing})(\text{lengths of tubing produced per day})] \\ &\quad \leq [\text{resources available per day}] \end{aligned}$$

In terms of the parameters and decision variables, the constraint for the polyethylene resource is

$$\begin{array}{ccccc} 2x_1 & + & 3x_2 & \leq & 12 \\ \begin{array}{c}[\text{cartons of} \\ \text{polyethylene} \\ \text{per day}]\end{array} & & \begin{array}{c}[\text{cartons of} \\ \text{polyethylene} \\ \text{per day}]\end{array} & & \begin{array}{c}[\text{cartons of} \\ \text{polyethylene} \\ \text{per day}]\end{array} \end{array} \tag{3-5}$$

The constraint on machine time is

$$6x_1 + 3x_2 \leq 24 \tag{3-6}$$

in which the units are machine hours per day for each term.

The final constraints require nonnegativity of the decision variables,

$$x_1 \geq 0, \qquad x_2 \geq 0 \tag{3-7}$$

That is, negative containers and tubing may not be produced.

In summary, the formulation is

$$\begin{aligned}
&\text{[daily profits]} \quad \text{maximize} \quad Z = x_1 + 6x_2 \\
&\text{s.t.} \quad \text{[daily use of polyethylene]} \quad 2x_1 + 3x_2 \leq 12 \\
&\qquad \text{[daily use of machine time]} \quad 6x_1 + 3x_2 \leq 24 \\
&\qquad \text{[nonnegativity]} \quad x_1, \quad x_2 \geq 0
\end{aligned} \tag{3-8}$$

where s.t. is the abbreviation for "subject to."

The company's resource allocation or activity analysis problem has been structured as a linear programming problem. The mathematical problem is to find feasible values for x_1 and x_2 that maximize the objective function. We have made a very important step by formulating the linear program. In practice, the formulation of the problem is the most difficult step. Solving linear problems has become quite straightforward since there are many good computer packages for finding the optimal solution.

3.4 THE GENERAL ACTIVITY ANALYSIS PROBLEM

The preceding sample problem is a simple example of the general activity analysis or resource allocation problem,

$$\begin{aligned}
\text{maximize} \quad & Z = c_1x_1 + c_2x_2 + \cdots + c_nx_n \\
\text{s.t.} \quad & a_{11}x_1 + a_{12}x_2 + \cdots + a_{1n}x_n \leq b_1 \\
& a_{21}x_1 + a_{22}x_2 + \cdots + a_{2n}x_n \leq b_2 \\
& \qquad \vdots \\
& a_{m1}x_1 + a_{m2}x_2 + \cdots + a_{mn}x_n \leq b_m \\
& x_1 \qquad\qquad\qquad \geq 0 \\
& \qquad x_2 \qquad\qquad \geq 0 \\
& \qquad\qquad \ddots \\
& \qquad\qquad\qquad x_n \geq 0
\end{aligned} \tag{3-9}$$

where x_j, $j = 1, 2, \ldots, n$ are potential activities (decision variables); c_j, $j = 1, 2, \ldots, n$ are the coefficients of the objective function—they indicate the payoff (profit) per unit of activity; b_i, $i = 1, 2, \ldots, m$, are the available resources; and a_{ij}, $i = 1, 2, \ldots, m$, $j = 1, 2, \ldots, n$ are the resources required per each unit of activity.

This problem can be rewritten with the use of the symbol $\sum$, which denotes summation. The notation $\sum_{j=1}^{n} x_j$ means $x_1 + x_2 + x_3 + \cdots + x_n$. The problem in (3-9) may be rewritten as

$$\begin{aligned} \text{maximize} \quad & Z = \sum_{j=1}^{n} c_j x_j \\ \text{s.t.} \quad & \sum_{j=1}^{n} a_{ij} x_j \leq b_i, \qquad i = 1, 2, \ldots, m \\ & x_j \geq 0, \qquad j = 1, 2, \ldots, n \end{aligned} \tag{3-10}$$

The activity analysis problem is equivalent to the resource allocation problem. Problems of this type are formulated when there is a set of finite resources (b_i, $i = 1, 2, \ldots, m$) that must be allocated among a set of activities (x_j, $j = 1, 2, \ldots, n$) so as to yield as high a payoff as possible (maximize Z). The coefficients a_{ij} represent production functions by requiring a certain amount of resource i to perform a unit of activity j.

The activity analysis formulation is in fact a general statement of the linear programming problem since any linear program can be modified to conform to this structure. Note that a minimization problem can be converted to a maximization by multiplying the objective function by -1. Similarly, the sign of an inequality constraint is changed by multiplication by -1. Also, equality constraints may be represented by two inequality constraints. Thus, the problem

$$\begin{aligned} \text{minimize} \quad & Z = -3x_1 + 2x_2 - 4x_3 \\ \text{s.t.} \quad & x_1 + x_2 + x_3 \geq 9 \\ & 2x_1 + 3x_2 + x_3 = 7 \\ & x_1, \quad x_2, \quad x_3 \geq 0 \end{aligned} \tag{3-11}$$

may be written in activity analysis form as

$$\begin{aligned} \text{maximize} \quad & (-Z) = 3x_1 - 2x_2 + 4x_3 \\ \text{s.t.} \quad & -x_1 - x_2 - x_3 \leq -9 \\ & 2x_1 + 3x_2 + x_3 \leq 7 \\ & -2x_1 - 3x_2 - x_3 \leq -7 \\ & x_1, \quad x_2, \quad x_3 \geq 0 \end{aligned} \tag{3-12}$$

3.5 GRAPHICAL SOLUTION OF LINEAR PROGRAMMING PROBLEMS

The set of feasible solutions to a linear program can be represented graphically when there are three or fewer decision variables. This is done in Fig. 3-1 for the Airtight Plastics problem formulated before in (3-8). The number of

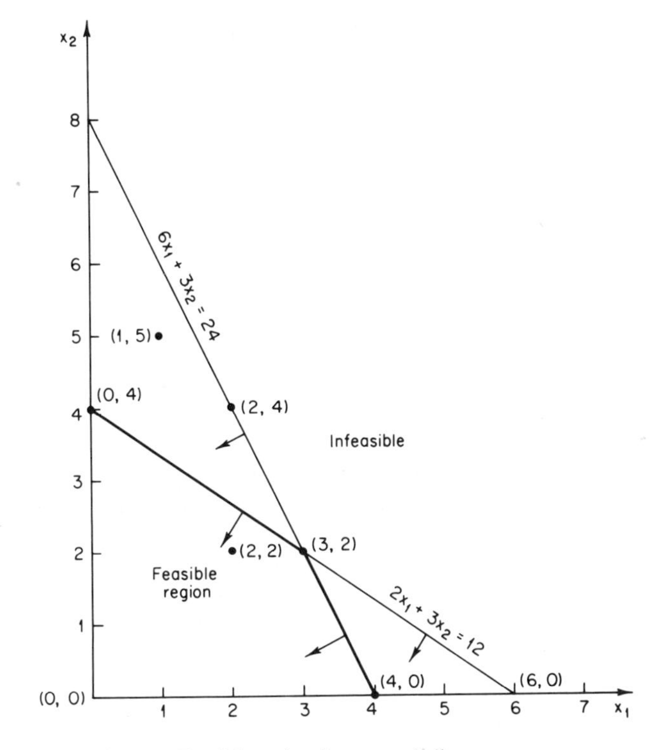

Fig. 3-1. Feasible region for a sample linear program.

axes in a plot of the feasible region equals the number of decision variables; for n variables the plot is n-dimensional.

Each constraint in (3-8) is represented by a straight line in Fig. 3-1, treating the function as if it were an equality. Thus the line $6x_1 + 3x_2 = 24$ (the machine-time constraint) is plotted as shown. The nonnegativity constraints are just the axes of the figure. To be feasible, solutions must lie on or on the appropriate side of each of the lines in Fig. 3-1. For the machine-time constraint, solutions that lie on the line $6x_1 + 3x_2 = 24$ or below it (since this is a *less*-than-or-equal-to constraint) are feasible. This is denoted in the figure by the arrows pointing toward the feasible side of the line. The feasible region is the collection of points that satisfy all of the constraints simultaneously. For example, the point (2, 2) is feasible because it lies below $2x_1 + 3x_2 = 12$ and $6x_1 + 3x_2 = 24$ and is nonnegative. The point (3, 2) is also feasible because it lies on the two resource constraints and above the two nonnegativity constraints. The feasible region is shown in Fig. 3-1. Points that violate any one of the constraints, such as (1, 5), which lies above $2x_1 + 3x_2 = 12$, are infeasible.

The boundary of the feasible region of a linear program is a series of straight lines that meet at corners called extreme points. The feasible region of a linear program has the property that the line connecting any two points in the region is also in the region. Such a region is said to be a convex set or, geometrically, a convex polyhedron. This is an important property in that it allows for a very efficient solution technique. This will be elaborated on later.

The objective function can be represented in the plot of the feasible region as a series of parallel lines, each of which is a contour that shows combinations of the decision variables that lead to a given, fixed value of the objective function. The value of the objective function increases as we move from one contour to another in a "northeasterly" direction. Contours for selected values of the Airtight objective function are shown in Fig. 3-2, in which the feasible region has been redrawn.

Six objective function contours, for $Z = 0, 6, 12, 18, 24, 30$, are drawn in Fig. 3-2 although there are an infinite number of them. A combination of values for x_1 and x_2, i.e., a solution, that lies on a contour would lead to profits equal to the value of the objective function on that contour. For

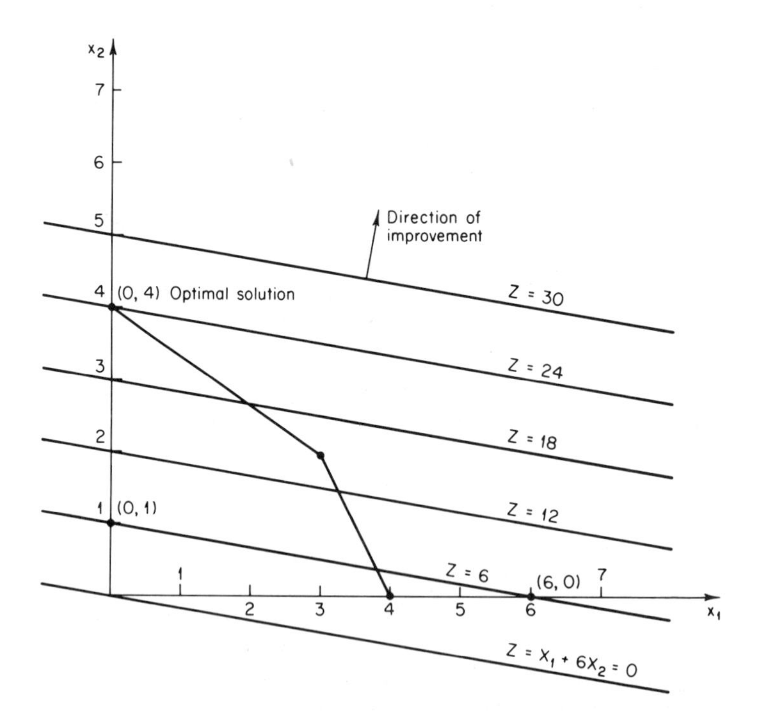

Fig. 3-2. Representation of the objective function for the sample linear program.

example, the solutions (0, 1) and (6, 0) both lie on the contour $Z = 6$ because they would lead to Airtight profits of $6.

The optimal solution to the Airtight problem can be read from Fig. 3-2 directly. The optimal solution to an optimization problem is that feasible solution that yields the highest value of the objective function. Graphically this means that we should look for the highest objective function contour that passes through at least one feasible point. In Fig. 3-2 such a contour is the line $Z = 24$, which touches the feasible region at the point (0, 4), i.e., $x_1 = 0$, $x_2 = 4$, from which it is concluded that Airtight should produce zero boxes of containers and four 100-ft lengths of tubing to earn maximum profits of $24. Note that higher contours do not intersect the feasible region, i.e., there are no feasible solutions that can yield higher profits. On the other hand, on the other side of $Z = 24$, contours do pass through the feasible region, but they yield lower profits; these feasible solutions therefore cannot be optimal. Notice also that only the slope of the objective function contours matters; i.e., it is the *relative* value of an activity that determines optimality. Thus, we can divide the objective function through by any positive number or add a constant to the objective function without affecting the relative desirability of solutions or optimality.

Graphical solution methods are applicable to problems with only two or three decision variables. Problems that are encountered in practice rarely have so few; they usually have hundreds or thousands. In order to develop the nongraphical solution method known as the simplex method, a few observations on the graphical representation are in order.

Notice that the optimal solution for the Airtight problem was an extreme point of the feasible region. The point (0, 4) is the "corner" where two constraints—the polyethylene and the nonnegativity of x_2 constraints—intersect. To be more precise, (0, 4) is an extreme point because there is no feasible line segment such that (0,4) lies in the middle (not on an end) of the segment. It is a general result for linear programs that the optimal solution will be found at an extreme point.

It is easy to see from Fig. 3-2 that the optimal solution will always be on the boundary of the feasible region because an interior point must always lie on an objective function contour that is lower than contours passing through at least one point on the boundary. (Keep in mind that there are an infinite number of objective function contours; only six are shown in Fig. 3-2.) Putting it another way, one can always find a feasible direction in which to move from an interior point that will improve the objective function. That the optimal solution must also be an extreme point of the feasible region can also be seen from Fig. 3-2. Recall that the object during a graphical solution is to push out along objective function contours until we can go no further. As long as the objective function is linear and the constraints form a convex feasible region the last point we will touch as we move out will be an extreme point. The only way that this cannot be true is by the introduction of nonlinear constraints or a nonlinear objective function. Experimentation with various feasible regions and objective functions by the reader may be useful to substantiate this fact.

To this point a very important result has been established: In solving linear programming problems we can restrict our attention to extreme points of the feasible region. This is important because there is a relatively small number of extreme points in comparison to the infinite number of feasible solutions. The simplex method exploits this characteristic of linear programming problems by searching only the extreme points.

There is an exception to the occurrence of the optimal solution of a linear program at a single extreme point: There may be alternative optimal solutions which are found at more than one extreme point and along the edges of the feasible region that connect the optimal extreme points. Suppose that Airtight begins to experience heavy competition from other

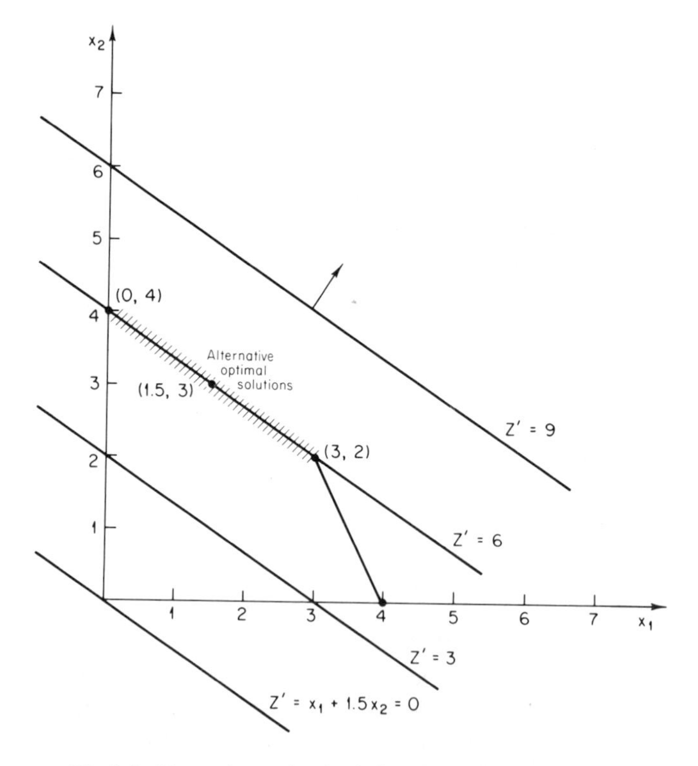

Fig. 3-3. Alternative optimal solutions in the sample linear program.

producers of plastic tubing that drives the price for tubing down and results in a decrease in Airtight's earnings from $6 to $1.50 per 100-ft length of tubing. There would now be a new objective function

$$Z' = x_1 + 1.5x_2 \tag{3-13}$$

which is to be maximized subject to the same constraints faced before by the company. The old feasible region and the new objective function contours are shown in Fig. 3-3.

The result from Fig. 3-3 is that the extreme points (3, 2) and (0, 4) and all of the points along the line segment that connects (3, 2) and (0, 4) are optimal solutions. An objective function contour last touches the feasible region at a boundary rather than at a single point. This means that Airtight can earn a maximum of $6 by producing no containers and four 100-ft lengths of tubing, three boxes of containers and two lengths of tubing, or any combination that lies along the line segment connecting these two alternatives, e.g., (1.5, 3).

Alternative optima are quite common in linear programming problems. They present no undue complexity, however, because we know that alternative optima occur only when more than one extreme point is found to be optimal. Thus, our attention can still be devoted exclusively to extreme points with the knowledge that there are no alternate optima when only one extreme point is found to be optimal.

In order to proceed toward an understanding of the simplex method for the solution of linear programs we must devise a mathematical characterization of extreme points. This is done in the next section.

3.6 MATHEMATICAL CHARACTERIZATION OF EXTREME POINTS

The characterization of extreme points rests on the notion of a *basis* of a linear program which will be defined below. First, the idea and use of *slack variables* will be discussed.

A slack variable is incorporated into a constraint in order to convert it from an inequality to an equality. For example, the polyethylene constraint

$$2x_1 + 3x_2 \leq 12 \tag{3-14}$$

is an inequality which says that no more than the available quantity of polyethylene may be used in the production of containers and tubing. If less than 12 units of polyethylene is used, then there is some left over. The excess polyethylene will be denoted x_3 and is equal to

$$x_3 = 12 - (2x_1 + 3x_2) \tag{3-15}$$

Rewriting this relationship so that all of the xs are on the left-hand side of the equality gives

$$2x_1 + 3x_2 + x_3 = 12 \tag{3-16}$$

Thus, x_3 is a slack variable equal to excess resources that has converted the polyethylene constraint to an equality constraint. Note that the slack variable must be nonnegative to maintain the original relationship in (3-14).

If a slack variable x_4 is also added to the machine-time constraint, a new formulation with four variables for the Airtight Plastic problem results:

$$\begin{aligned} \text{maximize} \quad & Z = x_1 + 6x_2 \\ \text{s.t.} \quad & 2x_1 + 3x_2 + x_3 = 12 \\ & 6x_1 + 3x_2 + x_4 = 24 \\ & x_1, \quad x_2, \quad x_3, \quad x_4 \geq 0 \end{aligned} \tag{3-17}$$

The slack variables do not appear in the objective function because Airtight earns nothing from excess resources. If any of the constraints had been of the greater-than-or-equal-to type, then *surplus variables*, which must also be nonnegative, would have been subtracted from the left-hand side.

The formulation in (3-17) is equivalent to the formulation in (3-8). Therefore the feasible regions in x_1–x_2 space for these two problems must be identical. Furthermore, the two problems share the same extreme points. That is, the feasible region in Fig. 3-1 is entirely applicable to the problem in (3-17). Now, in considering Fig. 3-1 we can think about the slack variables. When a constraint is satisfied as an equality, i.e., when a point is on a line in Fig. 3-1 that represents that constraint, the appropriate slack variable is equal to zero. For example, at the point (2, 4) the machine-time constraint is binding ($6x_1 + 3x_2 = 12 + 12 = 24$). Therefore there is no excess machine time and $x_4 = 0$. If, however, a constraint is not binding, then its corresponding slack variable must be nonzero. For example, at (2, 4) the polyethylene constraint is not an equality since $2x_1 + 3x_2 = 4 + 12 = 16 > 12$, which requires that $x_3 = -4 < 0$. This solution is also infeasible, as we know from an inspection of Fig. 3-1.

The two formulations—with and without slack variables—have the same feasible region and the same extreme points. The feasible extreme points are listed in Table 3-1, in which values of the decision and slack variables are given at each point. The values of x_3 and x_4 are computed from the relationships in (3-17). The list in Table 3-1 exhibits a very interesting property: At each extreme point there are exactly two variables that are zero and two variables that are strictly positive. This happens only at the extreme points; at any other feasible solution there are more than two nonzero variables. There is a general result here: An extreme point of a linear program

TABLE 3-1

LIST OF VALUES OF ALL DECISION VARIABLES AT EACH FEASIBLE EXTREME POINT FOR THE SAMPLE LINEAR PROGRAM[a]

x_1	x_2	x_3	x_4
0	0	12	24
4	0	4	0
3	2	0	0
0	4	0	12

[a] See Fig. 3-1.

is characterized by a number of nonzero variables (decision and slacks) which, it turns out, is equal to the number of constraints (not including non-negativity restrictions). In the example, m = number of constraints = 2, and every extreme point has two nonzero variables.

The above result is very useful in that now we know that we have an extreme point whenever exactly m variables are nonzero. The general problem in (3-9), which has n decision variables and m constraints, becomes, upon the addition of slack variables,

$$\begin{aligned} \text{maximize} \quad & Z = c_1x_1 + c_2x_2 + \cdots + c_nx_n \\ \text{s.t.} \quad & a_{11}x_1 + a_{12}x_2 + \cdots + a_{1n}x_n + x_{n+1} = b_1 \\ & a_{21}x_1 + a_{22}x_2 + \cdots + a_{2n}x_n + x_{n+2} = b_2 \\ & \qquad \vdots \\ & a_{m1}x_1 + a_{m2}x_2 + \cdots + a_{mn}x_n + x_{n+m} = b_m \end{aligned} \tag{3-18}$$

The $x_{n+1}, x_{n+2}, \ldots, x_{n+m}$ are the slack variables of which there are m, so that with the addition of the slack variables the general problem has $(n + m)$ variables. If we choose any m of the $(n + m)$ decision and slack variables to be nonzero, set the remaining $(n + m) - m = n$ variables to zero and solve the resulting equations, the solution will be an extreme point.

It is important to understand fully this mathematical scheme for the characterization of extreme points. To do this we will deal with a smaller problem that has three structural (or nonslack) variables and three constraints:

$$\begin{aligned} \text{maximize} \quad & Z = c_1x_1 + c_2x_2 + c_3x_3 \\ \text{s.t.} \quad & a_{11}x_1 + a_{12}x_2 + a_{13}x_3 \leq b_1 \\ & a_{21}x_1 + a_{22}x_2 + a_{23}x_3 \leq b_2 \\ & a_{31}x_1 + a_{32}x_2 + a_{33}x_3 \leq b_3 \\ & x_1, \quad x_2, \quad x_3 \geq 0 \end{aligned} \tag{3-19}$$

which becomes, upon introduction of slack variables,

$$\begin{aligned} \text{maximize} \quad & Z = c_1x_1 + c_2x_2 + c_3x_3 \\ \text{s.t.} \quad & a_{11}x_1 + a_{12}x_2 + a_{13}x_3 + x_4 = b_1 \\ & a_{21}x_1 + a_{22}x_2 + a_{23}x_3 + x_5 = b_2 \\ & a_{31}x_1 + a_{32}x_2 + a_{33}x_3 + x_6 = b_3 \\ & x_1, \; x_2, \; x_3, \; x_4, \; x_5, \; x_6 \geq 0 \end{aligned} \tag{3-20}$$

We can write the constraints of (3-20) in a more concise manner by defining $\mathbf{a}_j$ as the column of coefficients that multiply x_j in the constraint set:

$$\mathbf{a}_j = \begin{bmatrix} a_{1j} \\ a_{2j} \\ a_{3j} \end{bmatrix} \tag{3-21}$$

We can rewrite the constraints in (3-20) then as

$$\begin{bmatrix} a_{11} \\ a_{21} \\ a_{31} \end{bmatrix} x_1 + \begin{bmatrix} a_{12} \\ a_{22} \\ a_{32} \end{bmatrix} x_2 + \begin{bmatrix} a_{13} \\ a_{23} \\ a_{33} \end{bmatrix} x_3 + \begin{bmatrix} 1 \\ 0 \\ 0 \end{bmatrix} x_4 + \begin{bmatrix} 0 \\ 1 \\ 0 \end{bmatrix} x_5 + \begin{bmatrix} 0 \\ 0 \\ 1 \end{bmatrix} x_6 \geq \begin{bmatrix} b_1 \\ b_2 \\ b_3 \end{bmatrix} \tag{3-22}$$

or more simply as

$$\mathbf{a}_1x_1 + \mathbf{a}_2x_2 + \mathbf{a}_3x_3 + \mathbf{a}_4x_4 + \mathbf{a}_5x_5 + \mathbf{a}_6x_6 = \mathbf{b} \tag{3-23}$$

where $\mathbf{b}$ is a column of right-hand sides. The original constraints in (3-20) can be written quite compactly then as

$$\sum_{j=1}^{6} \mathbf{a}_jx_j = \mathbf{b} \tag{3-24}$$

Now let us return to extreme points and their mathematical characterization. Every extreme point for our three-constraint problem will have three variables that are nonzero and three variables that are zero. We find the values of the nonzero variables by setting $(6 - 3) = 3$ variables so chosen to be zero. For example, let us assume that we wish to find the extreme point associated with the variables x_2, x_4, x_5. We would set $x_1 = x_3 = x_6 = 0$. When this is done (3-23) reduces to

$$\mathbf{a}_2x_2 + \mathbf{a}_4x_4 + \mathbf{a}_5x_5 = \mathbf{b} \tag{3-25}$$

or

$$\begin{bmatrix} a_{12} \\ a_{22} \\ a_{32} \end{bmatrix} x_2 + \begin{bmatrix} 1 \\ 0 \\ 0 \end{bmatrix} x_4 + \begin{bmatrix} 0 \\ 1 \\ 0 \end{bmatrix} x_5 = \begin{bmatrix} b_1 \\ b_2 \\ b_3 \end{bmatrix} \tag{3-26}$$

which is a set of three simultaneous equations in three unknowns. In solving these equations we obtain values for x_2, x_4, and x_5. If all of these values are nonnegative, then we have obtained a feasible extreme point, i.e., a corner point of the feasible region.

There is another way to think of the process that we just went through. We collected the columns $\mathbf{a}_2, \mathbf{a}_4$, and $\mathbf{a}_5$, and discarded the other three of the original six columns of coefficients. The three columns that we collected are said to form a *basis* for the linear program. We shall denote this by $\mathbf{B}$ and for the example, $\mathbf{B} = [\mathbf{a}_2, \mathbf{a}_4, \mathbf{a}_5]$. When we form a basis and solve the resulting simultaneous equations so that the variables are all nonnegative, we obtain an extreme point. Accordingly, an extreme point is called a basic solution and the variables whose columns are in the basis (i.e., the variables are nonzero) are called basic variables. Variables associated with columns *not* in the basis (i.e., the variables are zero) are called nonbasic variables.

A basis for the general problem with m constraints will consist of m columns. Any columns may form a basis and may lead to a basic solution. It is possible, however, that some bases will lead to infeasible solutions. In the Airtight problem the choice of x_1 and x_4 as basic variables leads to the solution $x_1 = 6$, $x_2 = 0$, $x_3 = 0$, and $x_4 = -12$, which is an infeasible extreme point since x_4 is negative. This point appears in Fig. 3-1 as the intersection of the line $2x_1 + 3x_2 = 12$ with the x_1 axis.

Bases which lead to feasible points yield solutions that are called basic feasible solutions. These feasible extreme points are the solutions to which our attention is directed in solving linear programs since, as we know from the previous section, the optimal solution will occur at (at least) one feasible extreme point. In the next section we shall take up the simplex method, which uses the mathematical characterization of extreme points in a procedure for finding the optimal solution.

3.7 COMPUTATIONAL PROCEDURES FOR SOLVING LINEAR PROGRAMMING PROBLEMS

The most widely used approach to the solution of linear programming problems is the simplex method, which was developed through the work of Dantzig in 1947 and published in Koopmans (1951). The method is an iterative one that will find an optimal solution in a finite number of steps. Today, several programmed versions of the method are available on most computers. One of the most widely used versions is the Mathematical Programming System Extended (MPSX) (IBM, 1971) which can be used on IBM 360 and 370 series computers.

Linear programming codes such as MPSX have made model solution a relatively straightforward task. Indeed, analysts need not understand the simplex method and its procedures to use MPSX. After the model has been formulated the analyst need only organize the data in the appropriate format, read it into the computer, access the programmed simplex, and await the solution. Of course, it is suggested that users understand the method so that the results can be fully interpreted.

3.8 THE SIMPLEX METHOD

The simplex method is a procedure for solving linear programs that exploits the occurrence of the optimal solution at one or more (in the case of alternative optima) extreme points of the feasible region. The method starts at a basic feasible solution and moves from one adjacent extreme point to other adjacent extreme points until the objective function can no longer be improved.

The movement from one extreme point to an adjacent one is effected by replacing a column in the basis by a column currently not in the basis. An examination of Table 3-1 and Fig. 3-1 will show that adjacent extreme points have one basic variable in common in their corresponding solutions. For example, (0, 0) and (4, 0) both have x_3 as a basic variable; to move from (0, 0) to (4, 0), x_4 is replaced by x_1. In general, adjacent extreme points have $m - 1$ basic variables in common, where m is the number of constraints.

There are several questions that must be answered to develop the simplex method:

(1) At which extreme point do we begin?
(2) Given that the current basis is not optimal:
  (a) which column should leave the basis?
  (b) which column should enter the basis?
  (c) what value should the new basic variable have and what are the new values for the old basic variables?

(3) How can the optimal solution be detected, i.e., when should the algorithm terminate?

Each of these questions will be answered in the following sections, although the order in which they will be treated will vary from the order in which they are listed above. The three-constraint example in (3-20) will be used throughout the development of the method.

3.8.1 The Initial Basic Feasible Solution

An extreme point at which the simplex method begins is called an initial basic feasible solution. Any feasible extreme point may serve as the starting point. For simplicity, however, the columns corresponding to the slack variables are usually taken as the initial basis. This means that all of the structural (or nonslack) decision variables are initially set equal to zero. It is then a simple operation to find the values for the slack variables by solving

$$\sum_{j \in \mathbf{B}} \mathbf{a}_j x_j = \mathbf{b} \tag{3-27}$$

where $\mathbf{B}$ denotes the basis, $\in$ means "is an element of," and $j \in \mathbf{B}$ indicates that $\mathbf{a}_j$ is a basic column.

For the three-constraint problem introduced in the previous section the slack columns are $\mathbf{a}_4, \mathbf{a}_5, \mathbf{a}_6$. Equation (3-27) becomes, for the example,

$$\sum_{j=4}^{6} \mathbf{a}_j x_j = \mathbf{b} \tag{3-28}$$

or

$$\mathbf{a}_4 x_4 + \mathbf{a}_5 x_5 + \mathbf{a}_6 x_6 = \mathbf{b} \tag{3-29}$$

which from (3-22) becomes

$$\begin{bmatrix} 1 \\ 0 \\ 0 \end{bmatrix} x_4 + \begin{bmatrix} 0 \\ 1 \\ 0 \end{bmatrix} x_5 + \begin{bmatrix} 0 \\ 0 \\ 1 \end{bmatrix} x_6 = \begin{bmatrix} b_1 \\ b_2 \\ b_3 \end{bmatrix} \tag{3-30}$$

which gives the value of the basic variables directly as

$$\begin{aligned} x_4 \qquad &= b_1 \\ x_5 \quad &= b_2 \\ x_6 &= b_3 \end{aligned} \tag{3-31}$$

The reader can appreciate that the use of the slack variables as the initial basis saves a great deal of computational effort. In a realistic problem with, say, 1000 constraints the use of the structural (nonslack) variables in the starting basis would present the formidable numerical problem of solving 1000 simultaneous equations for the values of 1000 variables.

It should be pointed out that when some of the constraints are of the "$\geq$" or "$=$" type, the use of slack variables as the initial basis is not possible. Methods which use "artificial variables" to find an initial basic feasible solution have been devised. Hillier and Lieberman (1976, pp. 151–156) and Wagner (1969, pp. 111–113) discuss these methods. Those readers with a background in linear algebra will find a detailed discussion of methods for finding initial basic feasible solutions in Hadley (1962, pp. 116–121, 149–158).

3.8.2 The Simplex Tableau

Dealing with the remaining questions posed above will be facilitated by introducing a new way of representing a basic solution: the *simplex tableau.* Given a basis, it is possible to represent each nonbasic column as a linear function of the basic columns.

$$\mathbf{a}_j = \sum_{i \in \mathbf{B}} y_{ij} \mathbf{a}_i \qquad \text{for all} \quad j \notin \mathbf{B} \tag{3-32}$$

where the symbol $\notin$ means "not an element of," and the y_{ij}s are the coefficients in the linear functions. As an example assume that the current basis is the initial basis consisting of the slack variables, i.e., $\mathbf{B} = [\mathbf{a}_4, \mathbf{a}_5, \mathbf{a}_6]$. Then,

$$\mathbf{a}_j = \sum_{i=4}^{6} y_{ij} \mathbf{a}_i \qquad j = 1,2,3 \tag{3-33}$$

For the particular case of $j = 2$

$$\mathbf{a}_2 = \sum_{i=4}^{6} y_{i2} \mathbf{a}_i \tag{3-34}$$

or

$$\begin{bmatrix} a_{12} \\ a_{22} \\ a_{32} \end{bmatrix} = y_{42} \begin{bmatrix} 1 \\ 0 \\ 0 \end{bmatrix} + y_{52} \begin{bmatrix} 0 \\ 1 \\ 0 \end{bmatrix} + y_{62} \begin{bmatrix} 0 \\ 0 \\ 1 \end{bmatrix} \tag{3-35}$$

which is equivalent to

$$\begin{aligned} a_{12} &= y_{42} \\ a_{22} &= y_{52} \\ a_{32} &= y_{62} \end{aligned} \tag{3-36}$$

This is a special case in which the y_{ij}s are just the same as the original constraint coefficients, but it is a case that is observed at the initial basic feasible solution.

Any basic column can also be expressed as a linear function of the basic columns, but the result is a trivial one. Taking $[\mathbf{a}_2, \mathbf{a}_3, \mathbf{a}_4]$ as an example basis, the relationship for the column $\mathbf{a}_3 \in \mathbf{B}$ is

$$\mathbf{a}_3 = \sum_{i=2}^{4} y_{i3} \mathbf{a}_i \tag{3-37}$$

or

$$\mathbf{a}_3 = y_{23} \mathbf{a}_2 + y_{33} \mathbf{a}_3 + y_{43} \mathbf{a}_4 \tag{3-38}$$

The obvious solution is $y_{23} = y_{43} = 0$ and $y_{33} = 1$.

TABLE 3-2

A PARTIAL SIMPLEX TABLEAU FOR THE THREE-CONSTRAINT PROBLEM

		$\mathbf{a}_1$	$\mathbf{a}_2$	$\mathbf{a}_3$	$\mathbf{a}_4$	$\mathbf{a}_5$	$\mathbf{a}_6$
Basic columns	$\mathbf{a}_2$	y_{21}	1	0	0	y_{25}	y_{26}
	$\mathbf{a}_3$	y_{31}	0	1	0	y_{35}	y_{36}
	$\mathbf{a}_4$	y_{41}	0	0	1	y_{45}	y_{46}

With the linear relationships among columns, a convenient method of presentation is shown in Table 3-2, in which a partial simplex tableau is shown. A column is included in the table for each variable x_1 through x_6 (headed by their associated columns of coefficients) and there is a row for each column in the basis, which is taken to be $[\mathbf{a}_2, \mathbf{a}_3, \mathbf{a}_4]$ for this example tableau. For the general problem in (3-18) with $(n + m)$ variables and m constraints, the simplex tableau would have $n + m$ columns and m rows (since a basis for this problem would have m elements), plus the empty rows and column as in Table 3-2. The y_{ij}s in the linear relationships are the elements which appear in the table. Thus, y_{21} appears in the $\mathbf{a}_1$ column and the $\mathbf{a}_2$ row because that is the coefficient which multiplies $\mathbf{a}_2$ in the expression of $\mathbf{a}_1$ as a linear function of the basic columns. Notice that columns in the tableau which correspond to basic variables have zeros and ones as was established for $\mathbf{a}_3$ in (3-38) above. There are two empty rows and one empty column in Table 3-2. Other information will be added to the tableau in subsequent sections.

3.8.3 Selecting the Column to Leave the Basis and the Value of the Entering Variable

Given a basic feasible solution the algorithm proceeds to an adjacent extreme point (if the current one is nonoptimal) by removing one of the columns in the current basis and replacing it with a currently nonbasic column. In this section a method for determining the column that leaves the basis, i.e., has its associated variable driven to zero, and the value for the entering variable is presented.

Suppose that the current feasible basis is $[\mathbf{a}_2, \mathbf{a}_3, \mathbf{a}_4]$ and x_1 is the variable that will become basic. We know from the previous section that $\mathbf{a}_1$ can be expressed as a linear function of the current columns in the basis:

$$\mathbf{a}_1 = y_{21}\mathbf{a}_2 + y_{31}\mathbf{a}_3 + y_{41}\mathbf{a}_4 \tag{3-39}$$

The following relationship must also hold for the current basis to be feasible:

$$x_2\mathbf{a}_2 + x_3\mathbf{a}_3 + x_4\mathbf{a}_4 = \mathbf{b} \tag{3-40}$$

Upon introduction of $\mathbf{a}_1$ into the basis the current basic variables will assume new values denoted x'_2, x'_3, and x'_4, x_1 will assume a value that will be denoted θ. For feasibility, the following relationship must hold after $\mathbf{a}_1$ is introduced:

$$\theta\mathbf{a}_1 + x'_2\mathbf{a}_2 + x'_3\mathbf{a}_3 + x'_4\mathbf{a}_4 = \mathbf{b} \tag{3-41}$$

The left-hand sides of equations (3-40) and (3-41) must be equal so that

$$x_2\mathbf{a}_2 + x_3\mathbf{a}_3 + x_4\mathbf{a}_4 = \theta\mathbf{a}_1 + x'_2\mathbf{a}_2 + x'_3\mathbf{a}_3 + x'_4\mathbf{a}_4 \tag{3-42}$$

Substituting the expression in (3-39) for $\mathbf{a}_1$ and grouping terms gives

$$x_2\mathbf{a}_2 + x_3\mathbf{a}_3 + x_4\mathbf{a}_4 = (x'_2 + \theta y_{21})\mathbf{a}_2 + (x'_3 + \theta y_{31})\mathbf{a}_3 + (x'_4 + \theta y_{41})\mathbf{a}_4 \tag{3-43}$$

which implies that

$$x_2 = x'_2 + \theta y_{21}, \qquad x_3 = x'_3 + \theta y_{31}, \qquad x_4 = x'_4 + \theta y_{41} \tag{3-44}$$

or, in general,

$$x_i = x'_i + \theta y_{ij}, \qquad \forall i \in \mathbf{B} \tag{3-45}$$

where $\mathbf{a}_j$ is the entering column and the symbol $\forall$ means "for all."

The new values of the basic variables must be nonnegative so that

$$x'_2, x'_3, x'_4 \geq 0 \tag{3-46}$$

which implies from (3-44) that

$$x_2 - \theta y_{21} \geq 0, \qquad x_3 - \theta y_{31} \geq 0, \qquad x_4 - \theta y_{41} \geq 0 \tag{3-47}$$

or

$$\theta y_{21} \leq x_2, \qquad \theta y_{31} \leq x_3, \qquad \theta y_{41} \leq x_4 \tag{3-48}$$

which gives

$$\theta \leq x_2/y_{21}, \qquad \theta \leq x_3/y_{31}, \qquad \theta \leq x_4/y_{41} \tag{3-49}$$

which is true if

$$\theta \leq \min_{i=2,3,4} (x_i/y_{i1}) \tag{3-50}$$

Now we know that either x'_2, x'_3, or x'_4 must be driven to zero because one of these variables will become nonbasic. This means from (3-44) that

$$\theta = x_2/y_{21} \qquad \text{or} \qquad \theta = x_3/y_{31} \qquad \text{or} \qquad \theta = x_4/y_{41} \tag{3-51}$$

Therefore θ must simultaneously satisfy (3-50) and (3-51), which can be achieved only if

$$\theta = \min_{i=2,3,4} (x_i/y_{i1}) \tag{3-52}$$

This is the desired result: The entering variable will assume a value that is equal to the minimum of the ratios of the values of the old basic variables to the coefficients which express the entering column as a function of the old basic columns. The basic column that yields this minimum ratio is the leaving column. Note that we can only consider those i for which y_{i1} is nonnegative, since θ must be nonnegative for x_1 to assume a feasible value. Furthermore, $y_{i1} = 0$ will not be considered since θ would be undefined. Thus, y_{i1} must be strictly positive.

For a problem with m constraints in which j is the entering variable, (3-52) becomes

$$\theta = \min_{i \in \mathbf{B}, y_{ij} > 0} (x_i/y_{ij}) \tag{3-53}$$

That i which yields the minimum is called r. Thus, x_j replaces x_r, $x_j = \theta$, $x_r' = 0$, and the new values for the old basic variables x_i' that remain in the basis are computed from (3-45) as

$$x_i' = x_i - \theta y_{ij}, \qquad \forall i \in \mathbf{B} \tag{3-45}$$

The new basis is $\mathbf{B}'$, which has the same columns as $\mathbf{B}$ except that $\mathbf{a}_j$ replaces $\mathbf{a}_r$.

3.8.4 Determining Which Column to Enter the Basis and When to Stop

The remaining two questions of which variable should enter the basis and how to determine that the optimal solution has been obtained are answered here. The key to the answer is a quantity called a "reduced cost" that shows for each nonbasic variable the impact its introduction into the basis would have on the present solution. The mathematical development that follows leads to an expression for the reduced cost.

For a current basis $\mathbf{B}$ the value of the objective function is

$$Z = \sum_{i \in \mathbf{B}} c_i x_i \tag{3-55}$$

The summation is only over the variables in the basis because the nonbasic variables have a zero value. Of course, if there is a slack variable in $\mathbf{B}$, then its corresponding c_i is zero. For the three-constraint problem suppose $\mathbf{B} = [\mathbf{a}_2, \mathbf{a}_3, \mathbf{a}_4]$; then

$$Z = \sum_{i=2}^{4} c_i x_i = c_2 x_2 + c_3 x_3 + c_4 x_4 = c_2 x_2 + c_3 x_3 \tag{3-56}$$

since x_4 is a slack variable.

When the basis changes to $\mathbf{B}'$, the variables in the old basis $\mathbf{B}$ assume new values called x_i' and $x_r' = 0$ for the leaving variable x_r. The new basic variable is x_j with a value of θ. Thus, the new value of the objective function at $\mathbf{B}'$ is

$$Z' = \sum_{i \in \mathbf{B}} (c_i x_i') + c_j \theta \tag{3-57}$$

which becomes, after substitution for x_i' from (3-45),

$$Z' = \sum_{i \in \mathbf{B}} [c_i(x_i - \theta y_{ij})] + c_j \theta \tag{3-58}$$

Multiplying through in the term under the summation gives

$$Z' = \sum_{i \in \mathbf{B}} c_i x_i - \theta \sum_{i \in \mathbf{B}} (y_{ij} c_i) + c_j \theta \tag{3-59}$$

The first term of (3-59) is just the old value of the objective function as in (3-55). Noting this and factoring out θ in the last two terms of (3-59) gives

$$Z' = Z - \theta \left[\sum_{i \in \mathbf{B}} (y_{ij} c_i) - c_j \right] \tag{3-60}$$

The accepted notation of the simplex method is to denote the summation term in (3-60) by z_j, i.e.,

$$z_j = \sum_{i \in \mathbf{B}} y_{ij} c_i \tag{3-61}$$

and there is a z_j for every nonbasic variable x_j. Using z_j in (3-60) gives

$$Z' = Z - \theta(z_j - c_j) \tag{3-62}$$

The term $z_j - c_j$ in (3-62) is the reduced cost for variable x_j. It tells us by how much the objective function will change if x_j, which is currently nonbasic, is increased from zero to one. If, for example, $z_1 - c_1 = -2$, then making $x_1 = 1$ (which means $\theta = 1$) will *increase* Z by two units. If one considers the fact that the y_{ij}s are zero or one for a basic variable [see (3-38)], then from (3-61) $z_j - c_j = 0$ for all basic variables.

The reduced costs provide a stopping rule and a method for selecting a variable to enter the basis:

Given a basis with a current value of the objective function Z, which we are trying to maximize, the current solution is optimal if $z_j - c_j \geq 0$ for all j not in the basis.

Notice from (3-62) that if $z_j - c_j > 0$, then introducing x_j into the basis will decrease the objective function, which is undesirable. If the introduction of any nonbasic variable would lead to a decrease in Z, then we have the optimal solution. Recall that changing the basis through one column exchange is equivalent to moving to an adjacent extreme point. Thus, the stopping rule also states that if all adjacent extreme points lead to a decrease (when maximizing) or provide no improvement in the objective function, then the current extreme point is optimal. When the problem is a minimization, $z_j - c_j \leq 0$ $\forall j$ at optimality.

It may happen that $z_j - c_j = 0$ for some $j \notin \mathbf{B}$. This means that the introduction of these variables will lead to no change in the objective function, which is the situation for alternative optima that was demonstrated in Fig. 3-3. The reduced costs allow the simplex method to recognize the existence of alternative optimal solutions.

If all $z_j - c_j$ are not nonnegative, then those columns with negative reduced costs can be entered into the basis and the objective function will improve. This is the criterion for selecting the entering column. Actually, any variable with $z_j - c_j < 0$ may be chosen to enter; the choice of which one is arbitrary.

3.8.5 Summary of the Simplex Method

The following is the simplex algorithm for solving linear programming problems in which the objective function is to be maximized and all constraints are of the "$\leq$" type.

(1) Add slack variables to the constraints so that they are equalities.

(2) Choose the collection of slack columns as the initial basis and find an initial basic feasible solution.

(3) Compute the y_{ij}s which express the nonbasic columns as linear relationships of the basic columns. Use these coefficients to determine the reduced costs.

(4) If all reduced costs are nonnegative, then the current solution is the optimal solution. STOP. If not, continue.

(5) Choose a nonbasic column with a negative reduced cost to enter the basis at a value of θ. Call this column j. Find θ and the column to leave the basis (denoted r) from the expression in (3-53). Return to step (3).

3.8.6 Demonstration of Simplex Operations and Some Easy Rules for Performing Them

The Airtight Plastics problem will be used to demonstrate the simplex algorithm. In step (1) slack variables were added to the constraints to give

$$\begin{aligned} \text{maximize} \quad & Z = x_1 + 6x_2 \\ \text{s.t.} \quad & 2x_1 + 3x_2 + x_3 = 12 \\ & 6x_1 + 3x_2 + x_4 = 24 \\ & x_1, \quad x_2, \quad x_3, \quad x_4 \geq 0 \end{aligned} \tag{3-63}$$

In this problem there are now four variables with the following columns:

$$\mathbf{a}_1 = \begin{bmatrix} 2 \\ 6 \end{bmatrix}, \quad \mathbf{a}_2 = \begin{bmatrix} 3 \\ 3 \end{bmatrix}, \quad \mathbf{a}_3 = \begin{bmatrix} 1 \\ 0 \end{bmatrix}, \quad \mathbf{a}_4 = \begin{bmatrix} 0 \\ 1 \end{bmatrix} \tag{3-64}$$

In step (2) an initial basic feasible solution is formed by choosing the slack columns as the basis, i.e., $\mathbf{B}^1 = [\mathbf{a}_3, \mathbf{a}_4]$, where the superscript denotes that this is the first basis. At this point it is convenient to set up a simplex tableau similar to the one in Table 3-2. A general tableau for the current problem is shown in Table 3-3. The rows and columns that were empty in Table 3-2 are now filled. The second row shows the c_j or objective function coefficients for each variable. The last row shows the reduced costs for each variable. The last column, called $\mathbf{a}_0$, shows the values of the basic variables; nonbasic variables have the value zero.

TABLE 3-3

GENERAL SIMPLEX TABLEAU FOR THE AIRTIGHT PROBLEM

	$\mathbf{a}_1$	$\mathbf{a}_2$	$\mathbf{a}_3$	$\mathbf{a}_4$	$\mathbf{a}_0$
c_j	1	6	0	0	0
$\mathbf{a}_3$	y_{31}	y_{32}	y_{33}	y_{34}	$x_3 = y_{30}$
$\mathbf{a}_4$	y_{41}	y_{42}	y_{43}	y_{44}	$x_4 = y_{40}$
$z_j - c_j$	$z_1 - c_1$	$z_2 - c_2$	$z_3 - c_3$	$z_4 - c_4$	$Z = z_0 - c_0$

To continue the algorithm the elements of the tableau must be computed. Since $\mathbf{B}^1 = [\mathbf{a}_3, \mathbf{a}_4]$,

$$\mathbf{a}_j = \sum_{i=3,4} y_{ij}\mathbf{a}_i \qquad \text{for} \quad j = 1, 2, 3, 4 \tag{3-65}$$

For $j = 1$,

$$\mathbf{a}_1 = y_{31}\mathbf{a}_3 + y_{41}\mathbf{a}_4 \tag{3-66}$$

TABLE 3-4

SIMPLEX TABLEAU FOR THE INITIAL BASIS FOR THE AIRTIGHT PROBLEM

	$\mathbf{a}_1$	$\mathbf{a}_2$	$\mathbf{a}_3$	$\mathbf{a}_4$	$\mathbf{a}_0$
c_j	1	6	0	0	0
$\mathbf{a}_3$	2	3	1	0	12
$\mathbf{a}_4$	(6)	3	0	1	24
$z_j - c_j$	−1	−6	0	0	0

or

$$\begin{bmatrix} 2 \\ 6 \end{bmatrix} = y_{31} \begin{bmatrix} 1 \\ 0 \end{bmatrix} + y_{41} \begin{bmatrix} 0 \\ 1 \end{bmatrix} \tag{3-67}$$

which gives

$$2 = y_{31}, \qquad 6 = y_{41} \tag{3-68}$$

Recall that at the initial basic feasible solution, when the basis consists of slack columns only, the y_{ij}s are just the original coefficients a_{ij}. We can now fill in the y_{ij}s in Table 3-3 without further computation. The numerical values for Table 3-3 are shown in Table 3-4.

To compute the values for x_3 and x_4 we simply solve the constraint equations in (6-63) with $x_1 = x_2 = 0$. This is equivalent to

$$x_3 \mathbf{a}_3 + x_4 \mathbf{a}_4 = \mathbf{b} = \mathbf{a}_0 \tag{3-69}$$

or

$$x_3 \begin{bmatrix} 1 \\ 0 \end{bmatrix} + x_4 \begin{bmatrix} 0 \\ 1 \end{bmatrix} = \begin{bmatrix} 12 \\ 24 \end{bmatrix} \tag{3-70}$$

which gives

$$x_3 = 12, \qquad x_4 = 24 \tag{3-71}$$

If the last column of the tableau is thought of as corresponding to a variable x_0, then the computation of x_3 and x_4 is equivalent to computing y_{30} and y_{40} [compare the operations in (3-65)–(3-67) with (3-69) and (3-70)]. The column of right-hand sides, which gives the values of the basic variables, is therefore denoted as $\mathbf{a}_0$ with corresponding "variable" x_0 and is carried along in the tableau.

The remaining elements of Table 3-3 to be computed are the reduced costs. For x_1 the reduced cost is, using (3-61),

$$z_1 - c_1 = (y_{31}c_3 + y_{41}c_4) - c_1 \tag{3-72}$$

Since

$$c_3 = c_4 = 0 \tag{3-73}$$

$$z_1 - c_1 = -c_1 = -1 \tag{3-74}$$

Thus, at the initial basic feasible solution $z_j - c_j = -c_j$ since the objective function coefficients for all of the basic (slack) variables are zero. Therefore, we have that $z_2 - c_2 = -c_2 = -6$. Also, $z_3 - c_3 = z_4 - c_4 = 0$ because $\mathbf{a}_3$ and $\mathbf{a}_4$ are basic columns.

The element in the last row and last column of Table 3-3 is the current value of the objective function, which can be thought of as $z_0 - c_0$, i.e., the "reduced cost" of the "variable" x_0. Actually,

$$z_0 - c_0 = \sum_{i \in \mathbf{B}} c_i y_{i0} - c_0 \tag{3-75}$$

But $c_0 = 0$ and $y_{i0} = x_i$ for $i \in \mathbf{B}$, so that

$$z_0 - c_0 = Z \tag{3-76}$$

For the current basis,

$$z_0 - c_0 = c_3 y_{30} + c_4 y_{40} = 0 \tag{3-77}$$

The tableau with all of the elements filled in is shown in Table 3-4. In terms of Fig. 3-1 and 3-2, the current basis corresponds to the extreme point (0, 0).

In step (4) of the algorithm the reduced costs are examined. Since some of them are negative ($z_1 - c_1$ and $z_2 - c_2$), the current solution is not optimal and the algorithm continues to step (5). A column with a negative reduced cost is chosen to enter the basis. The variable x_1 will be chosen. To choose the column to leave and the value for x_1, (3-53) is used.

$$\theta = \min_{i=3,4} (x_i/y_{i1}) \tag{3-78}$$

or

$$\theta = \min(\tfrac{12}{2}, \tfrac{24}{6}) = \min(6, 4) \tag{3-79}$$

Therefore $x_1 = \theta = 4$ and since x_4/y_{14} gave the minimum, x_4 will become nonbasic: $\mathbf{a}_1$ replaces $\mathbf{a}_4$. The algorithm returns to step (3).

In step (3) all elements of the tableau are recomputed. This will be done, although for subsequent recomputations some simple rules will be followed for the transformation of the simplex tableau.

The basis is now $\mathbf{B}^2 = [\mathbf{a}_3, \mathbf{a}_1]$, so that

$$\mathbf{a}_j = y_{3j}\mathbf{a}_3 + y_{1j}\mathbf{a}_1 \tag{3-80}$$

For columns in the basis we know that $y_{33} = 1$, $y_{13} = 0$, and $y_{31} = 0$, $y_{11} = 1$. For $\mathbf{a}_2$,

$$\begin{bmatrix} 3 \\ 3 \end{bmatrix} = y_{32}\begin{bmatrix} 1 \\ 0 \end{bmatrix} + y_{12}\begin{bmatrix} 2 \\ 6 \end{bmatrix} \tag{3-81}$$

or

$$\begin{aligned} 3 &= y_{32} + 2y_{12} \\ 3 &= \qquad\quad 6y_{12} \end{aligned} \tag{3-82}$$

which gives

$$y_{12} = \tfrac{1}{2}, \qquad y_{32} = 2 \tag{3-83}$$

The computation for $\mathbf{a}_4$ gives

$$\begin{aligned} 0 &= y_{34} + 2y_{14} \\ 1 &= \qquad\quad 6y_{14} \end{aligned} \tag{3-84}$$

or

$$y_{14} = \tfrac{1}{6}, \qquad y_{34} = -\tfrac{1}{3} \tag{3-85}$$

and, for $\mathbf{a}_0$,

$$\begin{bmatrix} 12 \\ 24 \end{bmatrix} = y_{30}\begin{bmatrix} 1 \\ 0 \end{bmatrix} + y_{10}\begin{bmatrix} 2 \\ 6 \end{bmatrix} \tag{3-86}$$

Check the original problem in (3-63) to see that this is equivalent to setting $x_2 = x_4 = 0$ and solving for x_1 and x_3. The result is

$$y_{10} = x_1 = 4, \qquad y_{30} = x_3 = 4 \tag{3-87}$$

The reduced costs for x_1 and x_3 are zero because $\mathbf{a}_1, \mathbf{a}_3 \in \mathbf{B}^2$. The reduced cost for x_2 is

$$z_2 - c_2 = c_1 y_{12} + c_3 y_{32} - c_2 = 1(\tfrac{1}{2}) - 6 = -5\tfrac{1}{2} \tag{3-88}$$

A similar computation for x_4 gives $z_4 - c_4 = \frac{1}{6}$. $z_0 - c_0$ gives the new value for Z, which is 4. The new tableau is given in Table 3-5.

The motivation for the use of the simplex tableau is that all of the computations that were performed above can be summarized in two very simple operations. To state the rules for transforming tableaux a little more notation must be introduced. The row that corresponds to the leaving variable is called the *pivot row* and the column of the tableau that corresponds to the entering variable is called the *pivot column*. The element of the tableau in the pivot row and the pivot column is called the *pivot element*. The 6 in row $\mathbf{a}_4$ and column $\mathbf{a}_1$ in Table 3-4 is circled because $\mathbf{a}_4$ is to be replaced by $\mathbf{a}_1$ and it

TABLE 3-5

SIMPLEX TABLEAU FOR THE THIRD BASIS FOR THE AIRTIGHT PROBLEM

	$\mathbf{a}_1$	$\mathbf{a}_2$	$\mathbf{a}_3$	$\mathbf{a}_4$	$\mathbf{a}_0$
c_j	1	6	0	0	0
$\mathbf{a}_3$	0	(2)	1	$-\frac{1}{3}$	4
$\mathbf{a}_1$	1	$\frac{1}{2}$	0	$\frac{1}{6}$	4
$z_j - c_j$	0	$-5\frac{1}{2}$	0	$\frac{1}{6}$	4

is the pivot element. To transform a tableau (such as Table 3-4) into a new tableau (such as Table 3-5) the following rules apply:

(1) *Identify pivot element* Choose a column with a negative reduced cost; this is the pivot column. Take the minimum of the division of each element of the $\mathbf{a}_0$ column by each element of the pivot column (excluding the c_j and $z_j - c_j$ rows and negative elements in the pivot column). The row which yields the minimum is the pivot row.

(2) *Compute new elements of the tableau* (a) Subtract from each element y_{ij} of the old tableau the following quantity:

$$A = \frac{\begin{pmatrix}\text{the element in the pivot}\\ \text{row, in column } j\end{pmatrix} \times \begin{pmatrix}\text{the element in row } i \text{ and}\\ \text{the pivot column}\end{pmatrix}}{\text{pivot element}} \tag{3-89}$$

That is,

$$y'_{ij} = y_{ij} - A \tag{3-90}$$

where y'_{ij} is an element of the new tableau. This rule applies to all elements of the tableau except the pivot row and the c_j row.

(b) Divide each element of the pivot row by the pivot element.

The rules are more difficult to state than they are to apply. Compare Tables 3-4 and 3-5. The pivot row in Table 3-4 is $\mathbf{a}_4$, the pivot column is $\mathbf{a}_1$, and the circled 6 is the pivot element. The reader can obtain Table 3-5 by applying the rules to Table 3-4. The transformation of Table 3-5 to a new tableau in Table 3-6 will be done with the rules in detail below.

Note in Table 3-5 that $\mathbf{a}_2$ has a reduced cost of $-5\frac{1}{2}$, which means that the current extreme point (4, 0) in Fig. 3-1 is not optimal. The simplex algorithm

TABLE 3-6a

COMPUTATIONS TO OBTAIN THE SIMPLEX TABLEAU FOR THE THIRD BASIS

	$\mathbf{a}_1$	$\mathbf{a}_2$	$\mathbf{a}_3$	$\mathbf{a}_4$	$\mathbf{a}_0$
c_j	1	6	0	0	0
$\mathbf{a}_2$	0/2	2/2	1/2	$-\frac{1}{3}/2$	4/2
$\mathbf{a}_1$	$1 - \frac{0(\frac{1}{2})}{2}$	$\frac{1}{2} - \frac{2(\frac{1}{2})}{2}$	$0 - \frac{1(\frac{1}{2})}{2}$	$\frac{1}{6} - \frac{(-\frac{1}{3})(\frac{1}{2})}{2}$	$4 - \frac{4(\frac{1}{2})}{2}$
$z_j - c_j$	$0 - \frac{0(-5\frac{1}{2})}{2}$	$-5\frac{1}{2} - \frac{(-5\frac{1}{2})(2)}{2}$	$0 - \frac{(1)(-5\frac{1}{2})}{2}$	$\frac{1}{6} - \frac{(-\frac{1}{3})(-5\frac{1}{2})}{2}$	$4 - \frac{4(-5\frac{1}{2})}{2}$

continues. The variable x_2 will become basic at the next iteration, so $\mathbf{a}_2$ is the pivot column. The pivot row (the leaving variable) is determined by finding $\min(\frac{4}{2}, 4/\frac{1}{2})$, which is 2, so the $\mathbf{a}_3$ row is the pivot row. The pivot element in Table 3-5 is circled. The necessary computations for each element are shown in Table 3-6a and the transformed tableau is shown in Table 3-6b.

The new solution is the extreme point (3, 2), at which $Z = 15$ (check the last column of Table 3-6b and Fig. 3-1). The last row of Table 3-6b indicates that the current solution is not optimal and that x_4 should become basic, so $\mathbf{a}_4$ is the new pivot column. The determination of the pivot row is straightforward since $\frac{1}{4}$ is the only positive element in the $\mathbf{a}_4$ column of Table 3-6b. Thus, $\frac{1}{4}$ becomes the pivot element, and it is circled.

The final tableau is shown in Table 3-7. This is an optimal solution since all reduced costs are nonnegative. The optimal solution can be read from Table 3-7 as $x_1 = 0, x_2 = 4, x_3 = 0, x_4 = 12$, and $Z = 24$, which corresponds to the extreme point (0, 4). This was found to be the optimal solution when the problem was solved graphically in Fig. 3-2.

TABLE 3-6b

SIMPLEX TABLEAU FOR THE THIRD BASIS

	$\mathbf{a}_1$	$\mathbf{a}_2$	$\mathbf{a}_3$	$\mathbf{a}_4$	$\mathbf{a}_0$
c_j	1	6	0	0	0
$\mathbf{a}_2$	0	1	$\frac{1}{2}$	$-\frac{1}{6}$	2
$\mathbf{a}_1$	1	0	$-\frac{1}{4}$	$\textcircled{\frac{1}{4}}$	3
$z_j - c_j$	0	0	$\frac{11}{4}$	$-\frac{3}{4}$	15

TABLE 3-7

FINAL SIMPLEX TABLEAU FOR THE AIRTIGHT PROBLEM

	$\mathbf{a}_1$	$\mathbf{a}_2$	$\mathbf{a}_3$	$\mathbf{a}_4$	$\mathbf{a}_0$
c_j	1	6	0	0	0
$\mathbf{a}_2$	$\frac{2}{3}$	1	$\frac{1}{3}$	0	4
$\mathbf{a}_4$	4	0	-1	1	12
$z_j - c_j$	3	0	2	0	24

3.9 DUALITY, SENSITIVITY ANALYSIS, AND PARAMETRIC PROGRAMMING

The full power and richness of linear programming can be appreciated only by gaining an understanding of duality and by exploring the ways in which linear programming solutions can be interpreted. In this section the key concepts of duality are developed. A full treatment is presented in all of the linear programming references cited at the beginning of this chapter. Hillier and Lieberman (1967, Chapter 15) give a comprehensive presentation that does not rely on familiarity of linear algebra. The interpretation of the final simplex tableau and the generation of new solutions when parameters of the linear program are varied are also discussed below.

3.9.1 Dual Variables (Shadow Prices)

Duality is a complex and elegant body of mathematical theory. We shall concentrate, however, on the most important aspect of it for our later uses: dual variables or *shadow prices* as they are frequently called. It is shown below that each constraint of a linear program has associated with it a dual variable that is computed automatically by the simplex method. After discovering where the dual variables are in the simplex tableau, their economic interpretation as shadow prices will be discussed.

Recall that the reduced cost or $z_j - c_j$ indicates the magnitude of the change in the objective function if the nonbasic variable x_j is introduced into the basis at a value of one. A negative reduced cost indicates an increase in the objective function while a positive sign indicates a decrease. Consider Eq. (3-62) again to understand the implications of the sign of the reduced cost.

The reduced cost for a slack variable also indicates the desirability of including that variable in the basis. But what does the presence of a slack variable in the basis mean? Consider the constraint

$$\sum_{j=1}^{n} a_{ij}x_j \leq b_i \tag{3-91}$$

which, before solution, is converted to an equality constraint by the addition of the slack variable x_{n+i}:

$$\sum_{j=1}^{n} a_{ij}x_j + x_{n+i} = b_i \tag{3-92}$$

If x_{n+i} is *not* in the basis at the optimal solution, this means that $x_{n+i} = 0$ (and $z_{n+i} - c_{n+i} = z_{n+i} \geq 0$), which implies that (3-91) is binding and that all of the resources b_i is used in the optimal solution. If x_{n+i} were introduced into the basis at the level of one, the objective function would *decrease* by the amount of z_{n+i} (since the reduced cost is positive) and use of the resource b_i would be decreased by one unit. Thus, the reduced cost for the slack variable indicates the amount by which the objective function would decrease if the use of b_i were decreased by one unit.

Looking at it in another manner, z_{n+i} can be interpreted as the amount by which the objective function would *increase* if one more unit of b_i were available. This interpretation stems from the fact that

$$Z' = Z - \theta(z_{n+i} - c_{n+i}) \tag{3-93}$$

If θ could be made negative, i.e., if x_{n+i} were brought into the basis at a negative level, then the objective function would increase. But making x_{n+i} negative is equivalent to increasing the current use of resource i, $\sum_{j=1}^{n} a_{ij}x_j$, beyond the level b_i [check (3-92)]. Of course, the only feasible way to do this is by increasing b_i.

It was stated above that every constraint has associated with it a dual variable. Duality theory shows that these dual variables are the reduced costs of the appropriate slack variables. Thus, the dual variable for constraint i, denoted by λ_i, is just the reduced cost of x_{n+i}, i.e., $\lambda_i = z_{n+i}$, and can be found in the reduced-cost row of the final simplex tableau under the $\mathbf{a}_{n+i}$ column. The name "shadow price" for λ_i is motivated by the fact that λ_i (or z_{n+i}) tells us the value of an additional unit of resource i, as the above discussion indicated.

The interpretation of the dual variables can be enriched by considering another fundamental result from duality: complementary slackness. Every dual variable and its associated slack variable must satisfy the following relationship:

$$\lambda_i x_{n+i} = 0 \tag{3-94}$$

which requires that $\lambda_i = 0$ if $x_{n+i} > 0$ and $x_{n+i} = 0$ if $\lambda_i > 0$. The interpretation of complementary slackness is that the expansion of currently unused resources ($x_{n+i} > 0$) will not increase the objective function (λ_i must be zero since x_{n+i} is basic). Similarly if $\lambda_i > 0$, then all of the presently available resource must be used (x_{n+i} must be nonbasic to have a nonzero reduced cost). This is an intuitively appealing result. Table 3-7 shows that $\lambda_2 = z_4 = 0$ for the machine-time constraint since only 12 of the 24 hr available are used. More machine time will not improve the solution. On the other hand, all of the available polyethylene is used ($x_3 = 0$) and $\lambda_1 = z_3 = 2$, which indicates that increasing available polyethylene from 12 to 13 cartons would increase profits by $2.

Shadow prices are a significant output of the simplex algorithm. They indicate the marginal value of additional resources, which is a potentially important question in many resource allocation problems. As will be shown in subsequent chapters, the shadow prices can also be used to gain an understanding of the tradeoffs among objectives.

3.9.2 Sensitivity Analysis

Linear programming is applied extensively to real-world problems, not because the real world is linear, but because the technique is a very powerful one. A major aspect of the method's power is the information a single solution gives us about the impact on the solution of variations in the parameters. Thus, it is easy and inexpensive to conduct sensitivity analyses with a linear program.

Suppose we are interested in exploring what would happen to a current optimal solution if the objective function coefficient for variable x_e were to change. A little further analysis of the final simplex tableau will tell us the range of c_e that will maintain the optimality of the current basis. We know that the reduced cost for every variable must be nonnegative since we are maximizing and the current solution is optimal,

$$z_j - c_j = \sum_{i \in \mathbf{B}} c_i y_{ij} - c_j \geq 0, \qquad \forall j \tag{3-95}$$

Suppose further that x_e is a nonbasic variable. We would like to know by how much c_e can change without altering the optimality of the current basis, i.e.,

$$z_e - c_e = \sum_{i \in \mathbf{B}} c_i y_{ie} - c_e \geq 0 \tag{3-96}$$

which implies that

$$c_e \leq \sum_{i \in \mathbf{B}} c_i y_{ie} \tag{3-97}$$

We immediately have an upper bound for c_e since the values of c_i, the cost coefficients of the basic variables, and y_{ie}, the numbers in the simplex tableau under the $\mathbf{a}_e$ column, are known. Of course, there is no lower bound on c_e: Making c_e smaller will only make x_e more undesirable, but it is already nonbasic.

As an example we shall consider x_1 in the Airtight final tableau in Table 3-7. How large can c_1 become before the current basis is no longer optimal? Applying (3-97) to the example, we get

$$c_1 \leq c_2 y_{21} + c_4 y_{41} \tag{3-98}$$

Using Table 3-7 and noting that $c_4 = 0$, we get

$$c_1 \leq 6(\tfrac{2}{3}) = 4 \tag{3-99}$$

i.e., as long as Airtight earns $4 or less on containers, it is optimal to produce four lengths of tubing and zero boxes of containers.

A discussion of Fig. 3-4 will substantiate our result. As c_1 increases to 4, the objective function contours incline toward the x_1 axis. When $c_1 = 4$ the contours are parallel to the polyethylene constraint, as in Fig. 3-3. That is, $c_1 = 4$ and $c_2 = 6$ gives the same result of alternate optima as $c_1 = 1$ and $c_2 = 1.5$. Recall that we can divide an objective function through by a positive number without changing the problem. If c_1 is greater than 4, the solution at (0, 4) is no longer optimal. Figure 3-4 also shows why there is no lower bound on c_1. As c_1 decreases the objective function contour becomes

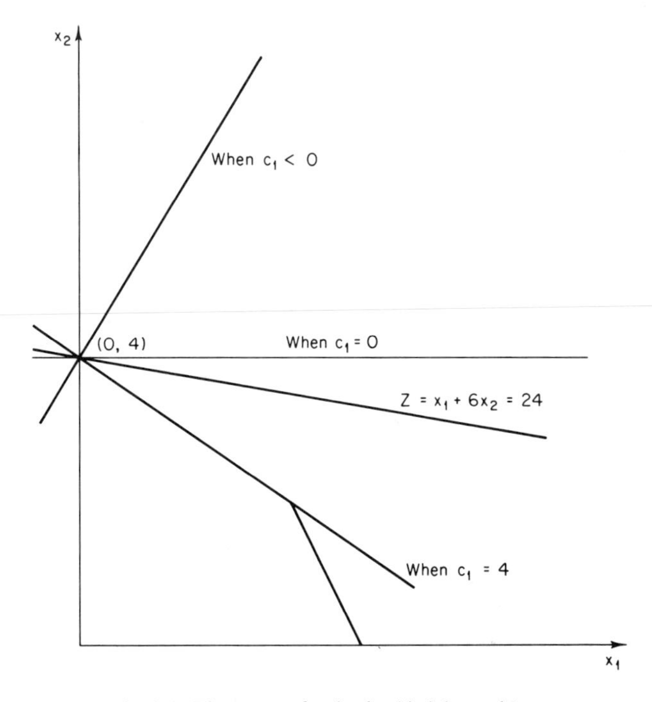

Fig. 3-4. The range of c_1 in the Airtight problem.

flatter until at $c_1 = 0$ the contour is a horizontal line. As c_1 becomes negative the contours continue to shift up toward the x_2 axis. Figure 3-4 shows that whenever $c_1 < 0$ while c_2 remains unchanged, the solution (0, 4) remains optimal.

The analysis is a bit different for changes in a basic variable's objective function coefficient. If x_e is basic, we must consider the reduced costs of all nonbasic variables. We shall rewrite (3-95) as

$$\sum_{\substack{i \in \mathbf{B} \\ i \neq e}} c_i y_{ij} + c_e y_{ej} - c_j \geq 0, \qquad \forall_j \notin \mathbf{B} \tag{3-100}$$

Notice that we need not consider (3-100) for $j \in \mathbf{B}$ since $y_{ej} = 0$ for all $j \in \mathbf{B}$.

We can solve (3-100) to get a series of inequalities:

$$y_{ej} c_e \geq c_j - \sum_{\substack{i \in \mathbf{B} \\ i \neq e}} c_i y_{ij}, \qquad \forall j \notin \mathbf{B} \tag{3-101}$$

Now we must be careful when we solve (3-101) for c_e since y_{ej} may have any sign. Therefore if $y_{ej} < 0$, (3-101) will be an upper bound on c_e, and if $y_{ej} = 0$, changes in c_e have no effect on $z_j - c_j$. Thus, we shall write (3-101) as

$$c_e \geq \frac{c_j}{y_{ej}} - \frac{1}{y_{ej}} \sum_{\substack{i \in \mathbf{B} \\ i \neq e}} c_i y_{ij}, \qquad \forall j \notin \mathbf{B},\ y_{ej} > 0 \tag{3-102a}$$

$$c_e \leq \frac{c_j}{y_{ej}} - \frac{1}{y_{ej}} \sum_{\substack{i \in \mathbf{B} \\ i \neq e}} c_i y_{ij}, \qquad \forall j \notin \mathbf{B},\ y_{ej} < 0 \tag{3-102b}$$

Consider the example in Table 3-7 for which bounds on c_2 are desired. Since y_{21} and y_{23} are both positive, we have

$$c_2 \geq \frac{c_1}{y_{21}} - \frac{1}{y_{21}}(c_4 y_{41}), \qquad c_2 \geq \frac{c_3}{y_{23}} - \frac{1}{y_{23}}(c_4 y_{43}) \tag{3-103}$$

which gives, noting that $c_4 = 0$,

$$c_2 \geq 1\tfrac{2}{3}, \qquad c_2 \geq 0 \tag{3-104}$$

Since the lower bound of zero is redundant, we simply get $c_2 \geq \frac{3}{2}$; i.e., as long as Airtight earns at least \$1.50 for each length of tubing, the current solution will remain optimal. In fact, the case of $c_2 = \frac{3}{2}$ was shown in Fig. 3-3 since it leads to alternate optima.

A similar analysis can be conducted for variations in the right-hand sides. For this case, however, the requirement is that the solution remain feasible.

That is, the question to be answered is how much can, say, b_i change before the current optimal basis is no longer feasible? As long as the basis remains feasible, it will also be optimal since the reduced costs do not depend on b_i. We will not go through this analysis here; the interested reader is referred to Hillier and Lieberman (1967, pp. 493–494).

Many commercial linear programming codes provide a user option by which information on the sensitivity of a solution to changes in the objective function coefficients and the right-hand sides may be obtained. With a single command the user obtains this information for all variables (basic and non-basic) and all constraints. The information can then be used to define those parameters that require further analysis or data collection.

3.9.3 Parametric Programming

The information developed above tells us over what range of the parameters a currently optimal basis will remain optimal or feasible. In many situations we would like to know, in addition, what the new optimal solution would be if the parameters change enough to render the current solution nonoptimal or infeasible. An obvious way to obtain the new solution is to change the parameters of interest and solve the linear program again. This is unnecessary, however, since we can exploit the fact that we already have a basis for the new linear program from which we can continue simplex iterations to find the new solution. This is called parametric programming, and as we shall see later, it is very useful for generating solutions to multi-objective problems.

Parametric programming is a standard option on many commercial linear programming codes. The user specifies which parameter should be varied, the range over which it should be varied, and the increments in that range at which new solutions should be reported. The code then goes through an analysis similar to our discussion of objective function coefficient variations. If the user-indicated change is enough to destroy the optimality or the feasibility of the current basis, then the code recommences simplex iterations until a new optimal solution is found. This is continued until the user-specified range for parameter variations is covered or until further changes in the parameter are known to cause no further change in the basis.

Suppose in an initial run of the Airtight problem we had $c_2 = 1$, i.e., the company could earn only \$1 for a length of tubing. We would have found from a simplex solution of the problem that $x_1 = 3$, $x_2 = 2$ was the optimal solution. (Consider Fig. 3-1 with $Z = x_1 + x_2$.) Now since there is a possibility that the true tubing earnings will be greater than \$1, we should like to vary c_2 from 1 to 10 in increments of 1. Using parametric programming on the objective function coefficients c_2, we would find a new optimal solution,

(0, 4), at $c_2 = 2$. For values of $c_2 > 2$ the solution would not change, and the linear programming code would tell us that by using results like (3-104).

Parametric programming will be used later in Chapter 6 on multiobjective generating techniques and in Chapter 9, in which some of those methods are applied.

*3.10 THE KUHN–TUCKER CONDITIONS†

In later chapters the Kuhn–Tucker conditions (Kuhn and Tucker, 1951) for the characterization of solutions to multiobjective problems will be presented and used to motivate techniques for finding these solutions. In this section the Kuhn–Tucker conditions for optimality for single-objective optimization problems are presented. In addition, a geometrical interpretation of the conditions is offered.

The single-objective or scalar optimization problem is

$$\begin{aligned} \text{maximize} \quad & Z(x_1, x_2, \ldots, x_n) \\ \text{s.t.} \quad & g_i(x_1, x_2, \ldots, x_n) \leq 0, \qquad i = 1, 2, \ldots, m \end{aligned} \tag{3-105}$$

where the $g_i(\quad)$ include nonnegativity restrictions. This may be rewritten with vector notation by denoting the n-dimensional vector of decision variables as $\mathbf{x}$. The problem in (3-105) becomes

$$\begin{aligned} \text{maximize} \quad & Z(\mathbf{x}) \\ \text{s.t.} \quad & g_i(\mathbf{x}) \leq 0, \qquad i = 1, 2, \ldots, m \end{aligned} \tag{3-106}$$

which can also be written as

$$\begin{aligned} \text{maximize} \quad & Z(\mathbf{x}) \\ \text{s.t.} \quad & \mathbf{x} \in \mathbf{X} \end{aligned} \tag{3-107}$$

where $\mathbf{X} = \{\mathbf{x}/g_i(\mathbf{x}) \leq 0, i = 1, 2, \ldots, m\}$.

The Kuhn–Tucker conditions will be presented in a manner similar to that in Hillier and Lieberman (1967, pp. 574–578), Wagner (1969, pp. 600–602), and Zangwill (1969, pp. 36–44). The conditions state that if $\mathbf{x}^*$ is an optimal solution to (3-107), then there exists $\mathbf{u}$ with $u_i \geq 0 \; \forall i = 1, 2, \ldots, m$ and

$$\mathbf{x}^* \in \mathbf{X} \tag{3-108}$$

$$u_i g_i(\mathbf{x}^*) = 0, \qquad i = 1, 2, \ldots, m \tag{3-109}$$

$$\nabla Z(\mathbf{x}^*) - \sum_{i=1}^{m} u_i \nabla g_i(\mathbf{x}^*) = 0 \tag{3-110}$$

† Readers without prior training in linear algebra and linear programming should consider skipping this section. The material is not essential for most of the material in later chapters.

where all functions are differentiable and the "constraint qualification" holds [see Zangwill (1969, pp. 39–40)].

Condition (3-108) requires $\mathbf{x}^*$ to be feasible, i.e., $\mathbf{x}^*$ must satisfy all constraints, while (3-109) is a statement of complementary slackness [see (3-94)]. Condition (3-110) relates the gradient of the objective function at $\mathbf{x}^*$ to the negative of the gradients of the binding constraints evaluated at $\mathbf{x}^*$, where

$$\nabla = \left[\frac{\partial}{\partial x_1}, \frac{\partial}{\partial x_2}, \ldots, \frac{\partial}{\partial x_n}\right].$$

The condition in (3-110) implies that movement from $\mathbf{x}^*$ along any direction that increases the value of the objective function must be infeasible and, further, that any move in a feasible direction cannot result in an increase in the objective function. That is, the direction of improvement must be opposite from the direction of feasibility. The minus sign on the second term of (3-110) results from our use of "$\leq$" constraints in (3-105). Thus, $-\nabla g_i(\mathbf{x})$ points toward feasibility.

The conditions in (3-108)–(3-110) are necessary for a constrained optimum. They are also sufficient if, in addition to the requirements mentioned above, the objective function is concave and $\mathbf{X}$ is a convex set. For a linear programming problem the conditions are necessary and sufficient.

The characterization of an optimal solution by the Kuhn–Tucker conditions and their interpretation can be demonstrated with an example. The Airtight example problem can be rewritten in the form of (3-106) as

$$\begin{aligned} \text{maximize} \quad Z(\mathbf{x}) &= x_1 + x_2 \\ \text{s.t.} \quad g_1(\mathbf{x}) &= 2x_1 + 3x_2 \leq 12 \\ g_2(\mathbf{x}) &= 6x_1 + 3x_2 \leq 24 \\ g_3(\mathbf{x}) &= -x_1 \leq 0 \\ g_4(\mathbf{x}) &= -x_2 \leq 0 \end{aligned} \tag{3-111}$$

where the constraint set $g_i(\mathbf{x})$, $i = 1, 2, 3, 4$, is represented by $\mathbf{X}$ so that a solution $\mathbf{x}$ that is feasible implies $\mathbf{x} = (x_1, x_2) \in \mathbf{X}$. Notice that a different objective function ($c_2 = 1$) is used in (3-111).

The constraint set and the optimal solution $\mathbf{x}^*$ are shown in Fig. 3-5. The optimal solution (3, 2) satisfies the first Kuhn–Tucker condition (3-108) because it is feasible. The third condition (3-110) can be written for the example as

$$\nabla Z(\mathbf{x}^*) - u_1 \nabla g_1(\mathbf{x}^*) - u_2 \nabla g_2(\mathbf{x}^*) - u_3 \nabla g_3(\mathbf{x}^*) - u_4 \nabla g_4(\mathbf{x}^*) = 0 \tag{3-112}$$

where $u_i \geq 0$ for $i = 1, 2, 3, 4$.

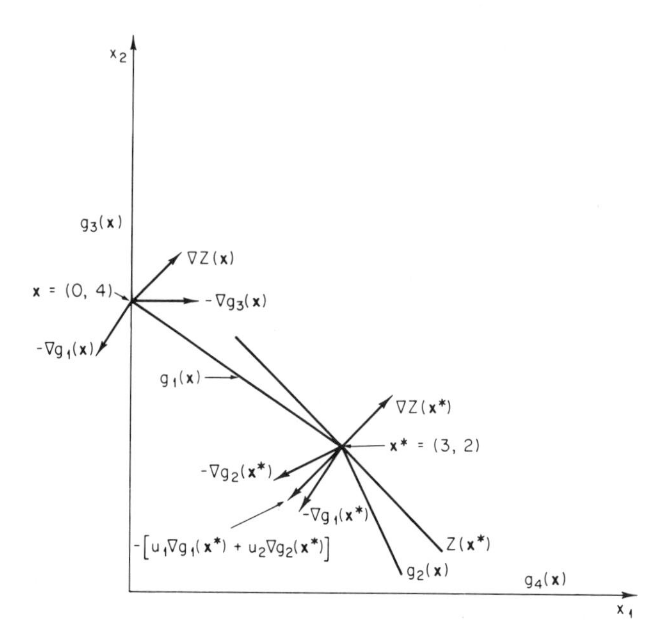

Fig. 3-5. The Kuhn–Tucker conditions for the Airtight problem.

Complementary slackness (3-109) implies that $u_i = 0$ when the ith constraint is not binding. Figure 3-5 shows that only $g_1(\mathbf{x})$ and $g_2(\mathbf{x})$ are binding at $\mathbf{x}^*$, so that

$$u_3 = u_4 = 0 \tag{3-113}$$

and (3-112) becomes

$$\nabla Z(\mathbf{x}^*) - [u_1 \nabla g_1(\mathbf{x}^*) + u_2 \nabla g_2(\mathbf{x}^*)] = 0 \tag{3-114}$$

The geometric interpretation of this derived condition is provided by Fig. 3-5. The condition implies that one can form a nonnegative linear combination of the negative of the gradients of the binding constraints at $\mathbf{x}^*$ that is equal in magnitude and opposite in direction to the gradient of the objective function. This linear combination is represented by the bracketed term in (3-114), where u_1 and u_2 are strictly positive, and is drawn in Fig. 3-5.

Using the data supplied in (3-111), Eq. (3-114) becomes

$$(1, 1)^t - u_1(2, 3)^t - u_2(6, 3)^t = 0 \tag{3-115}$$

which gives the two equations

$$2u_1 + 6u_2 = 1, \qquad 3u_1 + 3u_2 = 1 \tag{3-116}$$

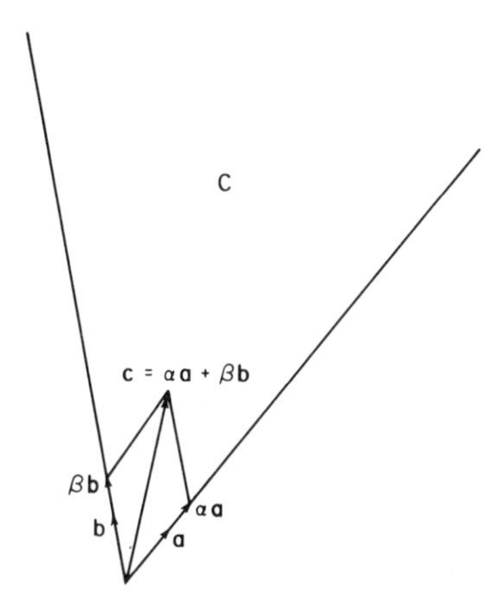

Fig. 3-6. Definition of a cone.

Solving (3-116) gives $u_1 = \frac{1}{4}$ and $u_2 = \frac{1}{12}$. This also shows that there exists a nonnegative linear combination that satisfies the third condition, (3-114).

It is interesting to consider a feasible point other than the optimal solution, such as $\mathbf{x} = (0, 4)$ in Fig. 3-5. At this point, a nonnegative linear combination of $-\nabla g_1(\mathbf{x})$ and $-\nabla g_3(\mathbf{x})$ that is equal and opposite to $\nabla Z(\mathbf{x})$ cannot be formed. Condition (3-110) does not hold here or at any other nonoptimal point.

For the subsequent development of the Kuhn–Tucker conditions for multiobjective problems it is worth introducing the geometrical concept of a cone. A cone generated by the vectors **a** and **b** is the set of all vectors such that

$$\mathbf{C} = \{\mathbf{c} \mid \mathbf{c} = \alpha\mathbf{a} + \beta\mathbf{b} \ \forall \alpha, \beta \geq 0\} \tag{3-117}$$

A cone **C** is shown in Fig. 3-6. A cone consists of the rays in the directions of **a** and **b** and all rays "in between" **a** and **b**.

Another interpretation of the third Kuhn–Tucker condition (3-110) can be stated with the definition of a cone as in Zangwill (1969, p. 41). At the optimal solution, the gradient of the objective function must lie in the cone generated by the gradients of the binding constraints. Or, put in the manner we have seen before, the negative of $\nabla Z(\mathbf{x}^*)$ must lie in the cone generated by the negative of the gradients of the binding constraints. This is just another way of saying that there exists a nonnegative linear combination of the binding constraint gradients equal and opposite to $\nabla Z(\mathbf{x}^*)$ since $\mathbf{c} \in \mathbf{C}$, from (3-117), is just a nonnegative linear combination of the cone-generating constraints.

CHAPTER 4

Formulation of the General Multiobjective Programming Problem

In this chapter, the general form of the multiobjective programming or vector optimization problem is introduced. The notation and terminology that are presented will be used throughout the remainder of the book.

4.1 FORMULATION OF THE PROBLEM

Multiobjective programming deals with optimization problems with two or more objective functions. The multiobjective programming problem differs from the classical (single-objective) optimization problem only in the expression of their respective objective functions. The general single-objective optimization problem (see Chapter 3) with n decision variables and m constraints is

$$\text{maximize} \quad Z(x_1, x_2, \ldots, x_n) \tag{4-1}$$

$$\text{s.t.} \quad g_i(x_1, x_2, \ldots, x_n) \leq 0, \qquad i = 1, 2, \ldots, m \tag{4-2}$$

$$x_j \geq 0, \qquad j = 1, 2, \ldots, n \tag{4-3}$$

The general multiobjective optimization problem with n decision variables, m constraints and p objectives is

$$\begin{aligned} \text{maximize} \quad & \mathbf{Z}(x_1, x_2, \ldots, x_n) \\ & = [Z_1(x_1, x_2, \ldots, x_n), \\ & \quad Z_2(x_1, x_2, \ldots, x_n), \\ & \quad \ldots, Z_p(x_1, x_2, \ldots, x_n)] \end{aligned} \tag{4-4}$$

$$\begin{aligned} \text{s.t.} \quad g_i(x_1, x_2, \ldots, x_n) \leq 0, & \qquad i = 1, 2, \ldots, m \\ x_j \geq 0, & \qquad j = 1, 2, \ldots, n \end{aligned} \tag{4-5}$$

where $\mathbf{Z}(x_1, x_2, \ldots, x_n)$ is the multiobjective objective function and $Z_1(\quad)$, $Z_2(\quad), \ldots, Z_p(\quad)$ are the p individual objective functions. Note that the individual objective functions are merely listed in (4-4); they are not added, multiplied, or combined in any way.

4.2 NONINFERIORITY

In single-objective problems the goal of solution is the identification of the optimal solution: the feasible solution (or solutions) that gives the best value of the objective function. Notice that even when there are alternative optima (as in Fig. 3-3) the optimal value of the objective function is unique. This notion of optimality must be dropped for multiobjective problems because a solution which maximizes one objective will not, in general, maximize any of the other objectives. What is optimal in terms of one of the p objectives is usually nonoptimal for the other $p - 1$ objectives.

Optimality plays an important role in the solution of single-objective problems. It allows the analyst and decision makers to restrict their attention to a single solution or a very small subset of solutions from among the much larger set of feasible solutions. A new concept called noninferiority will serve a similar but less limiting purpose for multiobjective problems. We will restrict our attention to noninferior solutions.

The idea of noninferiority is very similar to the concept of dominance. Noninferiority is called "nondominance" by some mathematical programmers, "efficiency" by others and by statisticians and economists, and "Pareto optimality" by welfare economists. Suppose three solutions in a two-objective problem are given as in Table 4-1. Alternative C is dominated by A and B because both of these alternatives give more of *both* objectives, Z_1 and Z_2. A solution that is dominated in this way is termed *inferior*. Solutions that are not dominated are called *noninferior*. Thus, for example, alternatives A and B

TABLE 4-1

AN EXAMPLE OF NONINFERIORITY

Alternative	Z_1	Z_2	
A	10	11	Noninferior
B	12	10	Noninferior
C	9	8	Inferior

in Table 4-1 are noninferior. To get a bit more formal, noninferiority can be defined in the following way:

> A feasible solution to a multiobjective programming problem is noninferior if there exists no other feasible solution that will yield an improvement in one objective without causing a degradation in at least one other objective.

This definition is easiest to understand graphically.

An arbitrary collection of feasible alternatives for a two-objective maximization problem is shown in Fig. 4-1. The area inside of the shape and its boundaries are feasible. Notice that the axes of this graph are the objectives Z_1 and Z_2. This plot is referred to as the *objective space* to distinguish from

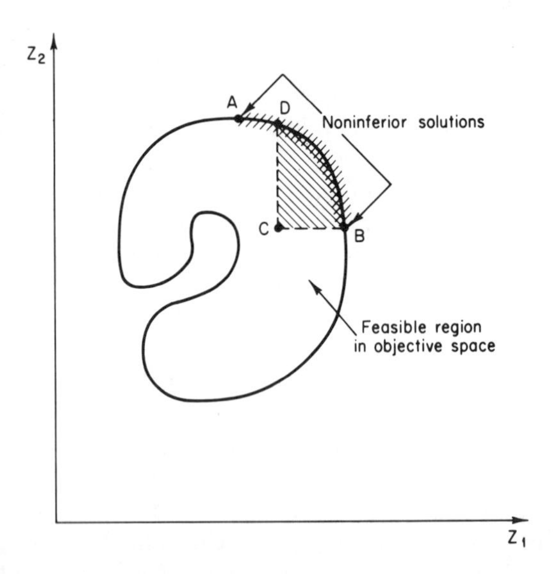

Fig. 4-1. Graphical interpretation of noninferiority for an arbitrary feasible region in objective space.

plots defined over the decision variables or *decision space*. The feasible area in Fig. 4-1 is called the feasible region in objective space.

Now the definition of noninferiority can be used to find noninferior solutions in Fig. 4-1. First, all interior solutions must be inferior for one can always find a feasible solution which leads to an improvement in both objectives simultaneously. Consider interior point C in Fig. 4-1, which is inferior, i.e., not noninferior. Alternative B gives more Z_1 than does C without decreasing the amount of Z_2. Similarly, D gives more Z_2 without decreasing Z_1. In fact, any alternative in the shaded area to the "northeast" of C in Fig. 4-1 dominates alternative C. There is a general rule here: When all objectives are to be maximized, a feasible solution is noninferior if there are no feasible solutions lying to the northeast, i.e., in an area such as the shaded part of Fig. 4-1. We will call this the *northeast rule*.

Applying the northeast rule to the rest of the feasible region in Fig. 4-1 leads to the conclusion that any point on the boundary that is not on the northeastern side of the feasible region is inferior. The noninferior solutions for the feasible region in Fig. 4-1 are found in the crosshatched portion of the boundary between points A and B.

Another set of feasible solutions is given in Table 4-2. The five alternatives A–E are the only feasible solutions; if there were more feasible alternatives, no conclusions could be drawn as to the noninferiority of the alternatives. An alternative must not be dominated by any feasible solution to be called noninferior. The value of each of three objectives, Z_1, Z_2, and Z_3, is given for each alternative in Table 4-2. To determine the noninferiority of an alternative, its values of the three objectives are compared with the objective values for each of the other alternatives. Consider alternative A: B and D give more Z_2 but less Z_1 and Z_3; C gives less of all three objectives; and E yields more Z_2 and Z_3 but less Z_1. Alternative A is therefore noninferior. Comparing each alternative with each of the others leads to the conclusion that C is inferior (it is dominated by A), and A, B, D, and E are noninferior. Notice that the graphical presentation for the two-objective problem in Fig. 4-1 allows a

TABLE 4-2

A THREE-OBJECTIVE EXAMPLE

Alternative	Z_1	Z_2	Z_3	
A	5	8	7	Noninferior
B	4	9	2	Noninferior
C	4	4	4	Inferior
D	3	10	6	Noninferior
E	2	9	8	Noninferior

more rapid comparison of alternatives in that the search over all feasible alternatives can be accomplished at a glance. The value of graphical presentations is diminished when there are more than two objectives.

4.3 TERMINOLOGY FOR MULTIOBJECTIVE PROGRAMMING

Noninferiority is the major new concept of multiobjective programming. There are some other terms, however, that will be used throughout the remainder of this book. These new terms will be introduced by way of an example.

The example problem we shall use here and in later chapters has two objectives and two decision variables:

$$\text{maximize} \quad \mathbf{Z}(x_1, x_2) = [Z_1(x_1, x_2), Z_2(x_1, x_2)] \tag{4-7}$$

where

$$Z_1(x_1, x_2) = 5x_1 - 2x_2$$
$$Z_2(x_1, x_2) = -x_1 + 4x_2$$

$$\text{s.t.} \quad -x_1 + x_2 \le 3, \qquad x_1 + x_2 \le 8$$
$$x_1 \le 6, \qquad x_2 \le 4,$$
$$x_1, x_2 \ge 0$$

The feasible region for this problem is drawn in Fig. 4-2. The axes of the graph in Fig. 4-2 are defined for x_1 and x_2, the decision variables. The area

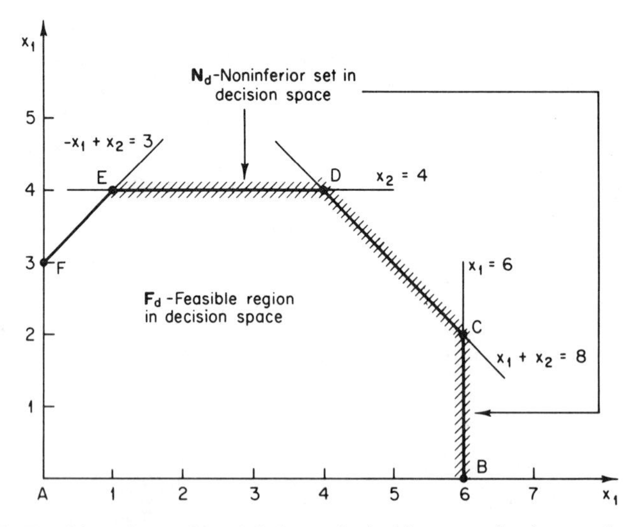

Fig. 4-2. Feasible region and noninferior set in decision space for the sample problem.

that is spanned by these axes will be called the decision space. The feasible region is the enclosed area labeled $\mathbf{F}_d$ in Fig. 4-2 (**F** for feasible, d for decision). $\mathbf{F}_d$ will be called the feasible region in decision space.

If this were a single-objective problem, we could proceed directly to the optimal solution by finding the feasible extreme point that gives the highest value of the objective function. This cannot be done for the current problem, however, because there are two objective functions. With which objective function should the desirability of an extreme-point solution be measured?

We shall concentrate on extreme points nevertheless. The six feasible extreme points of $\mathbf{F}_d$ are labeled A–F in Fig. 4-2. They are listed in Table 4-3 along with the value each extreme point yields for the two decision variables x_1 and x_2 and the two objectives Z_1 and Z_2. Notice that although the decision variables must be nonnegative, the objective functions may assume negative values.

The evaluation of Z_1 and Z_2 at an extreme point of the decision space yields an extreme point in a new space, the objective space. The objective space is defined by axes which correspond to the objectives. A plot in objective space is drawn in Fig. 4-3. The values of Z_1 and Z_2 at each of the extreme points A–F listed in Table 4-3 are plotted in Fig. 4-3. The plotted points in the objective space are images of corresponding points in decision space. Thus, point A in Fig. 4-2 leads to point A in Fig. 4-3 through the values of Z_1 and Z_2 that A yields. Notice that adjacent extreme points in Fig. 4-2 are still adjacent in Fig. 4-3.

The points A–F in Fig. 4-3 are connected by straight lines, which are images of the boundaries of the feasible region in decision space. Thus, the line which connects A and B in Fig. 4-2 becomes the line between A and B in Fig. 4-3 through the objective function evaluations. The result of connecting adjacent extreme points with lines leads to $\mathbf{F}_o$, the feasible region in objective space, in Fig. 4-3. It is important to understand the intimate relationship

TABLE 4-3

VALUES OF THE DECISION VARIABLES AND THE OBJECTIVE FOR THE SAMPLE PROBLEM

Extreme point	x_1	x_2	$Z_1(x_1, x_2) = 5x_1 - 2x_2$	$Z_2(x_1, x_2) = -x_1 + 4x_2$
A	0	0	0	0
B	6	0	30	−6
C	6	2	26	2
D	4	4	12	12
E	1	4	−3	15
F	0	3	−6	12

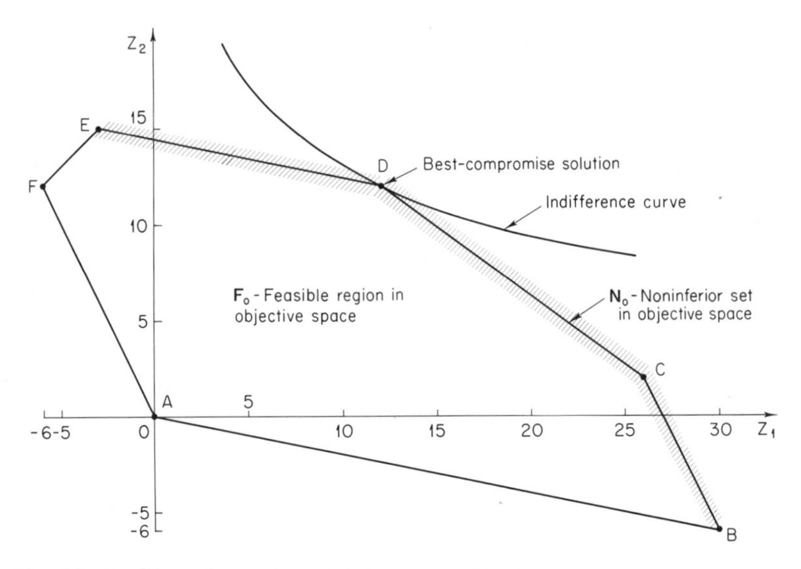

Fig. 4-3. Feasible region and noninferior set in objective space and a best-compromise solution for the sample problem.

between $\mathbf{F}_d$ and $\mathbf{F}_o$. The latter is a transformation of the former, where the particular shape of $\mathbf{F}_o$ depends on the objective functions since they serve as "mapping functions."

At this point the definition of noninferiority can be applied to $\mathbf{F}_o$ in order to find all noninferior solutions which we will call the noninferior set. Using the northeast rule, the noninferior set in objective space $\mathbf{N}_o$ is found to be the crosshatched portion of the boundary of $\mathbf{F}_o$ in Fig. 4-3. The extreme points B, C, D, and E and all of the solutions on the lines connecting them are noninferior. It follows directly from the relationship between $\mathbf{F}_d$ and $\mathbf{F}_o$ that the noninferior set in decision space $\mathbf{N}_d$ is the crosshatched portion of the boundary of $\mathbf{F}_d$ in Fig. 4-2. That is, if points B, C, D, and E are noninferior in objective space, they are still noninferior in decision space. A note of caution here: The northeast rule for finding noninferior solutions is applicable to the objective space *only*; the rule cannot be used in decision space.

The final bits of terminology are the notions of tradeoffs and the best-compromise solution. The noninferior set contains solutions that are not dominated by any other feasible solutions. Solutions in the noninferior set are not comparable. For example, point C gives 26 units of objective Z_1 and only two of objective Z_2, while D gives 12 units of each objective. Which is better? Is it worth giving up 14 units of Z_2 to gain 10 units of Z_1 in moving from D to C? The amount of one objective that must be sacrificed to gain an increase in the other objective is called a *tradeoff*. For the situations just cited, the tradeoff between Z_1 and Z_2 in moving from D to C is $\frac{14}{10}$ or $\frac{7}{5}$, i.e., $\frac{7}{5}$ units of Z_2 must be given up for every unit gained of Z_1. This could have been turned around. The tradeoff is also $\frac{5}{7}$ in that one unit of Z_2 must be given up for every $\frac{5}{7}$ of Z_1 gained. Alternatively, one could look at it in the other direction by moving from C to D and thereby sacrificing Z_1 to gain Z_2. The direction and the way in which the tradeoff is measured do not matter. It is only important to be clear and consistent when a tradeoff is stated.

It is characteristic of the noninferior set that the objectives must be traded off against each other in moving from one noninferior alternative to another. (The alternatives would not be noninferior if the tradeoff could be avoided.) The noninferior set and the tradeoffs are important information for decision making.

The noninferior set generally includes many alternatives, all of which obviously cannot be selected. The noninferior solution that is selected as the preferred alternative is called the *best-compromise solution.* Some would call this solution the optimal alternative, but this term will not be used here because the selected alternative is optimal on the basis of a single set of preferences: Another person would perhaps choose a different alternative. The term optimality will be reserved for that solution which is best in an objective manner (i.e., without the introduction of preferences) or as the result of solving an optimization problem.

How is the best-compromise solution chosen? One way is simply to consider a graphical representation of the noninferior set and then choose on the basis of the possibilities and the tradeoffs. This is a good approach. In fact, the methods in Chapter 6 are motivated by just this approach to decision making. The methods in Chapters 7 and 8 rely on a more formal characterization of preferences.

The two objectives in Fig. 4-3 can be thought of as commodities—apples and oranges or guns and butter. In this context, the noninferior set is then nothing more than a production possibility frontier and noninferior solutions are dominant in the same sense that efficient production alternatives dominate inefficient ones. (Indeed, the noninferior set is frequently referred to as the "net benefits transformation curve.") The selection of a combination of guns and butter or Z_1 and Z_2 depends on the preferences for the two commodities. The preferences can be represented as a family of indifference curves, each of which shows those combinations of Z_1 and Z_2 that lead to a given amount of utility. Indifference curves that yield higher levels of utility are found to the "northeast" in a plot of Z_1 and Z_2 since more of both objectives (commodities) is preferred to less. The preferred alternative or the best-compromise solution is the feasible solution that yields the greatest utility. The best-compromise solution is therefore found at the point where an indifference curve is tangent to the noninferior set (see Fig. 4-3). That the best-compromise solution is always noninferior follows from the nondominance of noninferior alternatives and from the desirability of having more of both objectives. We shall devote much more attention to utility functions and indifference curves in Chapter 7.

*4.4 MATHEMATICAL RESTATEMENT OF NONINFERIORITY[†]

In the discussion that follows, certain conventions are observed: Components of a vector are distinguished by subscripts and different vectors are distinguished by superscripts.

The single-objective or scalar (or unidimensional) optimization problem may be written as

$$\begin{aligned} \text{maximize} \quad & Z(\mathbf{x}) \\ \text{s.t.} \quad & g_i(\mathbf{x}) \leq 0, \qquad i = 1, 2, \ldots, m \\ & \mathbf{x} \geq 0 \end{aligned} \tag{4-8}$$

or

$$\begin{aligned} \text{maximize} \quad & Z(\mathbf{x}) \\ \text{s.t.} \quad & \mathbf{x} \in \mathbf{F}_\text{d} \end{aligned} \tag{4-9}$$

where

$$\mathbf{F}_\text{d} = \{\mathbf{x} \,|\, g_i(\mathbf{x}) \leq 0, i = 1, 2, \ldots, m; \mathbf{x} \geq 0\} \tag{4-10}$$

The multiobjective or vector optimization problem may be written as

$$\begin{aligned} \text{maximize} \quad & \mathbf{Z}(\mathbf{x}) = [Z_1(\mathbf{x}), Z_2(\mathbf{x}), \ldots, Z_p(\mathbf{x})] \\ \text{s.t.} \quad & \mathbf{x} \in \mathbf{F}_\text{d} \end{aligned} \tag{4-11}$$

in which the objective function is now a p-dimensional vector.

Scalar optimization problems are characterized by a complete ordering of their feasible solutions. Any two feasible solutions $\mathbf{x}^1$ and $\mathbf{x}^2$ to a unidimensional problem are comparable in terms of the objective function; i.e., either $Z(\mathbf{x}^1) > Z(\mathbf{x}^2)$, $Z(\mathbf{x}^1) = Z(\mathbf{x}^2)$, or $Z(\mathbf{x}^1) < Z(\mathbf{x}^2)$. This comparison can be made for all feasible solutions, i.e., all $\mathbf{x} \in \mathbf{F}_\text{d}$, and the solution $\mathbf{x}^*$ for which there exists no $\mathbf{x} \in \mathbf{F}_\text{d}$ such that $Z(\mathbf{x}) > Z(\mathbf{x}^*)$ is called an optimal solution.

Vector optimization problems are characterized by a *partial* ordering of alternatives. In general, it is not possible to compare all feasible solutions because the comparison on the basis of one objective function may contradict the comparison based on another objective function. Suppose there are two

[†] It is suggested that readers without prior training in linear algebra skip this and subsequent starred sections.

objective functions, $\mathbf{Z}(\mathbf{x}) = [Z_1(\mathbf{x}), Z_2(\mathbf{x})]$, and two solutions $\mathbf{x}^1$ and $\mathbf{x}^2$ are to be compared. That is, the two vector quantities $\mathbf{Z}(\mathbf{x}^1)$ and $\mathbf{Z}(\mathbf{x}^2)$ must be compared and if $\mathbf{Z}(\mathbf{x}^1) > \mathbf{Z}(\mathbf{x}^2)$, then $\mathbf{x}^1$ is better than $\mathbf{x}^2$. Now, since $\mathbf{Z}(\mathbf{x}^1) = [Z_1(\mathbf{x}^1), Z_2(\mathbf{x}^1)]$ and $\mathbf{Z}(\mathbf{x}^2) = [Z_1(\mathbf{x}^2), Z_2(\mathbf{x}^2)]$, $\mathbf{x}^1$ is better than $\mathbf{x}^2$ if and only if

$$Z_1(\mathbf{x}^1) > Z_1(\mathbf{x}^2) \qquad \text{and} \qquad Z_2(\mathbf{x}^1) \geq Z_2(\mathbf{x}^2)$$

or

$$Z_1(\mathbf{x}^1) \geq Z_1(\mathbf{x}^2) \qquad \text{and} \qquad Z_2(\mathbf{x}^1) > Z_2(\mathbf{x}^2)$$

because the comparison is of two vector quantities on a component-by-component basis. Suppose $Z_1(\mathbf{x}^1) > Z_1(\mathbf{x}^2)$ and $Z_2(\mathbf{x}^1) < Z_2(\mathbf{x}^2)$; then nothing can be said about the two solutions $\mathbf{x}^1$ and $\mathbf{x}^2$; i.e., they are incomparable. This is what is meant by a partial ordering: All solutions are not comparable on the basis of the objective functions.

Since a complete order is not available, the notion of optimality must be dropped. Of course, there may be some cases in which the solutions may be completely ordered. For example, given the three solutions $\mathbf{x}^1, \mathbf{x}^2, \mathbf{x}^3$ with the corresponding values of a two-dimensional objective function of $\mathbf{Z}(\mathbf{x}^1) = (5, 5)$, $\mathbf{Z}(\mathbf{x}^2) = (3, 7)$ and $\mathbf{Z}(\mathbf{x}^3) = (6, 8)$, then $\mathbf{x}^3$ is the optimal solution to this problem. In general, this is a very uncommon occurrence because most problems have conflicting objectives; otherwise, there would be little reason for formulating a multiobjective problem. In most problems, then, an optimal solution in the sense of a complete order on the basis of the objective functions will not be available.

The partial ordering in a vector optimization does allow some feasible solutions to be eliminated: Inferior solutions, those that are dominated by at least one feasible solution, may be dropped. Noninferior solutions are the alternatives of interest. A solution $\mathbf{x}$ is noninferior if there exists no feasible $\mathbf{y}$ such that

$$\mathbf{Z}(\mathbf{y}) \geq \mathbf{Z}(\mathbf{x}) \tag{4-12}$$

i.e.,

$$Z_k(\mathbf{y}) \geq Z_k(\mathbf{x}), \qquad k = 1, 2, \ldots, p \tag{4-13}$$

where (4-13) is satisfied as a strict inequality for at least one k. If such a feasible solution $\mathbf{y}$ exists, then $\mathbf{x}$ is inferior.

*4.5 KUHN–TUCKER CONDITIONS FOR NONINFERIOR SOLUTIONS

In addition to developing optimality conditions for scalar optimization problems, Kuhn and Tucker (1951) also stated conditions for noninferiority in vector optimization problems. In this section the conditions are presented.

A derivation of the third condition, which yields a graphical interpretation similar to the one offered in Chapter 3 for the scalar conditions, is given.

Given a vector maximization problem as in (4-11), if a solution $\mathbf{x}^*$ is noninferior, then there exist multipliers $u_i \geq 0$, $i = 1, 2, \ldots, m$ and $w_k \geq 0$, $k = 1, 2, \ldots, p$ and

$$\mathbf{x}^* \in \mathbf{F}_d \tag{4-14}$$

$$u_i g_i(\mathbf{x}^*) = 0, \qquad i = 1, 2, \ldots, m \tag{4-15}$$

$$\sum_{k=1}^{p} w_k \nabla Z_k(\mathbf{x}^*) - \sum_{i=1}^{m} u_i \nabla g_i(\mathbf{x}^*) = 0 \tag{4-16}$$

The Kuhn–Tucker conditions for noninferiority, (4-14)–(4-16) differ from the optimality conditions, (3-108)–(3-110), only in the last condition. The first term of (3-110) has been replaced by a nonnegative linear combination of the gradients of the p objective functions. The first condition, (4-14), requires feasibility and the second, (4-15), is a statement of complementary slackness.

The conditions in (4-14)–(4-16) are necessary for noninferiority. They are also sufficient if the $Z_k(\mathbf{x})$ are concave for $k = 1, 2, \ldots, p$, $\mathbf{F}_d$ is convex, and $w_k > 0$ for all k. This latter requirement for sufficiency ($w_k > 0$) will be discussed further in Chapter 6 when the weighting method is presented.

*4.5.1 Derivation and Interpretation of the Third Kuhn–Tucker Condition for Noninferiority

There are, in general, several noninferior solutions for a vector optimization problem. The noninferior set in decision space $\mathbf{N}_d$ is contained in the feasible region in decision space $\mathbf{F}_d$, which is in turn contained in the n-dimensional Euclidean vector space, i.e., $\mathbf{N}_d \subseteq \mathbf{F}_d \subseteq \mathbf{R}^n$. The sets $\mathbf{N}_d$ and $\mathbf{F}_d$ are mapped by the p-dimensional objective function into the noninferior set in objective space $\mathbf{N}_o$ and the feasible region in objective space $\mathbf{F}_o$, respectively. Furthermore, $\mathbf{N}_o \subseteq \mathbf{F}_o \subseteq \mathbf{R}^p$, where $\mathbf{R}^p$ is the p-dimensional Euclidean vector space.

Following Zadeh (1963), three disjoint subsets of the decision space $\mathbf{R}^n$ are defined relative to a feasible solution $\mathbf{x}$: $\mathbf{Q}^{>}(\mathbf{x})$, the set of all vectors in $\mathbf{R}^n$ that are superior to $\mathbf{x}$; $\mathbf{Q}^{\leq}(\mathbf{x})$, the set of all vectors in $\mathbf{R}^n$ inferior or equal to $\mathbf{x}$; and $\mathbf{Q}^{\sim}(\mathbf{x})$, the set of all vectors in $\mathbf{R}^n$ that are not comparable to $\mathbf{x}$ on the basis of the partial order implied by the objective function. Note that $\mathbf{x} \in \mathbf{Q}^{\leq}(\mathbf{x})$ only; it is not an element of either $\mathbf{Q}^{>}(\mathbf{x})$ or $\mathbf{Q}^{\sim}(\mathbf{x})$. Zadeh points out that since every vector in $\mathbf{R}^n$ must fall into one of the three subsets, the union of the subsets must equal the entire space

$$\mathbf{R}^n = [\mathbf{Q}^{>}(\mathbf{x})] \cup [\mathbf{Q}^{\leq}(\mathbf{x})] \cup [\mathbf{Q}^{\sim}(\mathbf{x})] \tag{4-17}$$

for any $\mathbf{x} \in \mathbf{R}^n$.

The definition of noninferiority can be restated in terms of these subsets. $\mathbf{x}^* \in \mathbf{F}_d \subseteq \mathbf{R}^n$ is noninferior if the intersection of $\mathbf{F}_d$ and $\mathbf{Q}^{>}(\mathbf{x}^*)$ is empty [where the set $\mathbf{Q}^{>}(\mathbf{x}^*)$ does not include $\mathbf{x}^*$ as an element]. In other words, a feasible solution is noninferior if there is no feasible solution that is superior to it.

For linear multiobjective programming problems, the objectives can be expressed in the form

$$Z_k(\mathbf{x}) = (\mathbf{c}^k)^t\mathbf{x} = \sum_{j=1}^{n} c_j^k x_j, \qquad k = 1, 2, \ldots, p \tag{4-18}$$

where the superscript t denotes transpose. Zadeh points out that in these cases, the subset $\mathbf{Q}^{>}(\mathbf{x})$ is the polar cone of the cone spanned by the objective function gradients $\mathbf{c}^k$ with origin at $\mathbf{x}$. The polar cone $\mathbf{A}^*$ of a cone $\mathbf{A}$ is the set of all vectors that make angles with all vectors of $\mathbf{A}$ of less than or equal to 90°. An example of a polar cone is shown in Fig. 4-4.

We will modify Zadeh's definition of $\mathbf{Q}^{>}(\mathbf{x})$ slightly. $\mathbf{Q}^{>}(\mathbf{x})$ will be defined as the *interior* of the polar cone of the cone spanned by the objective function gradients. More formally,

$$\mathbf{Q}^{>}(\mathbf{x}) = \{\mathbf{h} \mid \mathbf{h}^t\mathbf{c}^k > 0;\ k = 1, 2, \ldots, p\} \tag{4-19}$$

where $\mathbf{h}$ is an n-dimensional column vector with origin at $\mathbf{x}$.

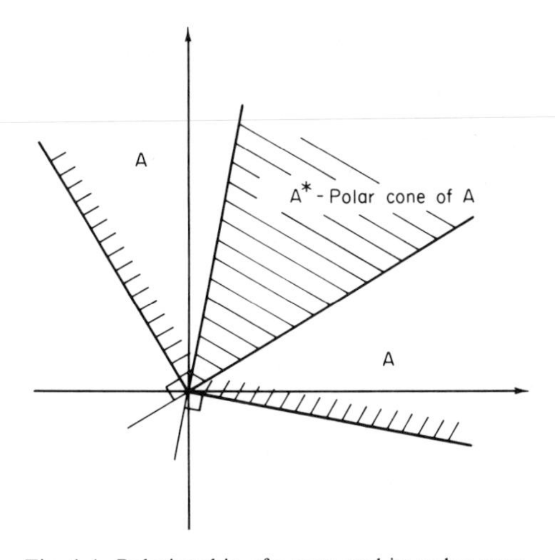

Fig. 4-4. Relationship of a cone and its polar cone.

The significance of the set $\mathbf{Q}^{>}(\mathbf{x})$ and its use in characterizing noninferior solutions can be demonstrated with the previous example problem given in (4-7).

First, notice that the set $\mathbf{Q}^{>}(\mathbf{x})$ indicates directions in which to move from $\mathbf{x}$ that lead to a simultaneous improvement of all objectives. When the objective function gradients are close together there are many simultaneously good directions, and $\mathbf{Q}^{>}(\mathbf{x})$ is relatively large. The size of $\mathbf{Q}^{>}(\mathbf{x})$ is a measure of the degree of conflict among the objectives: Small $\mathbf{Q}^{>}(\mathbf{x})$ means extensive conflict; large $\mathbf{Q}^{>}(\mathbf{x})$ means little conflict. The feasible region in decision space for the example of Fig. 4-2 is repeated in Fig. 4-5. Also shown is the cone $\mathbf{Q}^{>}(\mathbf{x}^*)$ with vertex at $\mathbf{x}^* = (1, 4)$, which was found by using the definition in (4-19) with gradients of the objective function $\mathbf{c}^1$ and $\mathbf{c}^2$.

By the definition of a noninferior point, $\mathbf{x}^* = (1, 4)$ is noninferior because the intersection of $\mathbf{Q}^{>}(\mathbf{x}^*)$ and $\mathbf{F}_\text{d}$ is empty, as shown in Fig. 4-5, since the cone does not include its vertex, i.e., $\mathbf{x}^* \notin \mathbf{Q}^{>}(\mathbf{x}^*)$. Zadeh also indicated that if $\mathbf{F}_\text{d}$ is a closed convex set, interior points of $\mathbf{F}_\text{d}$ cannot be noninferior since $\mathbf{Q}^{>}(\mathbf{x}) \cap \mathbf{F}_\text{d}$ is never empty for interior points in such a set. In a linear problem the shape of $\mathbf{Q}^{>}(\mathbf{x})$ does not change from one $\mathbf{x}$ to another; only its origin changes. Thus, one can find noninferior points at a glance by moving $\mathbf{Q}^{>}(\mathbf{x})$ to an $\mathbf{x}$ of interest. This applies, of course, to problems with only two decision variables.

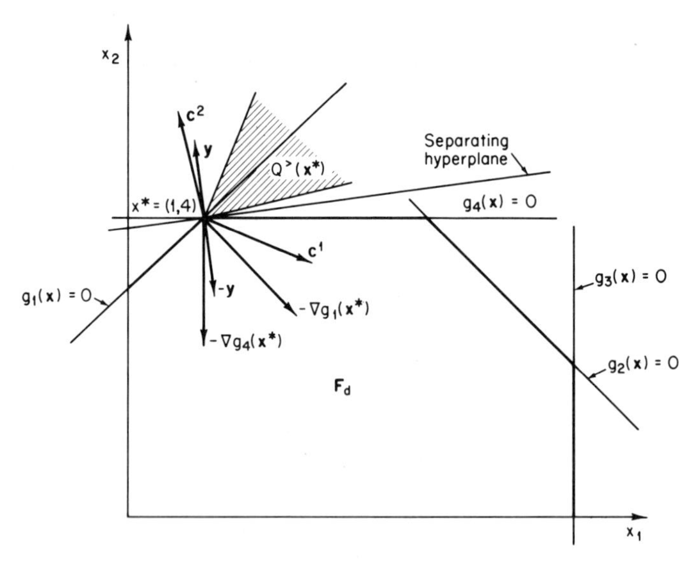

Fig. 4-5. Characterization of a noninferior solution for the multiobjective programming problem.

Further characterization of the noninferior solution, as shown in Fig. 4-5, can be developed. Zadeh (1963, p. 59) stated that when $\mathbf{F}_d$ is convex, noninferior solutions are those points $\mathbf{x}^*$ on the boundary of $\mathbf{F}_d$ through which hyperplanes that separate $\mathbf{F}_d$ and $\mathbf{Q}^>(\mathbf{x}^*)$ can be passed. Of course, this statement follows from the emptiness of the intersection of $\mathbf{F}_d$ and $\mathbf{Q}^>(\mathbf{x}^*)$. A hyperplane that separates $\mathbf{F}_d$ and $\mathbf{Q}^>(\mathbf{x}^*)$ and passes through the point (1, 4) is shown in Fig. 4-5. The normal to the separating hyperplane, which is directed away from the interior of $\mathbf{F}_d$, is the vector $\mathbf{y}$ in Fig. 4-5. Since the hyperplane separates $\mathbf{Q}^>(\mathbf{x}^*)$ and $\mathbf{F}_d$, $\mathbf{y}$ lies in the same half-space as $\mathbf{Q}^>(\mathbf{x}^*)$, which implies that $\mathbf{y}^t\mathbf{h} \geq 0$ for all $\mathbf{h} \in \mathbf{Q}^>(\mathbf{x}^*)$. This means that $\mathbf{y}$ is in the polar cone of $\mathbf{Q}^>(\mathbf{x}^*)$. Since $\mathbf{Q}^>(\mathbf{x}^*)$ is itself the polar cone of the cone formed by the $\mathbf{c}^k$, then $\mathbf{y}$ must belong to the cone spanned by the $\mathbf{c}^k$, because the polar of a polar cone is the original cone; i.e., $(\mathbf{A}^*)^* = \mathbf{A}$ (Philip, 1972, p. 209). This implies that $\mathbf{y}$ is a nonnegative linear combination of the $\mathbf{c}^k$, or for the problem in Fig. 4-5,

$$\mathbf{y} = w_1\mathbf{c}^1 + w_2\mathbf{c}^2 \tag{4-20}$$

where $w_k \geq 0$ for $k = 1, 2$.

The negative of the normal to the hyperplane must also lie in the cone generated by the negative gradients of the binding constraints, $-\nabla g_4(\mathbf{x}^*)$ and $-\nabla g_1(\mathbf{x}^*)$. If this were not true, then the hyperplane would not separate $\mathbf{Q}^<(\mathbf{x}^*)$ and $\mathbf{F}_d$. The implication of this is that $(-\mathbf{y})$ can be expressed as a nonnegative linear combination of the negative gradients.

$$-\mathbf{y} = u_1[-\nabla g_1(\mathbf{x}^*)] + u_4[-\nabla g_4(\mathbf{x}^*)] \tag{4-21}$$

where $u_i \geq 0$ for $i = 1, 4$.

Equating (4-20) and the negative of (4-21) gives

$$w_1\mathbf{c}^1 + w_2\mathbf{c}^2 = u_1 \nabla g_1(\mathbf{x}^*) + u_4 \nabla g_4(\mathbf{x}^*) \tag{4-22}$$

Noting that the $\mathbf{c}^k$ are the gradients of the objective functions and rearranging gives

$$w_1 \nabla Z_1(\mathbf{x}^*) + w_2 \nabla Z_2(\mathbf{x}^*) - u_1 \nabla g_1(\mathbf{x}^*) - u_4 \nabla g_4(\mathbf{x}^*) = 0 \tag{4-23}$$

which is the third Kuhn–Tucker necessary condition for noninferiority applied to the solution at (1, 4) in the example problem. Generalizing (4-23) to p objectives and m constraints gives

$$\sum_{k=1}^{p} w_k \nabla Z_k(\mathbf{x}^*) - \sum_{i=1}^{m} u_i \nabla g_i(\mathbf{x}^*) = 0 \tag{4-24}$$

which is the condition found by Kuhn and Tucker (1951) and Zadeh (1963).

The results in (4-24) along with feasibility and complementary slackness comprise the necessary and sufficient (when the Z_k are each concave and $\mathbf{F}_d$ is a convex set) conditions of noninferiority.

*4.5.2 Characterization of Some Special Cases

It is instructive to consider some special cases in which the objectives assume certain relationships. Understanding these cases will heighten insight and understanding.

Single Objective

If there is a single objective function, there is a single gradient ∇Z and a single vector in the cone of gradients. In this case the cone is a single ray and $\mathbf{Q}^{>}(\mathbf{x})$ is a halfspace defined by the hyperplane to which ∇Z is orthogonal (see Fig. 4-6). The hyperplane is really nothing more than a contour of the

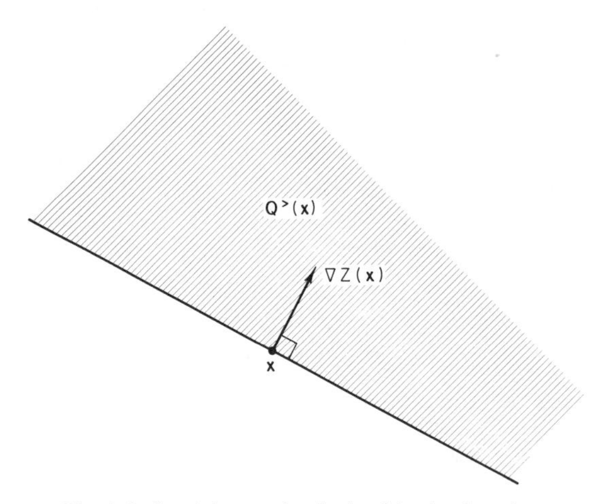

Fig. 4-6. Special case of a single objective function.

objective function. Recall that $\mathbf{Q}^{>}(\mathbf{x})$ does not include the boundary of the polar cone in (4-19). This is important, for if the hyperplane were in $\mathbf{Q}^{>}(\mathbf{x})$, alternate optima could not occur.

Collinear Objectives with No Conflict

In this case the objective function gradients point in the same direction, as in Fig. 4-7. $\mathbf{Q}^{>}(\mathbf{x})$ is a half-space defined by the hyperplane to which $\nabla Z_1(\mathbf{x})$ and $\nabla Z_2(\mathbf{x})$ are orthogonal. This situation is identical to the previous case, which is an intuitively appealing result: When the objectives do not conflict at all, the problem is equivalent to a single-objective problem.

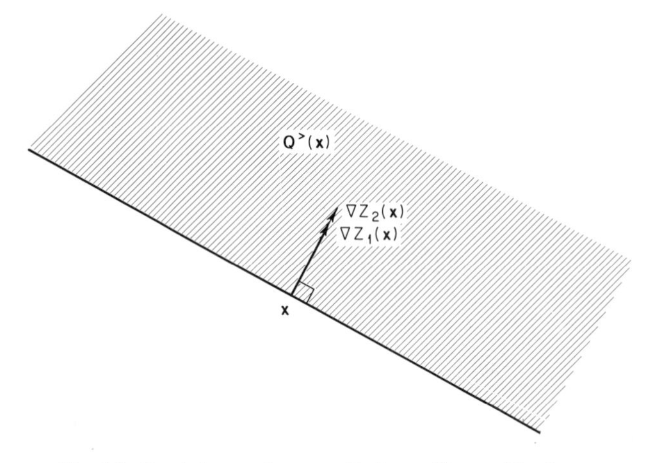

Fig. 4-7. Special case of nonconflicting collinear objectives.

Collinear Objectives with Complete Conflict

When the objectives are diametrically opposed as in Fig. 4-8, the set $\mathbf{Q}^>(\mathbf{x})$ is empty. This means, of course, that the intersection of $\mathbf{Q}^>(\mathbf{x})$ with $\mathbf{F}_\mathrm{d}$ is also empty for any $\mathbf{x} \in \mathbf{F}_\mathrm{d}$. The conclusion is that all feasible solutions are noninferior! A little thought will indicate the appeal of this result: When there is complete conflict, compromise is impossible. Holl (1973, p. 94) has proved this result.

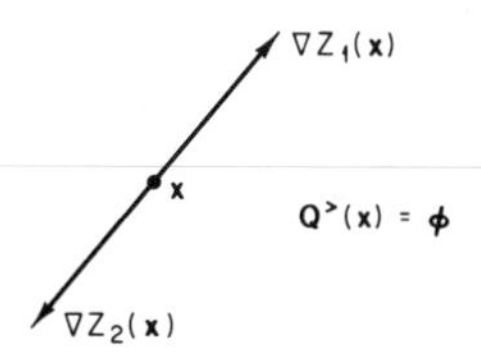

Fig. 4-8. Special case of completely conflicting objectives.

The last two cases can be generalized to the case of *dependent objectives*. If the gradient of an objective can be expressed as a linear combination of any of the remaining objectives,

$$\nabla Z_r(\mathbf{x}) = \sum_{\substack{k=1 \\ k \neq r}}^{p} \alpha_k \nabla Z_k(\mathbf{x}) \tag{4-25}$$

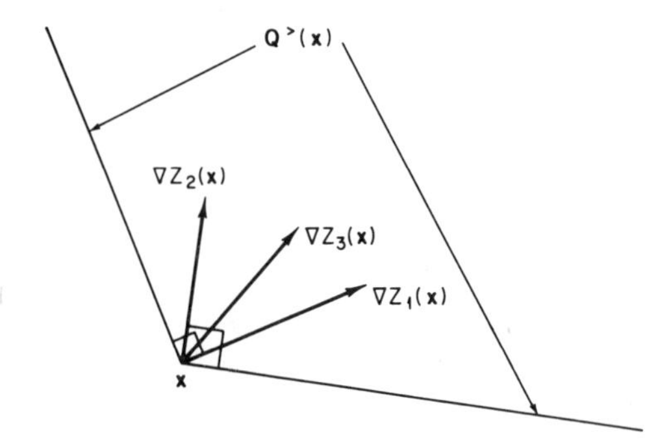

Fig. 4-9. Special case of dependent objectives: Z_3 may be eliminated since it is redundant for the definition of $\mathbf{Q}^{>}(\mathbf{x})$.

then one of the following is true:

(1) if $\alpha_k \geq 0$ for all k, then objective r can be eliminated from further consideration;

(2) if the summation in (4-25) is negative and $\alpha_k \neq 0$ for all $k \neq r$, then all feasible solutions are noninferior.

In the first result, the dependent gradient adds nothing to the definition of $\mathbf{Q}^{>}(\mathbf{x})$; see, for example, Fig. 4-9. In the second result the resultant vectors negate each other, as in Fig. 4-10, so that $\mathbf{Q}^{>}(\mathbf{x})$ is empty.

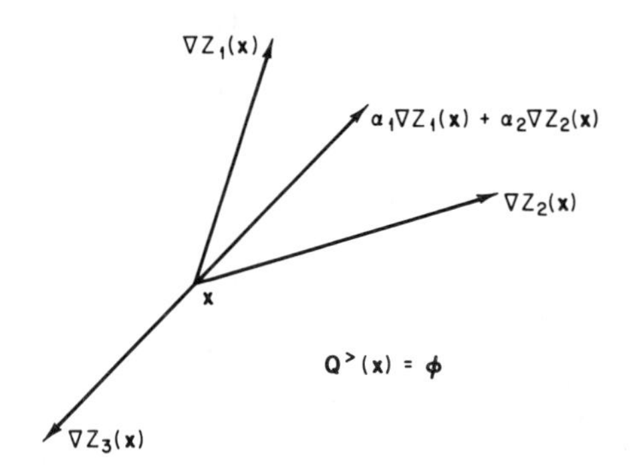

Fig. 4-10. Special case of dependent objectives: $\mathbf{Q}^{>}(\mathbf{x})$ is empty and all feasible solutions are noninferior.

CHAPTER 5

Classification of Multiobjective Programming Methods

Several different approaches to multiobjective programming and planning problems are presented in the next three chapters. In this chapter the techniques are categorized so that the presentation in Chapters 6–8 can proceed in an orderly fashion and so that the differences among the methods are appreciated. The basis for the classification is the role for the analyst in the planning process that is implied when a technique is used. After the techniques are categorized, the applicability of the methods in various decision-making contexts is discussed, computational requirements are assessed, and evaluation of the methods is considered.

5.1 A CATEGORIZATION OF TECHNIQUES

The characteristics of the decision-making process which will be used to categorize multiobjective programming methods are the information flows in the process and the decision-making context. Information flows are important because they determine the role that the analyst must play in the planning process. The decision-making context defines the goal of the analysis.

For our purposes it is sufficient to conceive of two types of information flows: from decision maker to analyst ("top–down") and from analyst to decision maker ("bottom–up"). The analyst–decision maker or bottom–up flow in this case contains results about the noninferior set—noninferior alternatives, their impacts on the objectives, and the tradeoffs among the objectives. The decision maker–analyst or top–down flow occurs when decision makers explicitly articulate preferences so that a best-compromise solution may be identified. There are many techniques that allow for only one of these types of information flows; some techniques (iterative methods) employ both types of flow during the solution process.

The decision-making context referred to above is the specific problem setting for the analysis. The two contexts that will be considered are multiobjective problems and multiple-decision-maker (conflict resolution) problems. The former setting is intended to include those situations in which there is a single decision maker, or a group of decision makers that share similar objectives and preferences, who must make a decision about a problem with many conflicting objectives. The latter situation is directed at those cases in which there are many decision makers or interest groups, each of which has its own conflicting objectives. In the first case of multiobjective decision making, the analytical goal is a best-compromise solution, which is found through the resolution of the decision maker's internal conflicts. In the second case, a particular decision maker must resolve internal conflicts among objectives and be aware of conflicts with other decision makers. The multiple-decision-maker setting may therefore require that prediction of preferences of other decision makers and their political reactions be an analytical goal.

It is difficult to develop a sharp distinction among decision-making contexts; indeed, the two cases discussed above are not mutually exclusive. It makes sense to treat many inherently multiple-decision-maker problems as more simplified multiobjective problems. Thus, a public decision maker may be simultaneously concerned with multiple objectives for clean air legislation precisely because that decision maker's constituency includes several interest groups: large urban areas, an active local chapter of an environmentalist group, and an economically important automobile industry. The decision maker can, perhaps, be served well by an analyst that shows the tradeoffs among urban air quality and costs to automobile manufacturers (a multiobjective analysis) without considering the political dynamics of the interest groups that stand for the objectives.

The range of analytical methods is segmented into three categories, depending on the decision-making setting for which they are best suited and

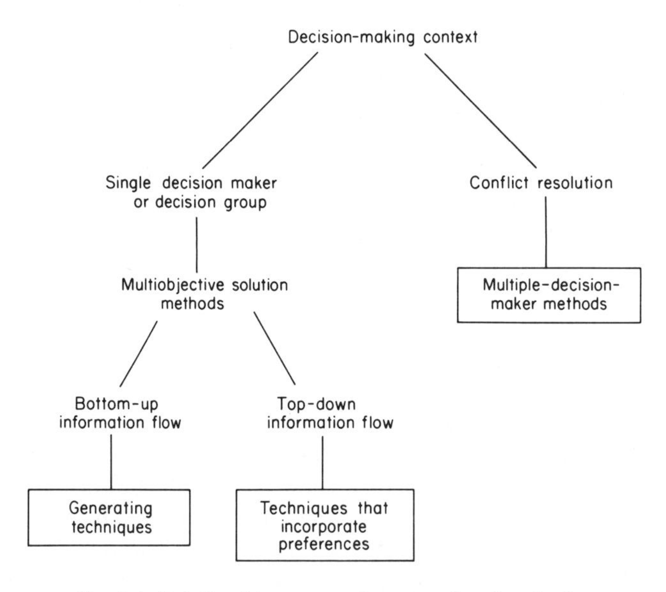

Fig. 5-1. Relationships among the categories of methods.

on the information flows that their use requires. The categories are generating techniques, methods that incorporate preferences, and multiple-decision-maker methods. Figure 5-1 explains the relationships of the methods. The decision-making context provides the first split into multiple-decision-maker methods and multiobjective solution techniques. The latter set of methods is then further categorized, on the basis of information flows, into generating techniques and techniques that incorporate preferences.

5.1.1 Generating Techniques

Generating techniques emphasize the development of information about a multiobjective problem that is presented to a decision maker in a manner that allows the range of choice and the tradeoffs among objectives to be well understood. Thus, the analytical goal is not political analysis or prediction, but the generation and evaluation of alternatives in terms of several objectives.

The information flow is of the bottom–up variety. Analysts apply a generating technique to find an exact representation or an approximation of the noninferior set in objective and decision spaces, as defined in Chapter 4. These results are then presented either graphically or in tabular form to decision makers who, based on this information, select a best-compromise solution. Notice that preferences need not be articulated explicitly by decision makers, although the process of choosing a best-compromise solution cannot be done without at least an implicit consideration of the decision makers' preferences for the objectives.

Several generating techniques are reviewed in Chapter 6: the weighting and constraint methods, the noninferior set estimation method, and the multiobjective simplex method. Approaches for displaying the results of a generating analysis are also presented.

5.1.2 Techniques That Incorporate Preferences

Techniques that incorporate preferences (*preference-oriented* methods) share the analytical goal of the generating methods: analysis of a multiobjective problem without explicit consideration of the political dynamics of the problem. Unlike the implicit treatment of preferences by the generating methods, however, preference-oriented techniques require that decision makers articulate their preferences and pass that information on to the analyst.

The articulation of preferences may be done in several ways; the particular approach used is the distinguishing feature of a specific preference-oriented method. A discussion of each of the specific treatments is provided in Chapter 7. For the present we shall further subdivide these methods into two classes: noniterative and iterative approaches.

The noniterative preference-oriented techniques require decision makers to articulate their preferences in advance of the analysis. Alternatively, the analyst may make assumptions about what form the preferences should assume. In either case, the best-compromise solution is defined without generating the noninferior set or portions of it. The noniterative techniques that are discussed in Chapter 7 are multiattribute utility functions and a method for finding the best-compromise solution when the utility function is previously defined; prior weighting of objectives; various geometrical notions of best, including goal programming; and the surrogate worth tradeoff method.

Iterative methods that incorporate preferences operate with local approximations of a decision maker's preferences. The locally approximated preference information is articulated by the decision maker in response to local information about the noninferior set generated by the analyst. The preference information is used by the analyst to find a new and better solution, which triggers new reactions from the decision makers. This iterative process continues until the decision maker is satisfied or until another termination condition becomes operative. The so-called step method and an iterative algorithm that includes local approximations of a multiattribute utility function are discussed in Chapter 7.

5.1.3 Techniques for Multiple-Decision-Maker Problems

This third category includes techniques that are consistent with an analytical goal very different from the focus of the first two classes. The methods in the multiple-decision-maker category are directed at the resolution of conflict among many interest groups or decision makers. In Chapter 8, we shall deal with conflict resolution in a public setting by turning to notions of the public interest—what it is and how it gets articulated.

The specific techniques that will be discussed in Chapter 8 are drawn from welfare economics and political science as well as from operations research. The history of welfare economics will be traced in order to understand how various schools of thought in that discipline have treated the problem of the public interest. Paretian analysis, a method for generating politically feasible alternatives, will be presented. Game theory and various voting models will also be discussed.

5.2 A VIEW OF PUBLIC DECISION MAKING

It is important to realize that many multiobjective programming techniques have been developed by theoreticians whose interest lies in the mathematical details of the multiobjective problem. The developers are, of course, interested in the application of their methods, but this is frequently left as an exercise for the practitioner. There are some very important issues that the practitioner must (and the theoretician should) consider in order to implement a technique.

A fundamental issue is the role that the analyst assumes in the planning process. Should he take a relatively passive position from which he acts as a provider of information? Or should the analyst go further and attempt to counsel the decision maker so that the latter can "discover" his preferences? Perhaps a particular decision maker's preferences are only a part of the analyst's task; should the analyst act as a predictor of an outcome from the decision-making process; i.e., should political feasibility be the goal of the analysis?

The particular role that the analyst or planner elects or is forced to play will depend on the nature of the decision-making process and where the analyst is located within that process. Therefore we must have in mind some notion of the ways in which public decisions are made. The reader is cautioned that the ideas advanced below are not well grounded in political theory. Rather, they are the views of a modeler who has collected empirical evidence to which common sense has been applied.

Our concern is with public decision-making processes, which have, of course, been widely studied by political scientists. It seems, however, that in spite of the efforts of political scientists (or because of them) there is no single theory that explains the process by which public decisions are made. Our lack of understanding in this area undoubtedly results from the complexity of the problem: Unlike natural and physical scientists who work with controllable systems, the political scientist must deal with the complicating and unpredictable influences of human behavior. Steiner (1969) presents a compact discussion of political decision-making theories, particularly as they relate to public investment decisions. Another interesting work is that of Greenberger *et al.* (1976), which considers the role of mathematical modeling in public decision-making processes.

Our specific concern is with those characteristics of decision-making processes that have implications for the role of the analysts in that process. There are many characteristics of public decision making that are relevant for analysis in general and modeling in particular. However, the discussion that follows will concentrate on those attributes relevant for the previously posed categorization of solution techniques. Thus, the attributes of the process that are important are the number of decision makers, their identifiability and accessibility, the structure of the decision-making process, and the analyst's institutional relationship to the process. Each of these characteristics is discussed more fully below in the context of alternative decision-making settings and as they relate to the applicability of multiobjective and multiple-decision-maker methods.

Decision-making frameworks vary considerably over branches and levels of government. It is sufficient for our purposes to consider two major types of settings: bureaucratic decision making and legislative decision making. While decision making in the judicial branch is becoming increasingly important as the courts take a more active role in setting public policy, such as in environmental protection, through their interpretation of existing laws, the most important arenas for the present are the executive and legislative branches.

5.2.1 Bureaucratic Decision Making

Our interest here is with the decision making of an agency at any level of government. The nature of such agencies varies widely in terms of centralization of power, the extent to which the agency must answer to other governmental institutions and nongovernmental interests, and thus in terms of all the attributes mentioned above.

There may be a single bureaucratic decision maker, e.g., a secretary of a multiagency department, an agency administrator, or the head of a section. On the other hand, more than one decision maker may be involved since there may be competing interests within the bureaucratic structure at several levels at which decisions are made. Similarly, interests exogeneous to the bureau may require attention from the analysts.

Identifying bureaucratic decision makers may be relatively easy, but even here the analyst must be careful. In any organization, particularly sprawling public institutions, there may be a real difference between the legal head of the organization and those who really make decisions. Playing organizational politics may seem distasteful, but it is a prerequisite of effective analysis. Again, one would expect large variations among institutions. Military and paramilitary organizations, e.g., police and fire departments, are at one extreme: The responsibility for decision making in particular problem areas is clearly defined and well known. Perhaps at the other extreme are massive federal departments, whose structure and relationships are difficult to fathom both for their complexity and temporal variability.

The accessibility of decision makers is always a problem for analysts, but the problem becomes more acute as one moves up the organizational chart. Thus, it is usually easier to see, say, an office head than it is to gain an audience with the Secretary of a federal department.

The structure of bureaucratic decision making and the analyst's position in that framework also vary considerably from one setting to another. The types of decision that an agency may be called on to make range from planning and design to regulation, either in response to legislation or to other executive units. The analyst, as an employee of the agency, must take into account the bureau's position as a responder to legislation or as a source of recommendations. The analyst's primary responsibility, however, is to the requirements of the agency's decision maker(s). For this reason, it seems reasonable to think of the bureaucratic decision-making structure as relatively well defined. Furthermore, the analyst's position will usually also be well defined, implying a reasonably well understood method by which analysis should be conducted.

5.2.2 Legislative Decision Making

Legislative bodies, as voting institutions, have a structure that requires attention to the political dynamics of the decision-making process. A particular decision maker's desires (or the desires of that legislator's constituency) can be implemented only if a prespecified minimum number of the other decision makers (usually a majority) agree with the alternative. Thus, the conflict among decision makers and its resolution through bargaining plays an important part in determining the analyst's role. This does not mean, however, that all legislative decision making is performed in a manner that demands the explicit consideration of conflict resolution. The decision maker must first define the desirable alternative for the constituency; this problem may be abstracted from the legislative decision-making context.

There are two institutional frameworks within which analysis may be performed to aid legislative decision making: The analyst may be an employee of a legislative committee or an employee of a specific legislator. In the first situation, the analyst serves many decision makers simultaneously, although the political party in the majority and therefore the chairman of the committee may control the committee's business. In the latter case, the analyst has a single boss who nevertheless must be responsive to multiple interest groups and other legislators.

The above discussion of decision-making structure indicates that, at least nominally, legislative decision making is characterized by multiple decision makers. For the purpose of some analyses, however, e.g., in the case of a legislative aide responding to a single legislator's requirements, one can characterize (a part of) the process as having a single decision maker.

The decision makers in a legislature are easily definable but rarely accessible. Try making an appointment with your United States Senator; the experiment will confirm the claim of inaccessibility. Of course, decision makers must be accessible to their analysts, but the requirements of the legislative process limit the amount of time that a legislator can devote to giving input to and considering results from analysis.

5.3 THE APPLICABILITY OF SOLUTION METHODS TO PUBLIC DECISION-MAKING PROBLEMS

In this section the three categories of methods proposed in Section 5.1 are evaluated in terms of their applicability to public decision-making problems. In going through this evaluation the rationale for the categorization and the importance for an awareness of the process by which decisions are made will be more fully appreciated.

It seems fair to say that every method considered in this book is applicable to at least one decision-making situation. It is not true, however, that every method is *best* for some situation or that the decision-making context to which a method is applicable is prevalent. The fact that this book could be written implies a range of problems and decision-making contexts that cannot be adequately captured by any one method. The lack of a universally applicable method makes this area of analysis interesting; it also places a burden on the analyst to choose the appropriate method for the problem at hand.

5.3.1 Applicability of Generating Techniques

Generating techniques put a relatively small burden on decision makers in terms of the supply of information. Because of the bottom-up information flow, decision makers need only react to the results of the solution process. The emphasis is on the demarcation of the range of choice, not on the explicit definition of preferences.

Generating methods, since they do not require inputs of preferences, are compatible with a wide range of decision contexts. The number of decision makers, their accessibility, and even their identifiability will not generally render a generating method inapplicable. If the model components are specified appropriately, then the analysis may proceed since the results—the noninferior set—may be displayed for any number of decision makers and the analyst need not gain access to the decision makers. Some of the decision makers may even be unknown to the analyst; yet the products of the analysis may still impact the decision-making process.

There are two reservations, however, regarding the applicability of generating methods. First, while a generating approach may proceed without complete identification of or access to the decision makers, conducting an analysis in such a void is not likely to succeed. That is, any attempt at analysis should strive to include decision makers to the maximum extent possible. Implementation is unlikely when decision maker–analyst interaction is not achieved.

Second, when there are multiple conflicting decision makers, generating methods may solve only a part of the problem. The methods aid each decision maker in assessing the alternatives and determining a best-compromise solution. The issue of conflict resolution, i.e., when the several best-compromise solutions are not the same, is not captured.

Generating methods are, with the qualifications mentioned above, applicable to a wide range of decision contexts. In the case of a single accessible decision maker, they offer a very useful way to proceed. The decision makers are offered insight into what is possible, what must be sacrificed, and what, therefore, is gained by achieving each noninferior alternative. Well-informed decision making is the goal.

In the case of multiple, inaccessible, partially unidentified decision makers, generating techniques may still play an important role. Each decision maker is well informed about the physical, economic, and social aspects of the problem (to the extent that the model captures these), but the political dimension may escape the analysis. Even in this case, however, the generation of alternatives in a multiobjective manner may aid conflict resolution. Decision makers who understand the alternatives and the tradeoffs may be more likely to compromise.

5.3.2 Applicability of Preference-Oriented Methods

Techniques that incorporate preferences place a relatively large demand on the decision maker in terms of information requirements. Preferences, articulated by the decision maker either prior to or as part of the solution process, are used to identify a best-compromise solution. Consequently, the range of decision contexts to which these techniques are applicable is rather narrow.

Obviously, the analyst must be able to identify the decision maker and have access to the decision maker in order to obtain an articulation of preferences as required by a top–down approach to planning. The requirement of accessibility may be particularly troublesome since some of the methods in this category demand a substantial amount of the decision maker's time to extract preference information, e.g., those methods based on multiattribute utility functions. Other methods require repeated interactions with the decision maker, e.g., the iterative techniques. If the decision maker is not identifiable or sufficiently accessible, the analyst may deal with a surrogate: People on the decision maker's staff or a panel of experts not associated directly with the true decision maker may be used. However, this approach is risky since a panel of experts and even a decision maker's staff person may not represent the decision maker's preferences correctly.

The number of decision makers is another complicating factor for preference-oriented techniques. You will note that in the preceding two paragraphs we dealt only with a single decision maker; when there are multiple decision makers, the obvious question is "whose preference is preponderant?" The question cannot be sidestepped for methods in this category, for preferences must be articulated for these techniques to work. As with the generating methods, one can claim that the definition of preferences may show the way to a compromise. This seems less likely for preference-oriented methods, however, since decision makers' preferences are frequently articulated in response to hypothetical situations that may have little to do with the problem at hand.

Solution techniques that incorporate preferences are applicable to those situations characterized by a single decision maker (or a single voice for a group of decision makers) who is willing to articulate preferences explicitly and audibly. This raises a question relative to the structure of the decision-making process. In many public decision-making settings decision makers may be reluctant to state, even privately to an analyst, that one part per million of a water or air quality indicator is worth $100,000 or that a human life is worth $350,000. These value judgments must be made if decisions are to be made, but the process by which these preferences are articulated may mean a great deal to decision makers. Generating methods allow decision makers to choose alternatives, thereby implying value judgments which need not be aired, while preference-oriented methods demand a statement, controversial and potentially politically explosive, that attaches values that are easily criticized.

5.3.3 Applicability of Multiple-Decision-Maker Methods

This category of methods is the most widely applicable in that it is directed at the most general decision context and subsumes simplified situations. That is, the case of a single decision maker is just a special case of a multiple-decision-maker problem. That statement is not meant to propose, however, that an analyst should use the methods in this category for single-decision-maker problems.

While multiple-decision-maker methods are the most widely applicable, they are also the least well developed. Most of the methods discussed later in Chapter 8 have been set forth theoretically or, at best, applied to hypothetical problems. The applicability of these methods must therefore be tempered by their rather primitive state of development.

Of the attributes of public decision making that we have discussed, the identifiability of decision makers is most critical for the multiple-decision-maker techniques. All of the techniques in this category require the explicit identification of decision makers. The methods then proceed to associate an index of political power or a utility function with each decision maker. When utility functions are used, the issue of accessibility may be crucial, since the difficulties associated with extracting a single decision maker's preferences mentioned in the previous subsection are magnified by the number of decision makers.

5.4 COMPUTATIONAL REQUIREMENTS

The previous discussion of the public decision-making process and the role which the analyst may play in it leaves out an important consideration in the analyst's choice of a solution method. In addition to the appropriateness of the method, the analyst must consider the computational requirements of the method. Analysis is conducted with a finite budget that may be exhausted prematurely by some methods, depending on the sizes of the problem and the budget.

Generating methods are generally more computationally intensive than are preference-oriented techniques. This should not be surprising in light of the greater amount of information supplied by a generating technique. Generating methods are used to find several noninferior solutions which comprise an exact representation or an approximation of the noninferior set.

For the most expensive methods in this category, each noninferior solution requires the solution of a linear program. If this linear program is large, which depends on the size of the problem under analysis, then it may be quite expensive to obtain a sufficiently accurate representation of the noninferior set. It is particularly expensive to achieve a given level of accuracy when there are several objectives, for the number of noninferior solutions required increases exponentially with the number of objectives.

Preference-oriented methods exploit the knowledge of preferences to avoid the computational costs associated with generating techniques. There are, however, different costs involved. Investments of time and money are required to obtain the preference information. It is generally true, nevertheless, that generating techniques are more expensive to implement than are methods that incorporate preferences.

It is difficult to assess the computational requirements of multiple-decision-maker methods. This category contains many techniques for which applicability to public decision-making problems of realistic dimensions is, at present, speculative at best. Furthermore, the specific procedures that are used vary considerably from one method to another. Some of the methods do not seem computationally feasible at this time, while others must await further testing before their computational requirements are fully understood.

5.5 EVALUATING THE METHODS

To this point, the applicability and computational requirements of the three categories of techniques have been assessed. It would be nice if the methods could now be evaluated and a best set of techniques chosen. This will not be done, however, because it cannot be done. The analyst also faces a multiobjective problem in the choice of a method. Some methods may seem most appropriate for some problems, but they may be too expensive to implement. Other techniques may be quite inexpensive, but not wholly appropriate from the perspective of the decision context. The selection of a method will depend on the problem, the decision-making context, the computational budget, and the analyst's preferences.

The bias of this author is toward generating methods. The rationale for this preference is based on the implied roles for the analyst and decision makers. Generating techniques place the analyst in the position of information provider, i.e., in the classical role of the analyst as a scientist; decision makers maintain complete control over the decision without abdicating any of that responsibility to the analyst. Generating methods are also quite flexible: They even may be applicable to many multiple-decision-maker situations.

In recognition of these biases, Chapters 6–8 are written as objectively as possible. No further evaluation of the methods is offered in those chapters. Rather, it is hoped that the reader will evaluate the methods in light of the discussion of the current chapter. Ask yourself questions about the applicability of the methods to various decision contexts and about computational requirements. Will these methods really work in a legislative or bureaucratic setting? What is required in the way of information from decision makers? How sensitive are the methods to the number of decision makers, their identifiability and accessibility? Above all, be critical.

CHAPTER 6

Techniques for Generating Noninferior Solutions

In this chapter methods for generating noninferior solutions from a previously formulated multiobjective optimization model are presented. Generating techniques, as they will be called, do not allow preferences to be incorporated into the solution process. Preferences are articulated, perhaps implicitly, by decision makers after an approximate or exact noninferior set is generated.

Generating techniques do not require prior statements about preferences, priorities, utilities, or any other value judgments about the objectives. The articulation of preferences is deferred until the range of choice, represented by the noninferior set, is identified and presented to decision makers. The emphasis is on tradeoffs among objectives over the entire range of feasibility.

After generation of the noninferior set the analyst may choose to pursue a structured or mathematical procedure for the articulation of preferences. It is sufficient in many applications simply to present the noninferior set and to explain the implications of the various alternatives. Generation alone is particularly effective when the noninferior set exhibits a "kinked" shape, as in Fig. 6-1. The best-compromise solution will usually be found in the vicinity of point B. On other parts of the curve a large amount of one objective must be sacrificed in order to gain relatively little of the other objective. For example, in going from B to A, $(Z_1^B - Z_1^A)$ of objective Z_1 is lost in order to gain only $(Z_2^A - Z_2^B)$ of objective Z_2. These "kinked" shapes will appear again in this chapter and in the case study discussion in Chapter 9.

An extreme shape of the noninferior set is not a prerequisite for the successful application of generating methods. A more "ordinary" curve may result in a higher computational burden and more difficult decisions, but the value of identifying the feasible range of choice is not diminished. The real gain associated with these techniques is the direction and structure that they bring to the planning process. Analysts are forced to concentrate on the formulation and evaluation of alternatives, and when results are reported they need not recommend a specific or best alternative. Analysts, instead, find themselves in the considerably more comfortable and defensible position of showing a range of alternatives. The responsibility for selection clearly rests with decision makers, which is as it should be. Models are less apt to be

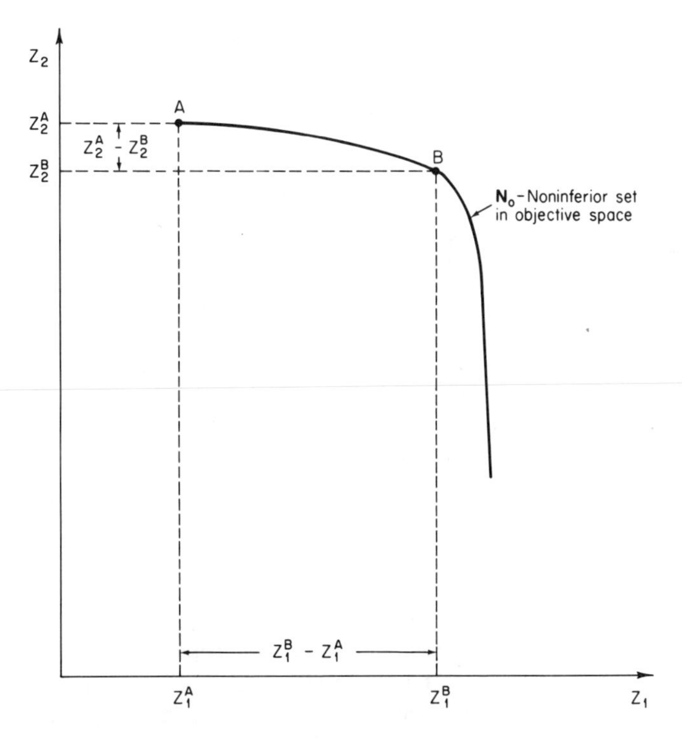

Fig. 6-1. A "kinked" noninferior set: Generating methods are particularly useful in these cases.

shelved when they are used in this way rather than in the traditional approach of finding a so-called "optimal" solution.

One more virtue that can be claimed for generating approaches is their generality relative to decision processes. They can be used in processes characterized by many decision makers, by relatively inaccessible decision makers, or by uncertainty as to who the decision makers are. Of course, the probability of successfully affecting policy may be diminished in any of these cases since interaction with decision makers is very useful. The techniques generate results that are of value without the formal incorporation of preferences. This permits the use of generating methods even in the most complex decision-making situations.

The major weakness of generating methods is a sensitivity to the number of objectives. Several objectives (usually more than three) cause two problems: high computational burden and complexity of result displays. High dimensionality creates difficulties for all methods in all categories, but the generating techniques exhibit the most sensitivity.

The following generating techniques are discussed in detail: the weighting method, the constraint method, the noninferior set estimation method, and the multiobjective simplex algorithm. Display mechanisms for multiobjective problems are also discussed. They are of obvious importance in the communication of results.

6.1 THE WEIGHTING METHOD

Weighting the objectives to obtain noninferior solutions is the oldest multiobjective solution technique. The method follows directly from the necessary conditions of noninferiority developed by Kuhn and Tucker (1951). This dependency is shown in a subsequent section. Gass and Saaty (1955) showed how noninferior solutions could be generated in two-objective problems by parametrically varying the objective function coefficients. Zadeh (1963) was the first to recommend the use of weights to approximate the noninferior set. Marglin (1967, pp. 23–24) and Major (1969) discussed the use of weighting in multiobjective public investment problems.

6.1.1 Motivation for Weighting Objectives

Suppose, for example, that we have a fire station location problem for which there are two objectives: maximize the property value (measured in dollars) within S miles of the facility and maximize the population within S miles of the facility. The property value and population objectives will be called Z_1 and Z_2, respectively. The two objectives conflict because commercial areas are characterized by high property value and low populations while residential areas have more people and lower property value. Since the fire station cannot be located such that the entire area is within S miles, the maxima of Z_1 and Z_2 cannot be obtained simultaneously. The objective function for this multiobjective location problem is

$$\text{maximize} \quad \mathbf{Z} = [Z_1, Z_2] \tag{6-1}$$

where the notation of previous chapters has been simplified by excluding the decision variables, in this case the location of the station.

Now, if someone were willing to articulate the value judgment that one person is worth w dollars, then the multiobjective problem could be reduced to a single-objective problem. The specification of w, which is called a *weight* on objective Z_2 (population), is equivalent to the identification of a desirable tradeoff between Z_1 and Z_2. Since we know the value of Z_2 in terms of Z_1, (6-1) can be rewritten as

$$\text{maximize} \quad Z(w) = Z_1 + wZ_2 \tag{6-2}$$

Notice that now the objective function has a single dimension and is denoted by $Z(w)$ to signify the dependence of the new function on the value of the weight w. Notice also that the units of the new objective function $Z(w)$ are dollars: Z_1 is measured in dollars and wZ_2 is (dollars/person)(persons) = dollars.

Given a single objective function such as (6-2) and a set of constraints, we can proceed directly to a solution by using well-known techniques such as the simplex method for linear problems. This solution would be the *best-compromise solution for the person who articulated the value of* w. To see this, suppose the feasible region in objective space $\mathbf{F}_0$ were as shown in Fig. 6-2. The noninferior set in objective space $\mathbf{N}_0$ is also indicated.

The optimization problem with (6-2) as the objective function can be solved graphically given $\mathbf{F}_0$ in Fig. 6-2. Contours of (6-2) in objective space are defined by the equation

$$Z_1 + wZ_2 = A \tag{6-3}$$

where A is an arbitrarily chosen constant. Rewriting (6-3) gives

$$Z_2 = (-1/w)Z_1 + (A/w) \tag{6-4}$$

so that the contour appears in objective space as a straight line with slope $(-1/w)$ and intercept A/w. This line is shown in Fig. 6-2.

The graphical solution of the problem is obtained by "pushing" the contour as far to the northeast as possible until it just touches the boundary of $\mathbf{F}_0$. This happens at point C in Fig. 6-2 when the constant in (6-3) is B. The slope of this line is $(-1/w)$, as was established in (6-4) above.

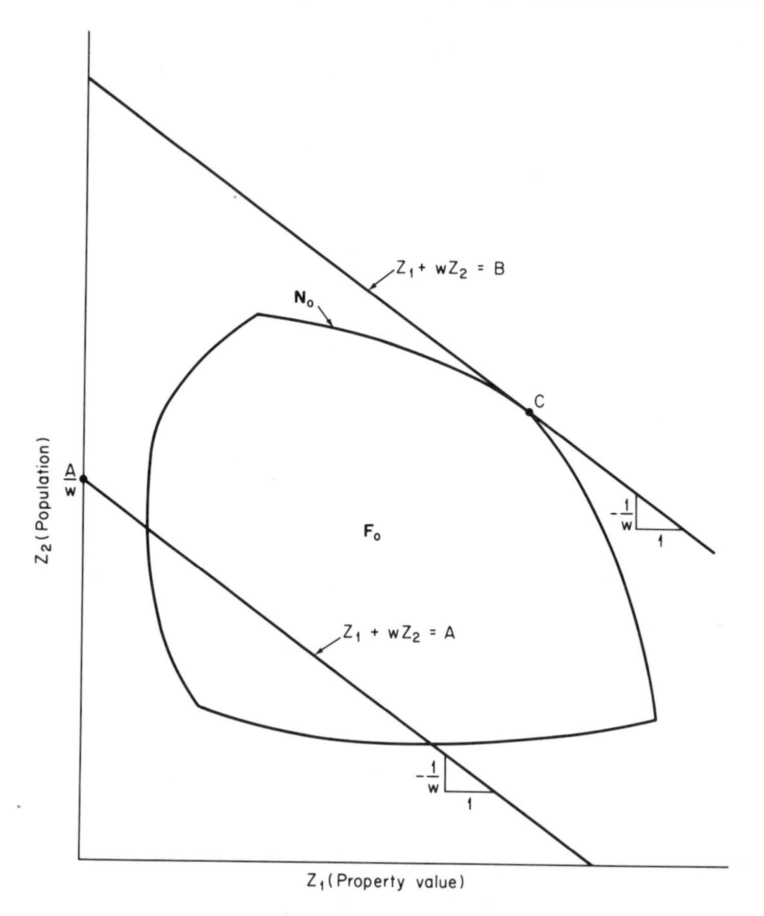

Fig. 6-2. Graphical interpretation of the weighting method: In objective space, a weighted objective function appears as a linear indifference curve.

Referring back to Fig. 4-3, the line expressed by (6-3) can be thought of as a linear indifference curve and point C is therefore the best-compromise solution. Thus, the specification of a desirable tradeoff between two objectives is equivalent to a statement of preferences that can be represented by linear indifference curves.

More insight into the role of weights and the sensitivity of the solution to their value can be gained by considering Fig. 6-3, in which $\mathbf{F}_0$ and the indifference curve from Fig. 6-2 are repeated. If w were increased in value what would happen to the best-compromise solution? As w increases to w' the slope of the indifference curves becomes less negative, which means that the indifference curves become flatter and the point at which the line touches $\mathbf{F}_0$

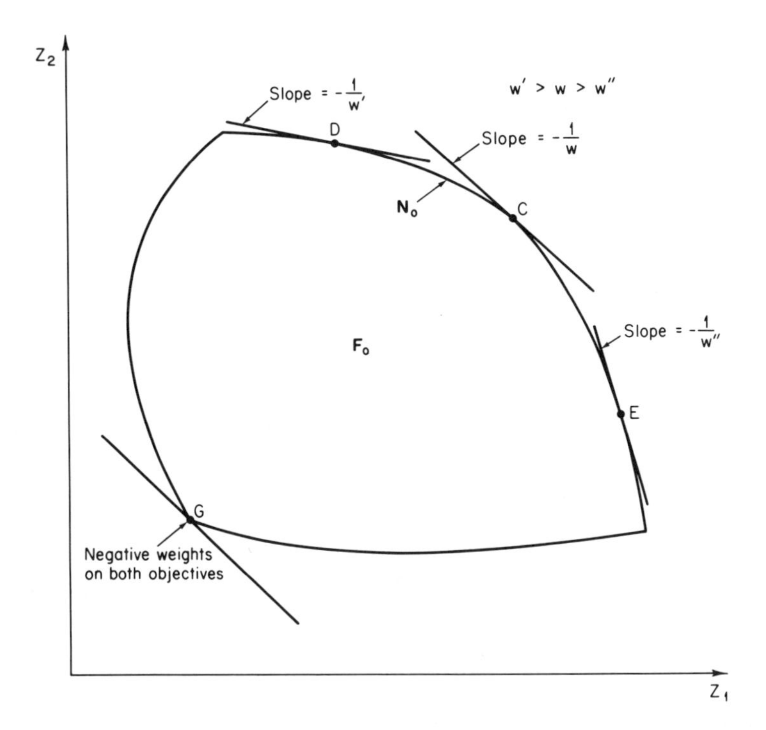

Fig. 6-3. The best-compromise solution varies as the weight w varies. Note: $w' > w > w''$.

shifts to the left from C to D in Fig. 6-3. Therefore, as more value is attached to population coverage relative to property value coverage (w increases) the best-compromise solution tends to favor population; i.e., more population and less property value are covered. Similarly when w is decreased to w'', less weight is put on population so that the best-compromise solution shifts to E in Fig. 6-3, where higher property value and lower population coverage are obtained.

6.1.2 Mathematical Approach

The use of weights can be generalized a bit. For a two-objective problem, both objectives can be weighted so that

$$\text{maximize} \quad \mathbf{Z} = [Z_1, Z_2] \tag{6-5}$$

becomes

$$\text{maximize} \quad Z(w_1, w_2) = w_1 Z_1 + w_2 Z_2 \tag{6-6}$$

Recall, however, from Chapter 3 that an objective function can be divided by a positive number without altering the solution. Dividing (6-6) by w_1 gives

$$\text{maximize} \quad Z(w_1, w_2) = Z_1 + (w_2/w_1)Z_2 \tag{6-7}$$

The quantity (w_2/w_1) can be redefined as w and we are back to the original formulation in (6-2). The purpose of this little exercise shows that what really matters is the *relative* weights on the objectives. In the future the formulation with weights on each objective as in (6-6) will be used with the understanding that the weight on one objective can be set to one or any arbitrary positive number.

The use of weights to find a best-compromise solution is contingent upon the willingness of a person or persons to specify the values of the weights for us. That is certainly a sticky problem for various reasons that will be discussed in the next chapter. We can use weights in a very constructive way nevertheless.

Notice that whether the weight in Fig. 6-3 is w, w', or w'', the solution is in the noninferior set $\mathbf{N}_o$. This is a very important observation since it implies that we can take a multiobjective optimization problem

$$\text{maximize} \quad \mathbf{Z}(x_1, \ldots, x_n) = [Z_1(x_1, \ldots, x_n), \ldots, Z_p(x_1, \ldots, x_n)] \tag{6-8}$$

$$\text{s.t.} \quad (x_1, \ldots, x_n) \in \mathbf{F}_\text{d} \tag{6-9}$$

where (6-9) means that solutions must satisfy the constraints which define $\mathbf{F}_\text{d}$, and transform it into

$$\text{maximize} \quad Z(x_1, \ldots, x_n; w_1, \ldots, w_p) = \sum_{k=1}^{p} w_k Z_k(x_1, \ldots, x_n) \tag{6-10}$$

$$\text{s.t.} \quad (x_1, \ldots, x_n) \in \mathbf{F}_\text{d} \tag{6-11}$$

where w_k is a weight on objective k.

The problem in (6-10) and (6-11) is a single-objective optimization problem for which solution methods exist. The optimal solution to the weighted problem in (6-10) and (6-11) is a noninferior solution for the multiobjective problem in (6-8) and (6-9) as long as all of the weights are nonnegative. This solution is also the best-compromise solution when the weights $(w_1, \ldots, w_p)$ express someone's preferences. We shall neglect this fact, however, since the weights in this case are parameters which are being used to find noninferior solutions. No value significance will be attributed to them.

The result that the optimal solution to the weighted problem is a noninferior solution to the original multiobjective problem follows from the fact that the optimal solutions to single-objective problems and noninferior solutions for a multiobjective problem occur on the boundary of the feasible region. The requirement that the weights be nonnegative insures that the optimal solution to the weighted problem will be on the right part, i.e., the

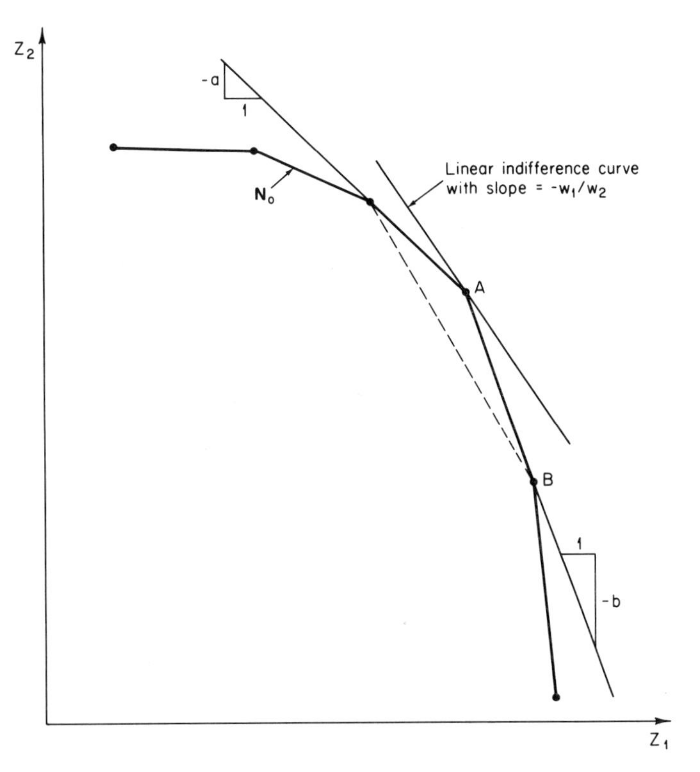

Fig. 6-4. Switching values for the weights in a linear problem.

noninferior portion, of the boundary. If the weights were chosen to be negative, then this is equivalent to transforming the original maximization to a minimization.

In Fig. 6-3, for example, an optimal solution to the problem with negative weights would be point G. Such a solution is obviously inferior.

The feasible region and noninferior set in Fig. 6-2 and 6-3 were drawn as smooth curves. In linear problems, with which we shall deal, the feasible region and noninferior set are actually a series of connected line segments as in Fig. 6-4. The points at which the line segments meet are noninferior extreme points. Notice that for point A in Fig. 6-4 there is a range of weights that will produce point A as the optimal solution to the weighted problem. In fact, as long as

$$-b \leq -(w_1/w_2) \leq -a \tag{6-12}$$

or

$$a \leq w_1/w_2 \leq b \tag{6-13}$$

point A will be the optimal solution to the weighted problem. The values of the weights at the extremes of the range are referred to as "switching values" (UNIDO, 1972). But if $w_1/w_2 > b$, then point B is optimal in the weighted problem. If $w_1/w_2 = b$, then A, B, and the points on the line segment joining them are alternative optima for the weighted problem.

The weighting method is generally used to approximate the noninferior set; it is not an efficient method for finding an exact representation of the noninferior set. A number of different sets of weights are used until an adequate representation of the noninferior set is obtained. Any sets of positive weights may be used in the weighting method but it makes sense to follow an orderly procedure. It seems reasonable to begin by optimizing each objective individually, i.e., by solving the weighted problem p times with $(w_1, w_2, \ldots, w_p)$ equal to $(1, 0, 0, \ldots, 0)$; $(0, 1, 0, \ldots, 0)$; $(0, 0, 1, \ldots, 0)$; $\ldots$; $(0, 0, 0, \ldots, 1)$. These solutions represent "end points" of the noninferior set. See, for example, the noninferior set in Fig. 6-5, in which the points A and B yield the optima of Z_1 and Z_2, respectively. When $w_1 = \alpha$ $(\alpha > 0)$ and $w_2 = 0$, we get

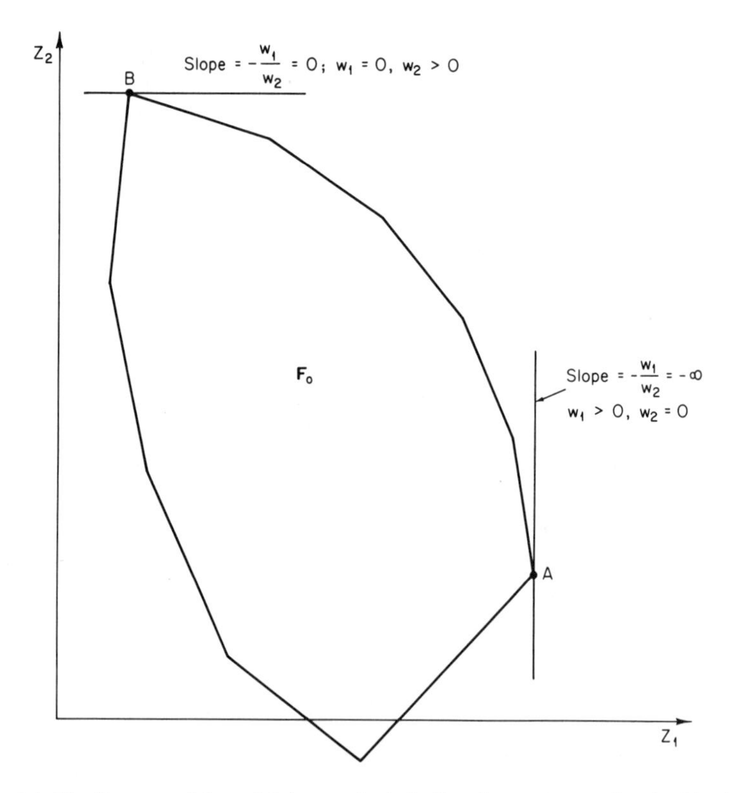

Fig. 6-5. The first step of the weighting method: finding the maximum of each objective.

$-w_1/w_2 = -\alpha/0 = -\infty$ so that the objective function for the weighted problem appears as a line with an infinite slope—a vertical line—in Fig. 6-5. When $w_1 = 0$ and $w_2 = \beta$ ($\beta > 0$), the weighted objective function appears as a line with a slope $-w_1/w_2 = -0/\beta = 0$. This is a horizontal line in Fig. 6-5 that gives point B as the optimal solution to the weighted problem.

After each objective is optimized individually, a systematic variation of the weights may be followed. That is, each weight may be varied from zero to some upper bound using a predetermined step size. The weighted problem is solved for each new set of weights that is generated in this manner. Suppose we had three objectives, Z_1, Z_2, and Z_3. Now, since relative values of the weights are of interest we can set one weight, say w_1, to a given value, say one, and vary only w_2 and w_3. The different weight sets that one would use if w_2 and w_3 were to be varied from zero to four in steps of one are shown in Table 6-1. There are 27 different weight sets, which means that the weighted problem would be solved 27 times to find, perhaps, 27 different noninferior solutions.

Every noninferior solution that is found with the weighting method requires the solution of a linear program. In practice, however, the weighting method can be accelerated through the use of parametric programming, which was discussed in Chapter 3. In this case the objective function is varied parametrically to capture the variations of the weights.

We must restrict ourselves to statements like "perhaps 27 different noninferior solutions" would be obtained because several weight sets may be within the switching values of one noninferior extreme point. On the other hand, the selected values for the weights may skip over many noninferior extreme points. In fact, this is generally the case in real-world problems for which the number of extreme points is enormous. It is worth getting more specific about the possible number of extreme points because this will give us a feeling for how good (or bad) the approximation of the noninferior set can be.

An upper bound on the number of feasible extreme points for a linear program with n variables (including slacks) and m constraints is

$$\left(\frac{n}{m}\right) = \frac{n!}{(n-m)!m!} \tag{6-14}$$

since an extreme point is formed by choosing m of the n variables to be positive. This is an upper bound because some of the extreme points may be infeasible. There are fewer noninferior extreme points than there are feasible ones, but it is not possible to generalize as to the number. It is safe to say that when (6-14) yields a mind-boggling number, as it normally will for real problems, the number of noninferior extreme points will also boggle the mind.

TABLE 6-1

TYPICAL SETS OF WEIGHTS FOR A PROBLEM WITH THREE OBJECTIVES

Weight set	w_1	w_2	w_3	
1	1	0	0	Initial phase
2	0	1	0	
3	0	0	1	
4	1	0	1	
5	1	0	2	
6	1	0	3	
7	1	0	4	
8	1	1	0	
9	1	1	1	
10	1	1	2	
11	1	1	3	
12	1	1	4	
13	1	2	0	
14	1	2	1	
15	1	2	2	
16	1	2	3	
17	1	2	4	
18	1	3	0	
19	1	3	1	
20	1	3	2	
21	1	3	3	
22	1	3	4	
23	1	4	0	
24	1	4	1	
25	1	4	2	
26	1	4	3	
27	1	4	4	

By an approximation of the noninferior set, which was stated to be the purpose of the weighting method, we mean that some extreme points are skipped over. This happens when weights that would lead to an extreme point are just not used. For example, if for the problem in Fig. 6-4 w_1/w_2 was never chosen to be in the range from a to b, then point A would not be found and the generated approximation of the noninferior set would follow the dashed line in Fig. 6-4. Of course, the analyst cannot usually see the actual noninferior set so that he may find it difficult to know just how good the current approximation is. There are no hard and fast rules that can be applied to assessing the sufficiency of an approximation. Generally, if there are no inordinately large gaps and the solutions give a reasonable account of the range of choice, then the approximation is adequate. The analyst can proceed by first using relatively large step sizes for the weights to identify a rough approximation of the noninferior set. If time allows, these results could then be used to identify a range of particular interest in which smaller step sizes would be used.

Before completing this discussion of the weighting method, one complication that may arise must be pointed out. For the special case in which one or more weights are set to zero, if there are alternative optima for the weighted problem, then some of these optima may be inferior. For a two-objective problem say that $w_1 = 0$ and $w_2 > 0$. This corresponds to the individual maximization of objective Z_2. These weights yield a horizontal indifference curve that touches $\mathbf{F}_o$ in Fig. 6-6 along the line segment AB. That is, all of the solutions which lie along AB are alternative optima for the weighted problem with $w_1/w_2 = 0$. Are all of the solutions on AB noninferior? Application of the definition of noninferiority tells us that only A is noninferior since A yields

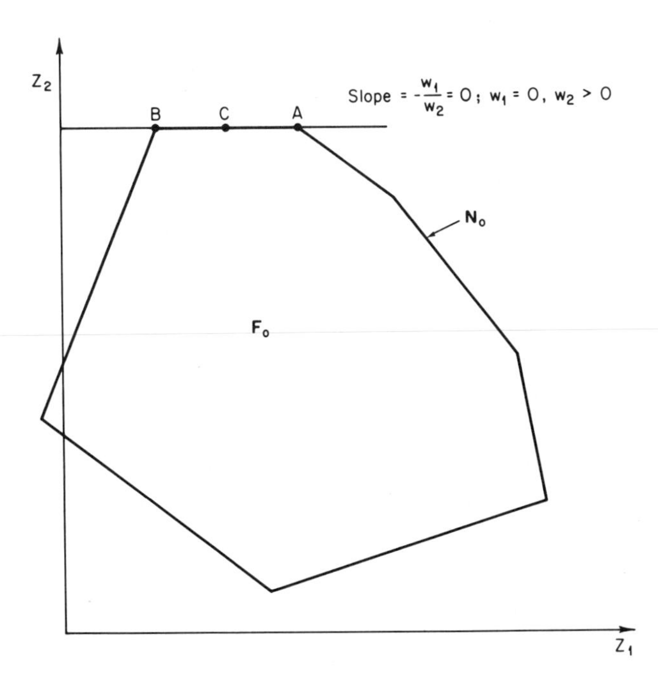

Fig. 6-6. Inferior alternate optimal solutions to the weighted problem when one weight equals 0.

the same amount for Z_2 and more Z_1 than point C or any other solution on AB. This demonstrates that when some of the weights are zero, inferior solutions may be obtained. Notice, however, that at least one of the alternative optima will be noninferior, as is point A in Fig. 6-6. Alternative optima for the weighted problem are noninferior when all of the weights are strictly positive. They may be noninferior even when some weights are zero, but the inferiority of the solutions must then be checked. A method for doing this is presented in the discussion of the constraint method in the next section.

6.1.3 Sample Application of the Weighting Method

The weighting method for finding noninferior solutions can be demonstrated with the example problem from Chapter 4. The strategy is to begin with a set of weights and then to vary them—each optimal solution of the transformed weighted problem yielding a noninferior solution to the original problem.

Applying weights to the objective functions in (4-7) gives

$$\text{maximize} \quad Z(x_1, x_2, w_1, w_2) = w_1 Z_1(x_1, x_2) + w_2 Z_2(x_1, x_2) \qquad (6\text{-}15)$$

or in terms of the decision variables,

$$\begin{aligned} \text{maximize} \quad & Z(x_1, x_2, w_1, w_2) = w_1(5x_1 - 2x_2) + w_2(-x_1 + 4x_2) \\ \text{s.t.} \quad & (x_1, x_2) \in \mathbf{F}_\mathrm{d} \end{aligned} \qquad (6\text{-}16)$$

where $\mathbf{F}_\mathrm{d}$ is the constraint set given in (4-7).

To begin, we optimize each objective individually. Two problems are solved:

$$\begin{aligned} \text{maximize} \quad & Z_1(x_1, x_2) = 5x_1 - 2x_2 \\ \text{s.t.} \quad & x_1, x_2 \in \mathbf{F}_\mathrm{d} \end{aligned} \qquad (6\text{-}17)$$

and

$$\begin{aligned} \text{maximize} \quad & Z_2(x_1, x_2) = -x_1 + 4x_2 \\ \text{s.t.} \quad & x_1, x_2 \in \mathbf{F}_\mathrm{d} \end{aligned} \qquad (6\text{-}18)$$

Contours for both of these objective functions are drawn in Fig. 6-7, in which a plot of the feasible region in decision space is drawn as in Fig. 4-2. There is a one-to-one correspondence between this plot and the objective space diagram in Fig. 6-8. An extreme point in Fig. 6-7 gives rise to a similarly labeled point in Fig. 6-8, as explained in Chapter 4. The objective function contours in Fig. 6-7 correspond to the linear indifference curves of Fig. 6-8, which are labeled similarly. The solution to problem (6-17) is $x_1 = 6, x_2 = 0$, and $Z_1 = 30$; Z_2 can be computed as -6. The point $(30, -6)$ corresponds to B in Fig. 6-8 where, not surprisingly, the linear indifference curve with a slope

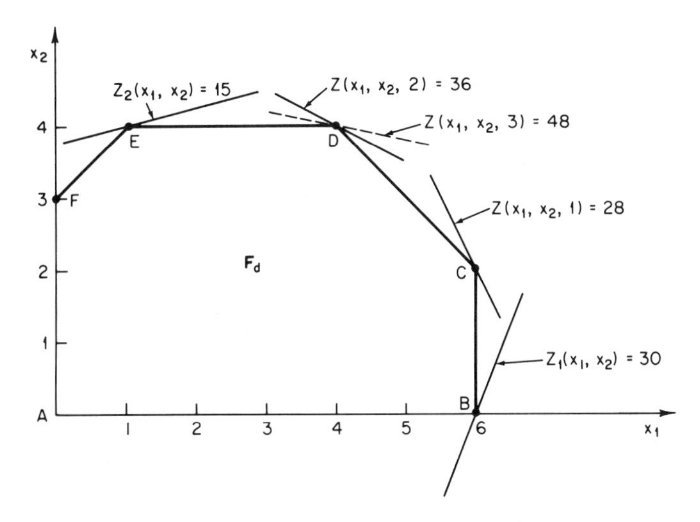

Fig. 6-7. Application of the weighting method to the sample problem: decision space.

of $(-w_1/w_2) = (-w_1/0) = -\infty$ (a vertical line in objective space) last touches $\mathbf{F}_o$. A similar result is obtained for problem (6-18) and points E in Fig. 6-7 and 6-8. For this problem the linear indifference curve is a horizontal line, i.e., a line with slope $= (-w_1/w_2) = (-0/w_2) = 0$. The individual problems have unique solutions so there is no problem with possibly inferior solutions that may arise when alternative optima are found.

As the weights are now varied, we know that only their relative values are of importance, i.e., it is the ratio w_1/w_2 that determines the slope of the linear indifference curve in objective space. For simplicity, the weight on objective Z_1 will be set to a value of one, $w_1 = 1$. The weighted problem in (6-16) becomes

$$\begin{aligned} \text{maximize} \quad & Z(x_1, x_2, w_2) = 5x_1 - 2x_2 + w_2(-x_1 + 4x_2) \\ \text{s.t.} \quad & (x_1, x_2) \in \mathbf{F}_d \end{aligned} \tag{6-19}$$

where w_1 is no longer among the arguments of the weighted objective function since w_1 will not vary during the rest of the procedure.

Now, it is decided arbitrarily that w_2 will be varied from zero to three in increments of one. That means that the weighted problem will be solved with $(w_1, w_2) = (1, 1)$, $(1, 2)$, and $(1, 3)$ since we have already solved the problem with $w_2 = 0$. For $(1, 1)$, the objective function of (6-19) becomes

$$\text{maximize} \quad Z(x_1, x_2, 1) = 5x_1 - 2x_2 + 1(-x_1 + 4x_2) \tag{6-20}$$

or

$$\text{maximize} \quad Z(x_1, x_2, 1) = 4x_1 + 2x_2 \tag{6-21}$$

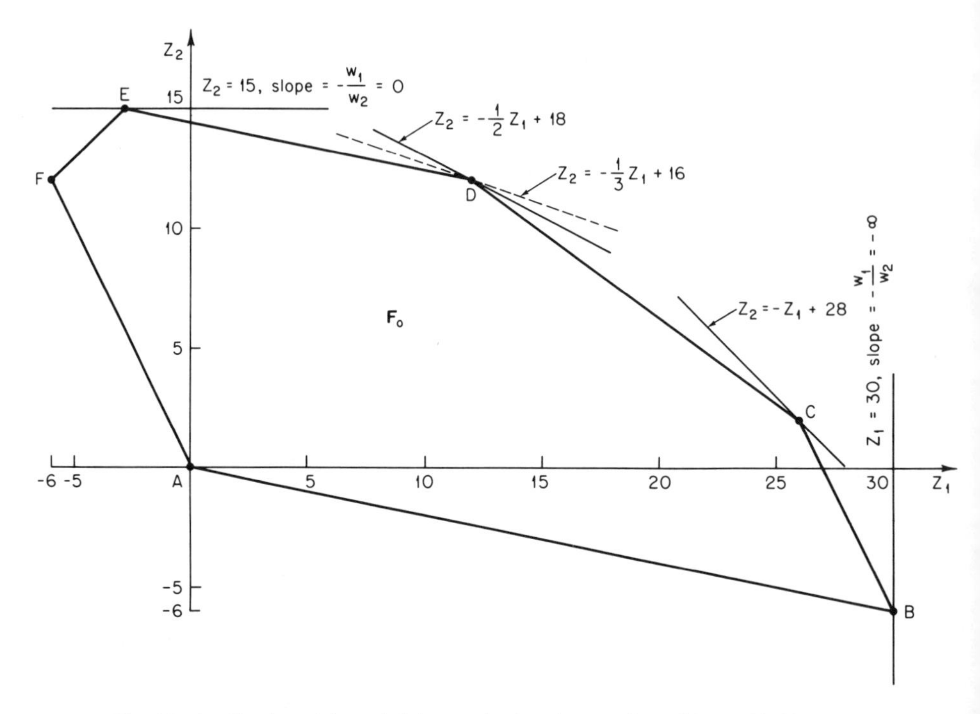

Fig. 6-8. Application of the weighting method to the sample problem: objective space.

The constraints are unchanged so the solution is found where a contour of (6-21) last touches $\mathbf{F}_d$ in Fig. 6-7. This occurs at point C, where $x_1 = 6$, $x_2 = 2$, $Z(x_1, x_2, 1) = 28$, $Z_1(x_1, x_2) = 26$, and $Z_2(x_1, x_2) = 2$. We can confirm this result by drawing a linear indifference curve with slope $= -1/w_2 = -1/1 = -1$ in Fig. 6-8 and observing where it last touches $\mathbf{F}_o$. The equation for the linear indifference curve that passes through C in Fig. 6-8 is, by analogy from (6-4),

$$Z_2 = -Z_1 + 28 \tag{6-22}$$

which is shown in Fig. 6-8.

When w_2 is increased to a value of two the weighted objective function becomes

$$Z(x_1, x_2, 2) = 5x_1 - 2x_2 + 2(-x_1 + 4x_2) \tag{6-23}$$

which is maximized at point D in Fig. 6-8, where $x_1 = x_2 = 4$, $Z(x_1, x_2, 2) = 36$, $Z_1(x_1, x_2) = Z_2(x_1, x_2) = 12$. The linear indifference curve corresponding to $w_2 = 2$ last touches $\mathbf{F}_o$ at D in Fig. 6-8.

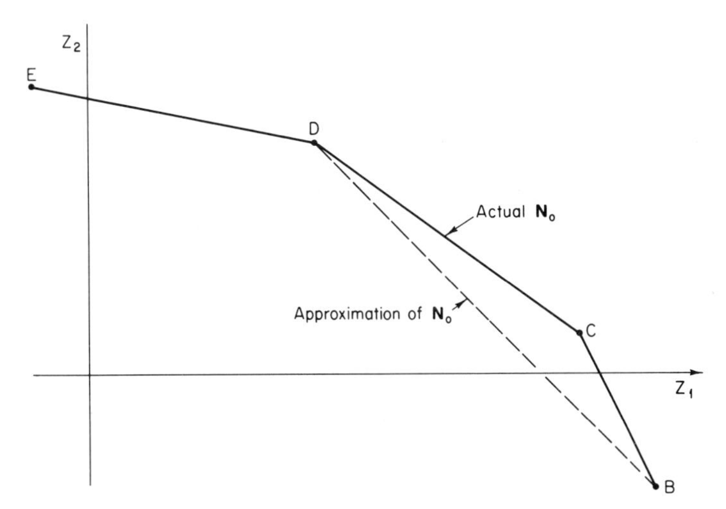

Fig. 6-9. An approximation of the noninferior set in objective space with the weighting method.

When w_2 is increased from two to three we get

$$Z(x_1, x_2, 3) = 2x_1 + 10x_2 \tag{6-24}$$

which also achieves its maximum at point *D*, as shown in Figs. 6-7 and 6-8.

Furthermore, the reader should check that as long as $\frac{5}{7} \leq w_2 \leq 5$, point *D* is the optimal solution to the weighted problem.

The values of the weights that were used above allowed all of the noninferior extreme points to be found. Suppose instead it was decided to vary w_2 from zero to four in increments of two, i.e., $(w_1, w_2) = (1, 2), (1, 4)$. In this case, point *C* would have been skipped since both weight sets are within the range defined by the switching values of point *D*. The noninferior set in objective space approximated with these weights appears in Fig. 6-9. Notice that point *C* is skipped and furthermore that the points on the line segment *BD* (the approximation) are feasible but inferior to points in the actual noninferior set.

*6.1.4 Further Characterization of the Weighting Method

Recall that the Kuhn–Tucker conditions for noninferiority state that a solution $\mathbf{x}$ is noninferior if there exist $w_k \geq 0$, $k = 1, 2, \ldots, p$, and $u_i \geq 0$, $i = 1, 2, \ldots, m$, and if conditions (4-14)–(4-16) are satisfied. If $\mathbf{x}$ satisfies these conditions, then it is also an optimal solution to

$$\text{maximize} \quad Z(\mathbf{x}, \mathbf{w}) = \sum_{k=1}^{p} w_k Z_k(\mathbf{x}) \tag{6-25}$$

$$\mathbf{x} \in \mathbf{F}_0 \tag{6-26}$$

since

$$\nabla Z(\mathbf{x}, \mathbf{w}) = \sum_{k=1}^{p} w_k \nabla Z_k(\mathbf{x}) \tag{6-27}$$

and the conditions for the single-objective problem, (3-108)–(3-110), are also satisfied.

The motivation for the weighting method comes from this observation: If one can find a weighted single-objective problem for which $\mathbf{x}$ is optimal, then $\mathbf{x}$ must also be noninferior since it satisfies both sets of conditions. The only qualification, in addition to the usual ones, is that the w_k must be strictly positive for sufficiency. If one or more $w_k = 0$ and alternative optima for (6-25), (6-26) are found, then some of the optimal solutions may be inferior.

The weighting method can also be tied to the graphical development of the Kuhn–Tucker conditions offered in Chapter 4. Figure 4-7 is repeated as Fig. 6-10 with the gradients of the weighted objective function contours drawn at the origin. The gradients for the two individual objectives are also drawn at the origin. Notice that as w_2 is increased from zero the weighted gradient sweeps across from $\mathbf{c}^1$ toward $\mathbf{c}^2$. The weighted gradient will always be between $\mathbf{c}^1$ and $\mathbf{c}^2$ since $w_k \geq 0 \; \forall_k$. This is another way of saying that the normal (gradient) to the separating hyperplane (weighted objective function contour) must lie in the cone generated by the individual objective function gradients.

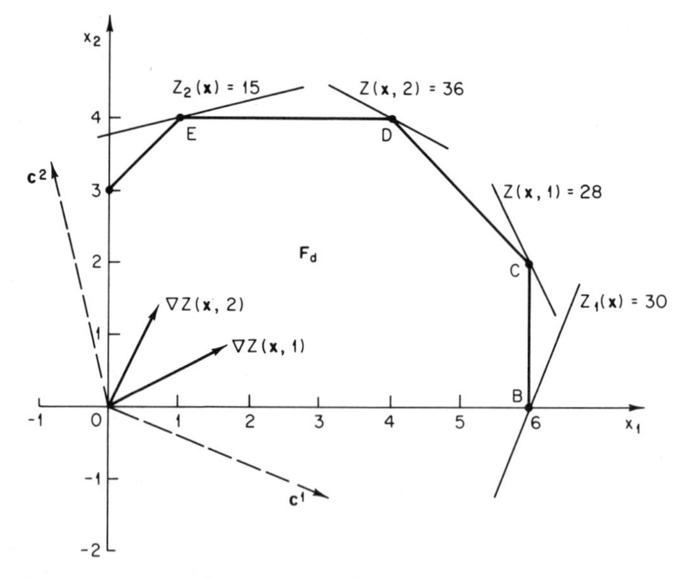

Fig. 6-10. Interpretation of the weighting method as a consequence of the Kuhn–Tucker conditions for noninferiority.

6.2 THE CONSTRAINT METHOD

The constraint method is perhaps the most intuitively appealing generating technique. It operates by optimizing one objective while all of the others are constrained to some value. Marglin (1967, pp. 24–25) appears to be the first to have suggested such an approach to multiobjective problems. In addition, Haimes (1973) presented the "ε-constraint" method, which is theoretically identical to the approach presented here, Cohon and Marks (1973) applied the constraint method to a water resource problem that is discussed in Chapter 9.

6.2.1 Motivation for the Constraint Method

In the fire station location problem of the previous section, suppose that instead of articulating a weight on population coverage our decision maker stated a constraint, e.g., at least L people shall be within S miles of the facility. The logical question becomes how much property value can we still cover when we are bound to cover at least L people? The answer to this question can be found in Fig. 6-11, in which $\mathbf{F}_0$ for the fire station problem has been redrawn.

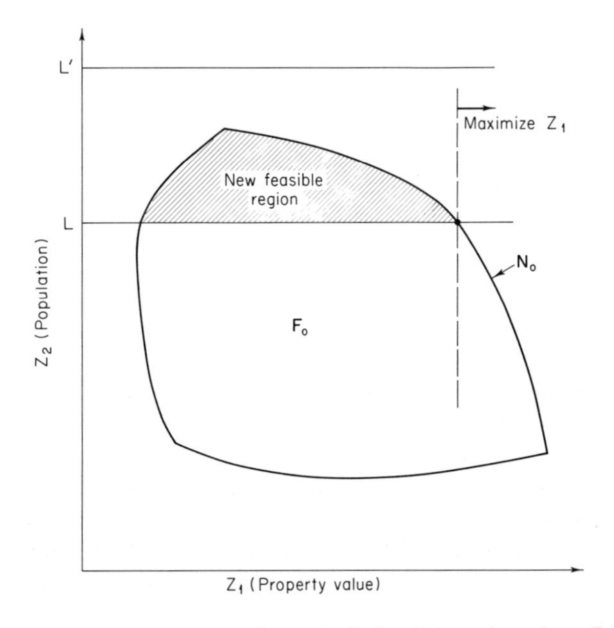

Fig. 6-11. Motivation for the constraint method: Specifying a bound on Z_2 (population) allows Z_1 (property value) to be maximized to find a noninferior solution.

The statement that "at least L people shall be within S miles" has been translated into the portion of $\mathbf{F}_0$ above the horizontal line in objective space, passing through point A and L in Fig. 6-11. The articulation of the constraint on population covered has reduced the size of the feasible region to the shaded area in Fig. 6-11. The maximization of Z_1 (property value covered) over the new feasible region leads to point A.

Notice in Fig. 6-11 that point A is a noninferior solution; it is in $\mathbf{N}_0$. We could change L to some other value, find a new maximum for Z_1, and this would yield a different noninferior solution to the original multiobjective problem. Proceeding in this manner, then, we can generate noninferior solutions by solving a series of single-objective problems. The bounds that are placed on the constrained objectives, i.e., the values of L, are considered parameters of the solution process; no value significance will be attached to them.

6.2.2 Mathematical Background

Given a multiobjective problem with p objectives

$$\begin{aligned}\text{maximize}\quad & \mathbf{Z}(x_1, x_2, \ldots, x_n)\\ &= [Z_1(x_1, x_2, \ldots, x_n), Z_2(x_1, x_2, \ldots, x_n)\\ &\quad, \ldots, Z_p(x_1, x_2, \ldots, x_n)]\end{aligned} \tag{6-28}$$

$$\text{s.t.}\quad (x_1, x_2, \ldots, x_n) \in \mathbf{F}_\text{d} \tag{6-29}$$

The constrained problem is

$$\text{maximize}\quad Z_h(x_1, x_2, \ldots, x_n) \tag{6-30}$$

$$\text{s.t.}\quad (x_1, x_2, \ldots, x_n) \in \mathbf{F}_\text{d} \tag{6-31}$$

$$\begin{aligned}&Z_k(x_1, x_2, \ldots, x_n) \geq L_k\\ &\qquad k = 1, 2, \ldots, h-1, h+1, \ldots, p\end{aligned} \tag{6-32}$$

where the hth objective was arbitrarily chosen for maximization. This formulation is a single-objective problem, so it can be solved by conventional methods, e.g., the simplex method for linear problems. The optimal solution to this problem is a noninferior solution to the original multiobjective problem if certain conditions, developed below, are satisfied.

The conditions that must be satisfied relate to the values for the L_k that are used in (6-32). Return to Fig. 6-11 and consider what would happen if a lower bound of L' were selected for Z_2. That is, we would have to satisfy the constraint

$$Z_2 \geq L' \tag{6-33}$$

Since $\mathbf{F}_0$ lies entirely below the $Z_2 = L'$ line in Fig. 6-11, there are no feasible solutions for which $Z_2 \geq L'$, i.e., the constrained problem is infeasible. The values for the L_k in (6-32) must be chosen so that feasible solutions to the resulting single-objective problem exist. This is a condition that we will attempt to satisfy in the algorithm discussed below.

Another condition that relates to our choice of the L_k is that all of the constraints on objectives should be binding at the optimal solution to the constrained problem. If this is not the case *and* if there are alternative optima to the constrained problem, then some of these optimal solutions may be inferior alternatives for the original multiobjective problem. This situation is equivalent to the case of zero weights in the weighting method. To see this we have to draw a correspondence between constraints on objectives and weights on those objectives.

Recall from Section 3.9 that the reduced cost for a slack variable (which converts a $\leq$ constraint to an equality) indicates the change in the objective function that would be observed if one more unit of the resource represented by the constraint were available. This reduced cost was called a dual variable or a shadow price for the constraint with which the slack variable is associated. Consider the constraint on the kth objective from the problem in (6-30)–(6-32),

$$Z_k(x_1, x_2, \ldots, x_n) \geq L_k \tag{6-34}$$

which can be converted to an equality by subtracting a surplus variable, a slack variable for $\geq$ constraints;

$$Z_k(x_1, x_2, \ldots, x_n) - x' = L_k \tag{6-35}$$

where x' is the surplus variable. When the constrained problem is solved, suppose (6-34) is satisfied as an equality (it is binding) so that $x' = 0$, i.e., the surplus variable is not in the basis. Its reduced cost, which will be called w_k, must be nonnegative since we are at the optimal solution of a maximization problem. This was shown in the development of the simplex method of Chapter 3.

If it were possible, we should like to make x' negative since this would increase the objective function $Z_h(x_1, x_2, \ldots, x_n)$. Making x' negative is equivalent to reducing L_k in (6-35), as can be seen by moving x' over to the right-hand side of (6-35). Thus, w_k tells us the amount by which $Z_h(x_1, x_2, \ldots, x_n)$ would increase for a decrease of one unit of $Z_k(x_1, x_2, \ldots, x_n)$. This shadow price is, therefore, equivalent to the tradeoff between the hth and kth objectives. It is also equivalent to a weight on the kth objective relative to the hth objective. This leads to the following observation: If L_k is used in the constrained problem (Z_h is being maximized) and the resulting shadow price is w_k, then w_k can be used in the weighted problem (with $w_h = 1$) to find the *same* noninferior solution. Thus, there is an intimate relationship between the weighting and constraint methods.

What can be concluded when the optimal solution to the constrained problem results in the satisfaction of (6-34) as a strict inequality? This means that $x' > 0$, i.e., the surplus variable is in the basis. We know from Chapter 3 that all basic variables have reduced costs equal to zero. Therefore w_k, the reduced cost for x' ,is equal to zero. If $w_k = 0$ were now used in the weighted problem and alternative optima were found, some of these solutions would possibly be inferior. Since nonbinding constraints in the constrained problem are equivalent to zero weights in the weighted problem, a similar caution on alternative optima, when there are nonbinding constraints, is required.

In discussing nonbinding constraints on objectives we have also discovered that the shadow price of such a constraint is the tradeoff between the constrained objective and the objective being optimized. Thus, in solving the constrained problem we get, as a by-product of the solution, the tradeoffs among the objectives at the resulting noninferior solution.

6.2.3 An Algorithm

The use of the weighting method was rather straightforward; a procedure was stated in just a few sentences. The application of the constraint method is a little more elaborate, so an algorithm will be stated in a formal manner.

Step 1 Construct a *payoff table.*

(a) Solve p individual maximization problems to find the optimal solution for each of the p objectives. Call the solution that maximizes objective k, $\mathbf{x}^k = (x_1^k, x_2^k, \ldots, x_n^k)$. If there are alternative optima for any of these problems, then choose those solutions from among the alternative optima that are noninferior.

(b) Compute the value of each objective at each of the p optimal solutions: $Z_1(\mathbf{x}^k), Z_2(\mathbf{x}^k), \ldots, Z_p(\mathbf{x}^k)$, $k = 1, 2, \ldots, p$. This gives us p values for each of the p objectives.

(c) Array the p values of each of the p objectives in a table in which the rows correspond to $\mathbf{x}^1, \mathbf{x}^2, \ldots, \mathbf{x}^p$ and the columns are labeled by the objectives (see Table 6-2).

(d) Find the largest number in the kth column; denote it by M_k. Find the smallest number in the kth column; denote it by n_k. Do this for $k = 1, 2, \ldots, p$.

Step 2 Convert a multiobjective programming problem such as (6-28) and (6-29) to its corresponding constrained problem as in (6-30)–(6-32).

TABLE 6-2

A PAYOFF TABLE FOR A PROBLEM WITH p OBJECTIVES

	$Z_1(\mathbf{x}^k)$	$Z_2(\mathbf{x}^k)$	$\cdots$	$Z_p(\mathbf{x}^k)$
$\mathbf{x}^1$	$Z_1(\mathbf{x}^1)$	$Z_2(\mathbf{x}^1)$	$\cdots$	$Z_p(\mathbf{x}^1)$
$\mathbf{x}^2$	$Z_1(\mathbf{x}^2)$	$Z_2(\mathbf{x}^2)$	$\cdots$	$Z_p(\mathbf{x}^2)$
$\vdots$	$\vdots$	$\vdots$	$\ddots$	$\vdots$
$\mathbf{x}^p$	$Z_1(\mathbf{x}^p)$	$Z_2(\mathbf{x}^p)$	$\cdots$	$Z_p(\mathbf{x}^p)$

Step 3 The n_k and M_k from step 1 represent a range for objective k in the noninferior set: $n_k \le Z_k \le M_k$. This range applies as well to L_k, the right-hand side of the constraint on objective k. Choose the number of different values of L_k that will be used in the generation of noninferior solutions. Call this r.

Step 4 Solve the constrained problem set up in step 2 for every combination of values for the L_k, $k = 1, 2, \ldots, h-1, h+1, \ldots, p$, where

$$L_k = n_k + [t/(r-1)](M_k - n_k), \qquad t = 0, 1, 2, \ldots, (r-1)$$

Since r values of each of the objectives (except objective h, which is in the objective function) will be used in step 4, there are r^{p-1} combinations of values of the L_k. Each of the r^{p-1} constrained problems that is feasible will yield a noninferior solution (if all objective constraints are binding). These solutions are the desired approximation of the noninferior set.

Each of the r^{p-1} constrained problems requires the solution of a linear program. In implementing the constraint method, some of the computational burden is saved through the use of parametric programming discussed in Section 3.9. In this case, the right-hand sides of the objective constraints are varied parametrically.

The payoff table introduced in step 1 is a device that is employed in the step method of Benayoun *et al.* (1971), which is discussed in Chapter 7. The payoff table provides a systematic way for finding a range of values for each of the L_k. The approach is predicated on the belief that the optima for the p individual problems represent "endpoints" of the noninferior set. This approach guarantees feasibility and noninferiority of the constrained problem for two-objective problems. Higher-dimensional problems will usually lead to some infeasible constrained problems. This is demonstrated below.

In a problem with two objectives, Z_2 achieves its minimum in the noninferior set at the solution that maximizes Z_1; a similar statement can be made for the minimum of Z_1 in the noninferior set. This result, which was proved

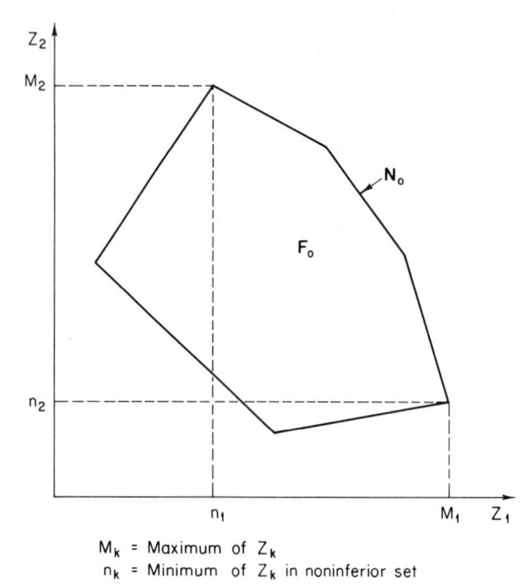

Fig. 6-12. The maximum for each objective in a two-objective problem represents an end-point of the noninferior set.

by Beeson (1971), is demonstrated in Fig. 6-12. Note that Z_1 does *not* achieve its *overall* minimum at maximum Z_2; it achieves its lowest value in the *noninferior set*.

The exactness of the approach with two objectives can also be demonstrated by constructing a payoff table, which is shown in Table 6-3. For the two-objective problem there are only two rows and two columns in the payoff table. Since there are only two elements in each column, one of them must be M_k, the maximum for objective k, and the other must be the minimum. Obviously, $Z_1(\mathbf{x}^1)$ and $Z_2(\mathbf{x}^2)$ are M_1 and M_2, respectively, so that $Z_2(\mathbf{x}^1) = n_2$ and $Z_1(\mathbf{x}^2) = n_1$. This states mathematically the verbal claim made above.

This whole discussion about values for L_k seems rather trivial for two-objective problems. For higher-dimensional problems, however, infeasibilities

TABLE 6-3

A Payoff Table for a General Two-Objective Problem

	$Z_1(\mathbf{x}^k)$	$Z_2(\mathbf{x}^k)$
$\mathbf{x}^1$	M_1	n_2
$\mathbf{x}^2$	n_1	M_2

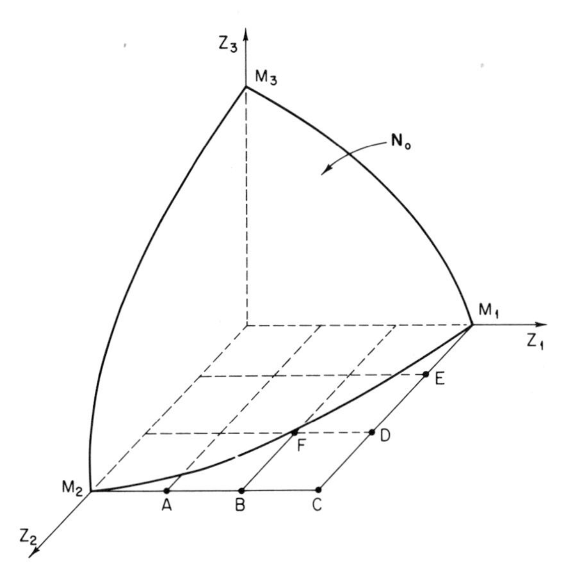

Fig. 6-13. Applying the constraint method to a three-objective problem: Some of the constrained problems are infeasible.

can be expected. Consider the three-dimensional noninferior set in Fig. 6-13. The maximum and minimum for each objective are shown; the minimum for each objective is at the origin for graphical simplicity. It has been assumed that Z_3 will be maximized while Z_1 and Z_2 will be constrained. Notice that even with the use of the payoff table, six of the 16 combinations of L_1 and L_2 will lead to infeasible problems. These combinations are labeled *A–F* in Fig. 6-13. Without the use of the payoff table or some other systematic procedure the number of infeasible constrained problems might be much greater.

Figure 6-13 is also useful in demonstrating the philosophy of the constraint method. In effect, the analyst places a grid over objective space, one that is the smallest possible that still contains $\mathbf{N}_o$. An attempt is then made to find a noninferior solution at each point on the grid.

6.2.4 A Method for Dealing with Alternative Optima in the Weighted and Constrained Problems

Alternative optima for both the weighted and constrained problems can cause difficulties in that some of the optimal solutions may be inferior. All alternative optima are noninferior in the weighted problem if all weights are strictly positive, and in the constrained problem if all constraints on objectives are binding. It is only in the case of some zero weights or some nonbinding objective constraints that some alternate optima *may* be inferior. A method for checking the noninferiority of alternative optima in the appropriate situations is presented below.

When some alternative optima may be inferior we want to search among all of them for the noninferior solution(s) and discard the inferior ones. In the case of the weighting method, suppose that $w_k > 0$ for $k = 1, 2, \ldots, h$ and $w_k = 0$ for $k = h + 1,\ h + 2, \ldots, p$ and that in the solution to the weighted problem of (6-10) and (6-11) we found alternative optima which gave $Z_k(x_1, x_2, \ldots, x_n) = Z_k^*$ for $k = 1, 2, \ldots, h$. Then we can solve a new problem to find the noninferior alternative optima.

$$\text{maximize} \quad \sum_{k=h+1}^{p} w_k Z_k(x_1, x_2, \ldots, x_n) \tag{6-36}$$

$$\text{s.t.} \quad (x_1, x_2, \ldots, x_n) \in \mathbf{F}_\mathrm{d} \tag{6-37}$$

$$Z_k(x_1, x_2, \ldots, x_n) = Z_k^*, \qquad k = 1, 2, \ldots, h \tag{6-38}$$

In (6-36), $w_k > 0$ for $k = h + 1, h + 2, \ldots, p$. The values of w_k are chosen in various combinations to find an approximation of that portion of the noninferior set which lies among the alternative optima.

To understand the procedure, a problem with three objectives will be considered. Suppose we began the weighting method by maximizing Z_1 individually, i.e., $w_1 > 0$ and $w_2 = w_3 = 0$. Alternative optima were found to exist since the feasible region in objective space is as shown in Fig. 6-14a. All of the solutions on the crosshatched face of $\mathbf{F}_\mathrm{o}$ in Fig. 6-14a yield the maximum of objective Z_1, called Z_1^*. We proceed then by solving

$$\text{maximize} \quad \sum_{k=2,3} w_k Z_k(x_1, x_2, \ldots, x_n) \tag{6-39}$$

$$\text{s.t.} \quad (x_1, x_2, \ldots, x_n) \in \mathbf{F}_\mathrm{d} \tag{6-40}$$

$$Z_1(x_1, x_2, \ldots, x_n) = Z_1^* \tag{6-41}$$

By adding the constraint in (6-41) to the original feasible region $\mathbf{F}_\mathrm{o}$, we have created a new feasible region $\mathbf{F}_\mathrm{o}'$, which is just the face of $\mathbf{F}_\mathrm{o}$ that gives Z_1^*. This face is redrawn in Fig. 6-14b. All of the feasible solutions of $\mathbf{F}_\mathrm{o}'$ yield Z_1^*; those that are also noninferior are shown as the crosshatched portion of $\mathbf{F}_\mathrm{o}'$ in Fig. 6-14b. These are found by using combinations of strictly positive values for w_2 and w_3 in (6-39). The reader will note that the procedure is equivalent to the weighting method as applied to a subset of the original feasible region.

For alternative optima in the constrained problem we proceed in a similar fashion but, instead of defining the set of objectives $Z_1, Z_2, \ldots, Z_h$

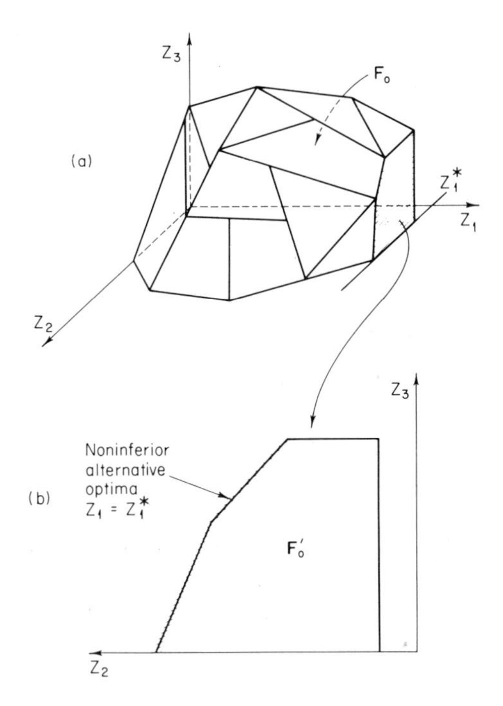

Fig. 6-14. Dealing with alternative optima when some of them may be inferior: (a) three-dimensional noninferior set; (b) face that maximizes Z_1.

as those with strictly positive weights, we define them as those objectives that had binding constraints and that objective selected for optimization. Furthermore, the set of objectives $Z_{h+1}, Z_{h+2}, \ldots, Z_p$ are those that had nonbinding constraints. The procedure is then pursued in an identical manner.

6.2.5 Application of the Constraint Method to the Sample Problem

In this section the algorithm presented in Section 6.2.3 is applied to the sample problem from Chapter 4.

Step 1 is to construct a payoff table. First, each objective is optimized individually. The maximization of $Z_1(x_1, x_2)$ leads to the unique solution (no alternative optima) $\mathbf{x}^1 = (x_1^1, x_2^1) = (6, 0)$ with $Z_1(\mathbf{x}^1) = 30$ and $Z_2(\mathbf{x}^1) = -6$. The maximization of $Z_2(x_1, x_2)$ gives the unique solution $\mathbf{x}^2 = (x_1^2, x_2^2) = (1, 4)$ with $Z_1(\mathbf{x}^2) = -3$ and $Z_2(\mathbf{x}^2) = 15$. The payoff table is shown in Table 6-4. The table shows that $M_1 = 30$, $n_1 = -3$, $M_2 = 15$, $n_2 = -6$.

TABLE 6-4

PAYOFF TABLE FOR THE SAMPLE PROBLEM

	$Z_1(\mathbf{x}^k) = 5x_1^k - 2x_2^k$	$Z_2(\mathbf{x}^k) = -x_1^k + 4x_2^k$
$\mathbf{x}^1 = (6, 0)$	30	-6
$\mathbf{x}^2 = (1, 4)$	-3	15

In step 2 the constrained problem is set up:

$$\text{maximize} \quad Z_1(x_1, x_2) = 5x_1 - 2x_2 \tag{6-42}$$

$$\text{s.t.} \quad (x_1, x_2) \in \mathbf{F}_\mathrm{d} \tag{6-43}$$

$$Z_2(x_1, x_2) = -x_1 + 4x_2 \geq L_2 \tag{6-44}$$

where objective Z_1 has been chosen (arbitrarily) for maximization.

In step 3 the range for the right-hand side of (6-44) is established as $n_2 \leq L_2 \leq M_2$ or $-6 \leq L_2 \leq 15$. The value of r, the number of different values for L_2 to be used in generating noninferior solutions, is set. Choose $r = 4$. In step 4 the problem in (6-42)–(6-44) is solved $r = 4$ times. Using (6-36), the values for L_2 can be computed from

$$L_2 = -6 + \tfrac{1}{3}t[15 - (-6)], \qquad t = 0, 1, 2, 3 \tag{6-45}$$

i.e.,

$$L_2 = -6 + 7t, \qquad t = 0, 1, 2, 3 \tag{6-46}$$

The constrained problem will be solved four times with $L_2 = -6, 1, 8, 15$.

The feasible regions in decision space and objective space for the *original*, not the constrained problem, are presented in Fig. 6-15 and 6-16. The constrained problem has an additional constraint, (6-44), which results in a reduced version of $\mathbf{F}_\mathrm{d}$ and $\mathbf{F}_\mathrm{o}$. There are really four different feasible regions in the two spaces; one for each of the four values of L_2. The lines corresponding to (6-44) for four different values of L_2 are drawn in Figs. 6-15 and 6-16. Considering Fig. 6-15, when $L_2 = -6$, the new feasible region is just $\mathbf{F}_\mathrm{d}$; i.e., the line labeled $Z_2(x_1, x_2) = -6$ "cuts through" $\mathbf{F}_\mathrm{d}$ at the point (6, 0). $Z_1(x_1, x_2)$ achieves its maximum on the new feasible region at the point (6, 0). As L_2 is increased from -6 to 1, the line corresponding to (6-44) shifts up to the northwest. At $L_2 = 1$ the feasible region is reduced to that part of $\mathbf{F}_\mathrm{d}$ which lies above $Z_2(x_1, x_2) = 1$. $Z_1(x_1, x_2)$ is optimized on this new feasible region at (6, 1.75). Other noninferior solutions are found at (4.8, 3.2) when $L_2 = 7$ and at (1, 4) when $L_2 = 15$.

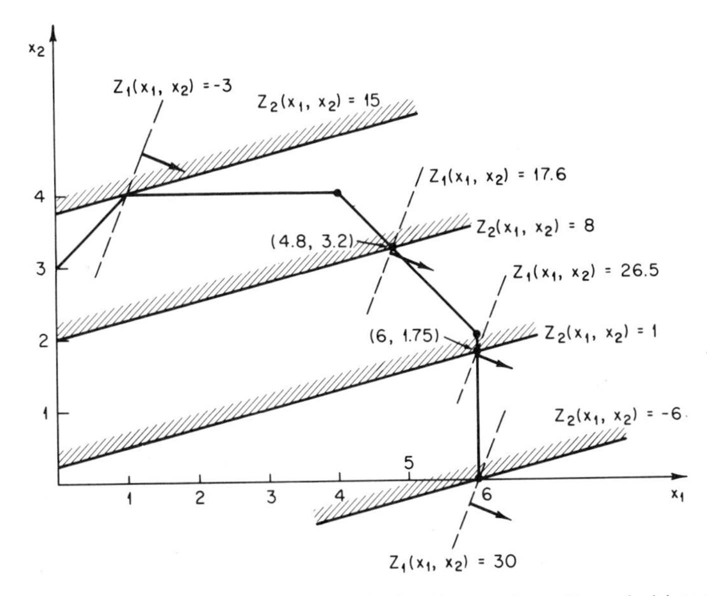

Fig. 6-15. Application of the constraint method to the sample problem: decision space.

The progression of the constraint method can be followed in the objective space diagram of Fig. 6-16. The new constraint in (6-44) now appears as a horizontal line. The maximum of Z_1 for the constrained problem is found where the appropriate horizontal line representing the new constraint touches the noninferior set. Thus, for example, the horizontal line passing through $Z_2 = 1$ crosses the noninferior set at $Z_1 = 26.5$, which is, of course, the solution found by the analysis in decision space.

A few additional remarks about this example are in order. Notice that solving the constrained problem at $L_2 = -6$ and $L_2 = 15$, i.e., at n_2 and M_2, is redundant and need not be done; these two solutions were found by the individual maximization of the objectives. This is the case for all two-objective problems. For problems with more than two objectives it is worth solving the constrained problem at the minima and maxima of the objectives.

Consider Fig. 6-16 again, in which $\mathbf{F}_o$, $\mathbf{N}_o$, and the approximation of the noninferior set found with the constraint method are drawn. The approximation looks a lot like the weighting method approximation in Fig. 6-9, but there is a difference. The constraint method found noninferior solutions that are not noninferior extreme points. The constraint method will usually identify these nonextreme noninferior points because the original feasible region is modified, as are some of the extreme points; new ones are created and old ones are destroyed. In the weighting method, the feasible region is never altered.

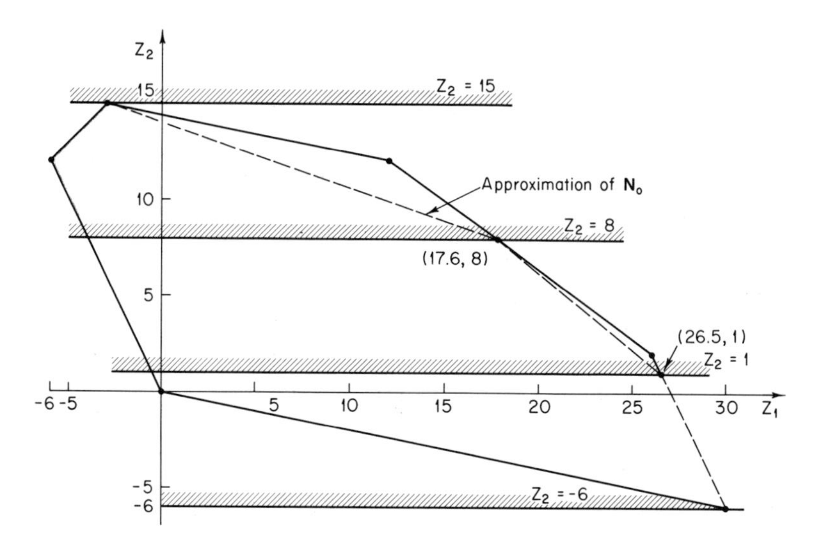

Fig. 6-16. Application of the constraint method to the sample problem: objective space.

*6.2.6 Theoretical Basis for the Constraint Method from the Kuhn–Tucker Conditions

The third Kuhn–Tucker condition in (4-16) can be rewritten as

$$w_h Z_h(\mathbf{x}^*) + \sum_{\substack{k=1 \\ k \neq h}}^{p} w_k \nabla Z_k(\mathbf{x}^*) - \sum_{i=1}^{m} u_i \nabla g_i(\mathbf{x}^*) = 0 \tag{6-47}$$

where $\mathbf{x}^*$ is a noninferior solution, $w_k \geq 0$, $k = 1, 2, \ldots, p$, and $u_i \geq 0$, $i = 1, 2, \ldots, m$. If we assume that $w_h > 0$, which is not a restrictive assumption, particularly if our interest lies in sufficiency, then we can interpret (6-47) as the third condition for optimality for the single-objective problem

$$\text{maximize} \quad w_h Z_h(\mathbf{x}) \tag{6-48}$$

$$\text{s.t.} \quad g_i(\mathbf{x}) \leq 0 \quad \forall i \tag{6-49}$$

$$Z_k(\mathbf{x}) \geq L_k \quad \forall k \neq h \tag{6-50}$$

Now, w_h is just a positive number, so (6-48) can be divided by it without altering the problem:

$$\text{maximize} \quad Z_h(\mathbf{x}) \tag{6-51}$$

which gives, with (6-49) and (6-50), the constrained problem in (6-30)–(6-32). This, then, provides a theoretical basis for the constraint method.

The sense of the constraints in (6-50) comes from the fact that the second term in (6-47) is preceded by a plus sign. We could have written (6-50) as

$$-Z_k(\mathbf{x}) \leq -L_k \qquad \forall k \neq h \tag{6-52}$$

so that in (6-47) we would get

$$-\sum_{\substack{k=1 \\ k \neq h}}^{p} w_k[-\nabla Z_k(\mathbf{x}^*)] \tag{6-53}$$

for the second term.

The right-hand sides L_k in (6-50) seem to have appeared magically which, of course, they did. They are not at issue in the third condition, for which any right-hand side would do. They are, of course, important for the first condition, (4-14), which requires feasibility and for the second condition, (4-15), which states complementary slackness.

6.3 THE NONINFERIOR SET ESTIMATION (NISE) METHOD

The noninferior set estimation (or NISE) method was developed by Cohon *et al.* (1978) to converge quickly on a good approximation of the noninferior set. In addition, the accuracy of the approximation can be controlled in the NISE method through a predetermined error criterion, which is compared to the maximum possible error at every iteration of the method. The method is related to a problem discussed by Zeleny (1974a, pp. 149–158).

The NISE method is developed below for problems with two objectives. Modifications in the algorithm for its application to higher-dimensional problems are discussed at the end of this section. It is assumed throughout that the feasible region is a convex set and that the objective functions are linear.

6.3.1 Mathematical Background

The NISE method operates by finding a number of noninferior extreme points and evaluating the properties of the line segments between them. Suppose two noninferior extreme points have been found; then the line segment between them is feasible and it may or may not be noninferior. If the line segment is noninferior, then moving in a direction out from the line segment is infeasible. If the line segment is inferior, then there are noninferior points in the outward direction. Figure 6-9 demonstrates both cases. The line segment between E and D is noninferior and any movement to the northwest is toward infeasibility. The line segment between D and B is inferior; there are points in the northwesterly direction that are noninferior, e.g., point C.

Noninferior points are found in the NISE method through the use of the weighted problem, which was discussed in Section 6.1. In the weighting method one can use any nonnegative weights. In the NISE method the values of the weights are chosen so that the next noninferior point is the feasible solution farthest out in a direction perpendicular to the line segment connecting two adjacent previously found points.

The method begins by optimizing each objective individually, yielding points A and B in Fig. 6-17. The next solution should be that feasible solution farthest out along the indicated direction. This point is found by solving the weighted problem as in (6-10) and (6-11) with weights w_1 and w_2 that satisfy

$$-w_1/w_2 = (\text{slope of line } AB) \tag{6-54}$$

The slope is an easy quantity to compute and we can set one of the weights to an arbitrary value such as 1. Therefore, the weights can be completely specified for each step of the algorithm. For the problem depicted in Fig. 6-17, the next noninferior solution that we would find is point C.

The accuracy of the approximation of the noninferior set is controlled by the maximum allowable error, which is preset by the analyst. The algorithm terminates when the maximum possible error does not exceed the maximum allowable error. The maximum possible error can be determined from what is known about noninferiority and the feasible region.

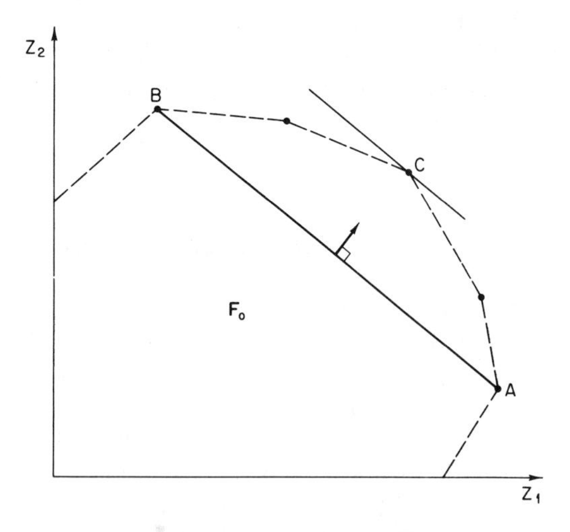

Fig. 6-17. Noninferior set estimation (NISE) method: Noninferior points are found by pushing out line segments defined by previously identified noninferior solutions.

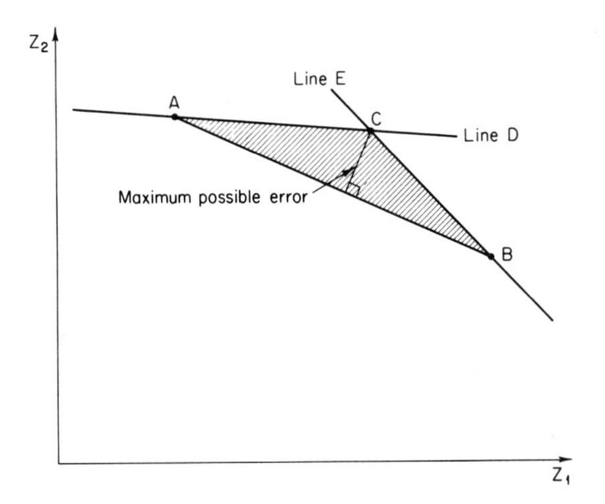

Fig. 6-18. Maximum possible error for a given set of previously identified noninferior solutions.

Consider Fig. 6-18 in which two noninferior points *A* and *B* are drawn. It is known that they are noninferior because they were found by solving the weighted problem. The weights used in the weighted problem that resulted in A can be used to compute the slope of a linear indifference curve corresponding to the weighted objective function (see Section 6.1). This linear indifference curve is shown as line *D* in Fig. 6-18. Line *E* in Fig. 6-18 is the linear indifference curve passing through *B* that corresponds to the weights used in the weighted problem that yielded *B*. We know that there can be no feasible solutions that lie above lines *D* or *E*; if there were, then *A* and *B* could not be optimal solutions to a weighted problem. The lines *D* and *E* therefore represent an upper bound to the noninferior set; i.e., the actual noninferior set cannot lie above these lines.

A lower bound for the actual noninferior set can also be identified. The line segment *AB* in Fig. 6-18 is such a lower bound because the feasible region has been assumed to be a convex set; i.e., if two points are feasible, than all of the solutions on the line segment that joints them are also feasible. That means that points which lie below the line *AB* must be inferior.

We can conclude from these observations that the actual noninferior set must lie in the shaded area of Fig. 6-18. The approximation of the noninferior set is taken as the points *A* and *B* and the line segment between them. Thus, the maximum possible error, given our approximation, is the longest line perpendicular to *AB* that can be drawn between the lower and upper bounds. The maximum possible error is shown in Fig. 6-18 as the perpendicular from *AB* that is drawn to *C*, the point of intersection of lines *D* and *E*.

6.3.2 The NISE Algorithm for Two-Objective Problems

An algorithm for the application of the NISE method to problems with two objectives is presented in this section; modifications required to extend the method to higher-dimensional problems are discussed in Section 6.3.4. The NISE algorithm begins by maximizing each objective individually, which yields two points in objective space. The slope of the line segment connecting these two points is used to compute weights for use in the weighted problem. When the new problem is solved, the resulting noninferior solution is used in the computation of the maximum possible error. The algorithm continues in this fashion until the maximum possible error in all parts of the noninferior set is less than or equal to the maximum allowable error.

Some new notation must be introduced before the algorithm can be stated. We also require a technique for ordering noninferior extreme points so that adjacent points can be determined. Define P_t as the tth noninferior extreme in objective space generated with the algorithm. Since the method begins by optimizing the objectives individually, P_1 is the solution that maximizes objective Z_1 and P_2 is the solution that maximizes objective Z_2. Note that as subsequent noninferior points in objective space are generated, P_t and P_{t+1} will not in general be adjacent.

The ordering scheme rests on the definition of S_i, which is defined as that noninferior point with the ith highest value of Z_2. Three noninferior solutions P_1, P_2, and P_3, are shown in Fig. 6-19. P_1 and P_2 maximize Z_1 and Z_2, respectively. P_3 was found by solving the weighted problem with weights corresponding to the slope of the line passing through P_1 and P_2. These noninferior solutions are relabeled with appropriate S_is in Fig. 6-19. $P_2 = S_1$ because it gives the highest value of Z_2. Since Z_2 always decreases as one moves from S_1 through the approximation of the noninferior set, S_i and S_{i+1} will always be adjacent points in a given approximation of the noninferior set. Note that S_i and S_{i+1} are not in general adjacent extreme points of the feasible region; they are only adjacent in a current approximation.

The computation of the weights to be used in the weighted problem is based on the slope of the line segment connecting two adjacent points S_i and S_{i+1} in the current approximation. Call this slope α; then

$$\alpha = \frac{Z_2(S_i) - Z_2(S_{i+1})}{Z_1(S_i) - Z_1(S_{i+1})} \tag{6-55}$$

where $Z_k(S_i)$ is the value of objective k at the point S_i. The slope α is always negative since $Z_2(S_i) > Z_2(S_{i+1})$ and $Z_1(S_i) < Z_1(S_{i+1})$ by the definition of S_i and the nature of the noninferior set.

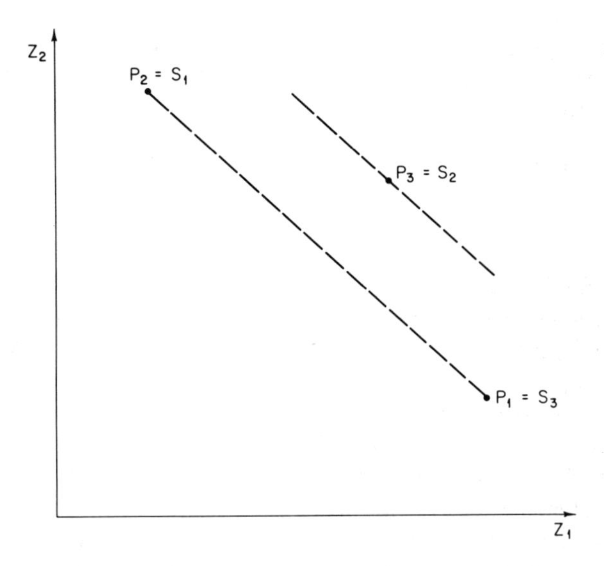

Fig. 6-19. The ordering scheme for the NISE method.

The weights in the weighted problem should be chosen so that

$$w_1/w_2 = -\alpha \tag{6-56}$$

We will follow the convention here that $w_2 = 1$, so that w_1 can be computed, by substituting (6-55) into (6-56) and noting the sign change, from

$$w_1 = \frac{Z_2(S_i) - Z_2(S_{i+1})}{Z_1(S_{i+1}) - Z_1(S_i)} \tag{6-57}$$

which is always positive.

The desired weighted objective function is

$$\text{maximize} \quad Z(x_1, \ldots, x_n; w_1, w_2) = w_1 Z_1(x_1, \ldots, x_n) + w_2 Z_2(x_1, \ldots, x_n) \tag{6-58}$$

which, upon substitution of $w_2 = 1$ and (6-57), becomes

$$\text{maximize} \quad Z(x_1, \ldots, x_n; i, i+1) = \left[\frac{Z_2(S_i) - Z_2(S_{i+1})}{Z_1(S_{i+1}) - Z_1(S_i)}\right] Z_1(x_1, \ldots, x_n) + Z_2(x_1, \ldots, x_n) \tag{6-59}$$

where i and $(i + 1)$ are included in the argument of Z to indicate the dependency of the weighted objective function on the choice of the line segment in the current approximation for which the approximation will be improved. (6-59) can be rewritten by multiplying through by the denominator of the bracketed term:

$$\begin{aligned} \text{maximize } & Z(x_1, \ldots, x_n; i+1) \\ &= [Z_2(S_i) - Z_2(S_{i+1})]Z_1(x_1, \ldots, x_n) \\ &\quad + [Z_1(S_{i+1}) - Z_1(S_i)]Z_2(x_1, \ldots, x_n) \end{aligned} \tag{6-60}$$

The weighted objective function in (6-60) is the equation of the line in objective space that goes through points S_i and S_{i+1}. Thus, when we solve this weighted problem, if the solution is on the line between S_i and S_{i+1}, then the old approximation is in fact noninferior. If it is above the line, then the old approximation is inferior. We can get this in another way. A value for $Z(x_1, \ldots, x_n; i, i+1)$ at S_i or S_{i+1} can be computed before solution of the weighted problem. We will call this value $B_{i,i+1}$. Upon solution of the weighted problem, $Z(x_1, \ldots, x_n; i, i+1) \geq B_{i,i+1}$. If $Z(x_1, \ldots, x_n; i, i+1) = B_{i,i+1}$, then the line segment between S_i and S_{i+1} is noninferior. If

$$Z(x_1, \ldots, x_n; i, i+1) > B_{i,i+1},$$

then a new noninferior solution will have been found, and the line segment between S_i and S_{i+1} is inferior.

The final bits of terminology are the definitions of the maximum possible and maximum allowable errors. We will define $\Psi_{i,i+1}$ as the maximum possible error in the line segment connecting S_i and S_{i+1}. It is measured as the maximum distance between the lower bound and the upper bound between S_i and S_{i+1}, measured in a direction perpendicular to the lower bound. T is the maximum allowable error; its value is preset by the analyst. The index n is used in the algorithm to indicate the number of noninferior solutions currently generated.

The NISE Algorithm follows.

Step 1 Maximize the objectives individually. The image in objective space of the optimum for objective Z_1 is P_1 and for objective Z_2 is P_2. (If there are alternative optima for either of these two problems, then the noninferior points are selected by using a method such as the one presented in Section 6.2.4). Let $S_1 = P_2$, $S_2 = P_1$, and $n = 2$. Compute Ψ_{12}.

Step 2 If $\Psi_{i,i+1} \leq T$ for $i = 1, 2, \ldots, n-1$, then STOP; a sufficiently accurate approximation consisting of the points S_i, $i = 1, 2, \ldots, n$, and the line segments between adjacent points has been found. Otherwise, go to step 3.

Step 3 Search the $\Psi_{i,i+1}$, $i = 1, 2, \ldots, n-1$, for the largest value. For the i, $i+1$ that yields the largest maximum possible error solve the weighted problem, using the objective function in (6-60). Compute $B_{i,i+1}$ at the weighted objective function evaluated at S_i or S_{i+1}. If the resulting objective function value $Z(x_1, \ldots, x_n; i, i+1) = B_{i,i+1}$, then set $\Psi_{i,i+1} = 0$ and return to step 2. If $Z(x_1, \ldots, x_n; i, i+1) > B_{i,i+1}$, designate the new noninferior solution as P_{n+1}. Go to step 4.

Step 4 Reorder the points P_t, $t = 1, 2, \ldots, n+1$. Note that i is the highest value of t such that $Z_2(S_t) \geq Z_2(P_{n+1})$, i.e., the new solution gives a value of Z_2 that is next highest after $Z_2(S_i)$. This allows the use of the following reordering scheme:

$$S'_t = S_t, \qquad t = 1, 2, \ldots, i \tag{6-61}$$

$$S'_{i+1} = P_{n+1} \tag{6-62}$$

$$S'_{t+1} = S_t, \qquad t = i+1, \ldots, n \tag{6-63}$$

The Ψ' terms must also be relabeled:

$$\Psi'_{t,t+1} = \Psi_{t,t+1}, \qquad t = 1, 2, \ldots, i-1 \quad (\text{if } i \geq 1) \tag{6-64}$$

$$\Psi'_{t+1,t+2} = \Psi_{t,t+1}, \qquad t = i+1, \ldots, n-1 \quad (\text{if } i \leq n-2) \tag{6-65}$$

Compute $\Psi'_{i,i+1}$ and $\Psi'_{i+1,i+2}$. Increment n by one and return to step 2.

The computation of $\Psi_{i,i+1}$ in steps 1 and 4 is a straightforward geometric–trigonometric problem. The upper and lower bounds always form a triangular area, as in Fig. 6-18. The vertices A and B in Fig. 6-18 are always known. They correspond to S_i and S_{i+1}. The vertex C can be found as the intersection of two lines, lines D and E in Fig. 6-18, with known equations; these equations are just the weighted objective functions with the appropriate vertical intercept values. The computation of $\Psi_{i,i+1}$ therefore becomes the determination of the altitude of a triangle with vertices A, B, and C and base AB. This procedure and the entire algorithm are demonstrated with an example in the next section.

6.3.3 Application of the NISE Algorithm to the Sample Problem

The sample problem, from Chapter 4, with two objectives and two decision variables will be used to demonstrate the NISE algorithm. A value for T, the maximum allowable error, must be selected prior to the start of the algorithm. There is a minor difficulty here in that distance in objective space, which has axes measured in different, noncommensurable units, is not terribly meaningful. A useful way to proceed therefore is to designate T as a percentage of Ψ_{12} computed in step 1. We will choose $T = 0.50$ (Ψ_{12} from step 1).

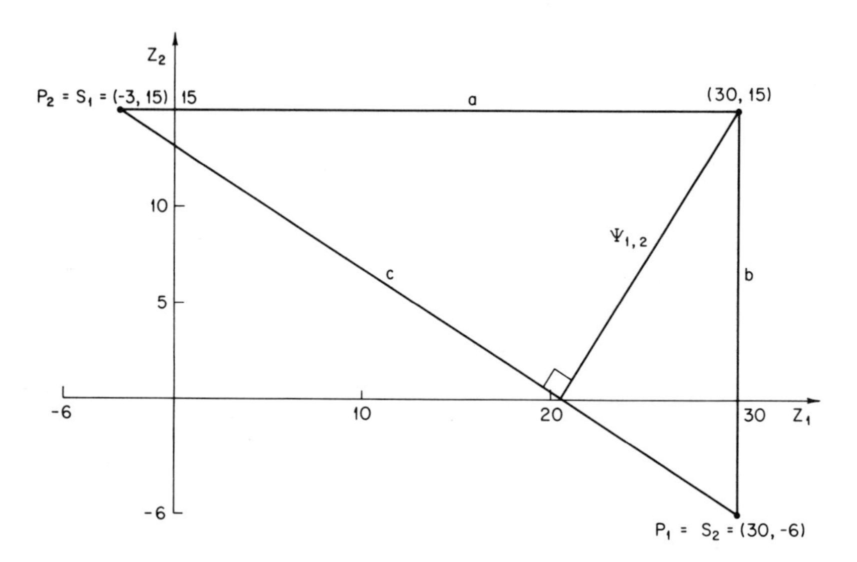

Fig. 6-20. Application of the NISE method to the sample problem: step 1 and the computation of the maximum possible error.

In step 1 objectives Z_1 and Z_2 are optimized individually. We know from before that this gives for objective Z_1 the point (6, 0) in decision space and the point (30, -6) in objective space; i.e., $P_1 = (30, -6)$. The optimization of objective Z_2 gives $(x_1, x_2) = (1, 4)$ and the point $P_2 = (-3, 15)$ in objective space. We set $S_1 = P_2$ and $S_2 = P_1$ since P_2 gives the highest value of Z_2 and P_1 gives the second highest value of Z_2. The two points are shown in Fig. 6-20. The lower bound and the upper bounds and the location of the actual noninferior set are also shown in Fig. 6-20. Recall that the upper bounds are the linear indifference curves corresponding to the weighted objective function used in the generation of a point in the current approximation of the noninferior set. In this case, weights used to generate P_2 were $w_1 = 0$ and $w_2 > 0$, i.e., Z_2 was optimized individually. Thus, the upper bound passing through P_2 is a line with slope $= -w_1/w_2 = 0$, which is the horizontal line passing through P_2 in Fig. 6-20. A similar line of reasoning leads to the other upper bound—the vertical line passing through P_1.

The value for Ψ_{12} is computed in step 1. Figure 6-20 shows that Ψ_{12} is the altitude of the triangle with vertices at P_1, P_2, and the point of intersection of the two upper bounds and a base equal to the length of the lower bound. The point of intersection in this case is (30, 15). We know for a triangle with sides a, b, and c and altitude d_c (with c as the base) that

$$\mathscr{A} = \tfrac{1}{2}cd_c \tag{6-66}$$

where $\mathscr{A}$ is the area. We know c for our triangle and would like to find d_c so that the area must be computed. One formula for the area of a triangle is

$$\mathscr{A} = [s(s - a)(s - b)(s - c)]^{1/2} \tag{6-67}$$

where s, known as the semiperimeter, can be computed from

$$s = \tfrac{1}{2}(a + b + c) \tag{6-68}$$

The sides of the triangle in Fig. 6-20 have been labeled a, b, and c. We can compute their lengths from the coordinates of the vertices of the triangle. In a Euclidean space, such as our objective space, the distance between two points q and r with coordinates (q_1, q_2) and (r_1, r_2) is

$$\text{distance from } q \text{ to } r = [(q_1 - r_1)^2 + (q_2 - r_2)^2]^{1/2} \tag{6-69}$$

Using this formula, we get

$$a = \{[30 - (-3)]^2 + (15 - 15)^2\}^{1/2} \tag{6-70}$$

which gives $a = 33$. A similar computation gives $b = 21$ and

$$c = \{(-3 - 30)^2 + [15 - (-6)]^2\}^{1/2} \tag{6-71}$$

which gives $c = 39.1$.

The formula in (6-68) can be used to compute the semiperimeter, which gives $s = 46.6$. Equation (6-67) is then used to compute the area of the triangle. This yields $\mathscr{A} = 348.9$. Finally, (6-66) can be used to compute $\Psi_{12} = d_c = 17.8$.

The last part of step 1 is to set $n = 2$. We can also now set T as 50% of Ψ_{12}, i.e., $T = 8.9$.

In step 2 we check to see if the error criterion is met. Obviously $\Psi_{12} < T$ so the algorithm continues to step 3.

There is only one Ψ_{12} so the segment connecting S_1 and S_2 is used to compute weights for the weighted problem. Using (6-60), the weighted objective function is

$$\begin{aligned} \text{maximize} \quad & Z(x_1, x_2, 1, 2) \\ & = [15 - (-6)](5x_1 - 2x_2) + [30 - (-3)](-x_1 + 4x_2) \end{aligned} \tag{6-72}$$

Carrying out the multiplications and grouping terms in (6-72) gives the weighted problem to be solved:

$$\text{maximize} \quad Z(x_1, x_2, 1, 2) = 72x_1 + 90x_2 \tag{6-73}$$

$$\text{s.t.} \quad (x_1, x_2) \in \mathbf{F_d} \tag{6-74}$$

where $\mathbf{F_d}$ is the feasible region in decision space defined by the constraints given in (4-7). The $(\ldots, 1, 2)$ in the argument of the weighted objective refers to the use of the segment between S_1 and S_2 for the definition of the weights.

The reader can draw the objective function contour for (6-73) in the decision space of Fig. 6-10 to find that the optimal solution for this weighted problem is $x_1 = 4$, $x_2 = 4$, which gives $Z_1 = 12$, $Z_2 = 12$, and $Z(x_1, x_2, 1, 2) = 648$. To check whether this solution lies on or above the line segment between S_1 and S_2 in Fig. 6-20, we compute $B_{1,2}$—the value of $Z(x_1, x_2, 1, 2)$ for any point on the line segment between S_1 and S_2. $B_{1,2}$ is computed by using the values for x_1 and x_2 that lead to a point on the line segment between S_1 and S_2 in (6-74). Any point will do so we shall use S_1. $(x_1, x_2) = (1, 4)$ leads to S_1, so $B_{1,2} = 432$. A check that the weights used to derive (6-74) are correct is to compute $B_{1,2}$ at another point, say S_2. S_2 comes from $(x_1, x_2) = (6, 0)$, which also gives 432.

Since $B_{12} < 648$, the new solution lies above the line segment between S_1 and S_2, from which one can conclude that this line segment is inferior. Step 3 then tells us to designate the new point P_{n+1}. Since $n = 2$, $P_{n+1} = P_3 = (12, 12)$.

In step 4 the points are reordered. Since $i = 1$, (6-61) gives $S_1' = S_1 = P_2$, (6-62) gives $S_2' = P_3$, and (6-63) gives $S_3' = S_2 = P_1$. Also, since $i = 1$, the expressions in (6-65) and (6-66) are not used. Instead, the values for $\Psi'_{1,2}$ and $\Psi'_{2,3}$ must be computed. This can be done with the help of Fig. 6-21.

The three noninferior points found by this stage of the algorithm are shown in Fig. 6-21. The lower bounds, which comprise the current approximation

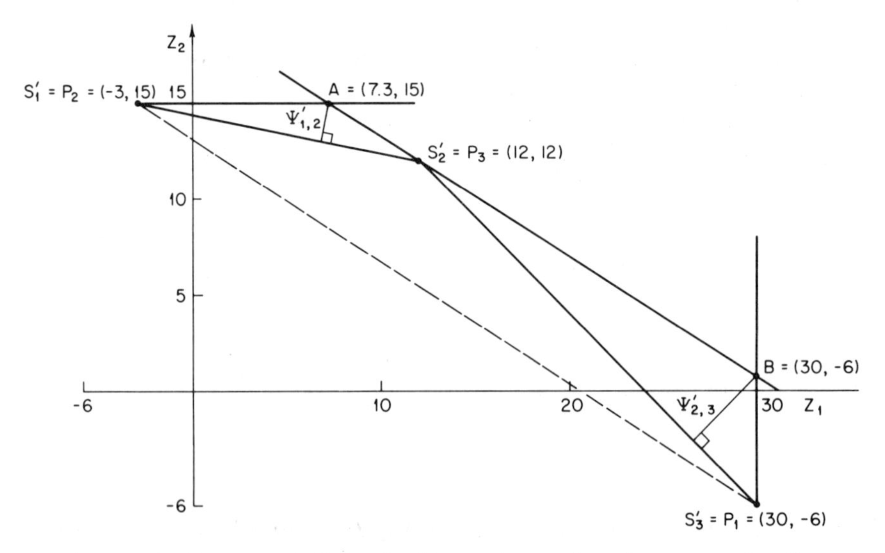

Fig. 6-21. Application of the NISE algorithm to the sample problem: second iteration.

of the noninferior set, consist of the line segments connecting adjacent points. The upper bound consists of the linear difference curves passing through the points. The vertical line for P_1 and the horizontal line for P_2 are shown in Fig. 6-21. The linear indifference curve passing through P_3 is also shown in Fig. 6-21; it is a line parallel to the line connecting P_1 and P_2, which was our old lower bound.

The computation of the maximum possible error requires first that the points of intersection of the linear indifference curves be found. This can be done by solving simultaneous equations. Consider point A in Fig. 6-21. It is the point of intersection of the line $Z_2 = 15$ and the line for which an equation can be derived from (6-72). This equation is equivalent to

$$21Z_1 + 33Z_2 = 648 \tag{6-75}$$

where the right-hand side is the value for the weighted objective at P_3. Since $Z_2 = 15$, (6-75) can be used to find $Z_1 = 7.3$. The sides of the triangle with vertices at P_2, P_3, and A can now be computed, and the formulas in (6-66)–(6-68) can be used to compute $\Psi'_{1,2} = 2.1$.

Using (6-75) with the equation $Z_1 = 30$ gives point $B = (30, 0.5)$. The maximum possible error in this segment can then be computed as $\Psi'_{2,3} = 4.6$.

The final part of step 4 is to increase n to a value of 3 and then return to step 2. In step 2 we compare $\Psi'_{i,i+1}, i = 1, 2$ to $T = 8.9$. We find that $\Psi'_{1,2} = 2.1 < T$ and $\Psi'_{2,3} = 4.6 < T$; the algorithm terminates with a sufficiently accurate approximation of the noninferior set, given the maximum allowable error. The final approximation of $\mathbf{N}_o$ is shown in Fig. 6-22. The approximation

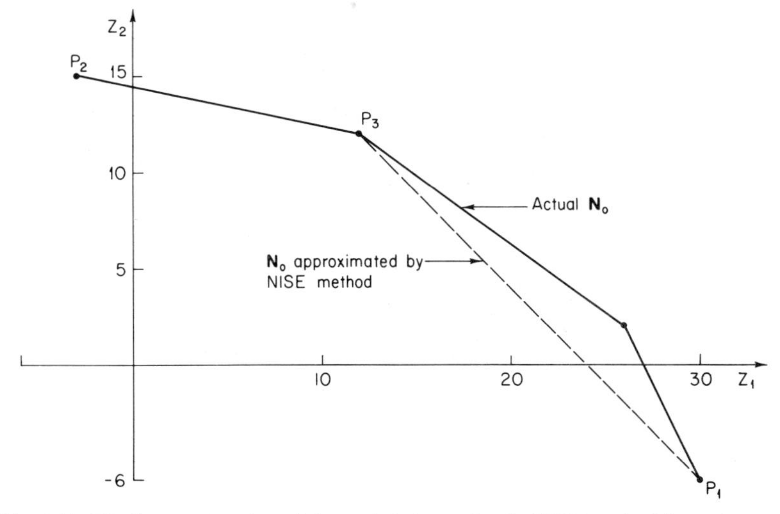

Fig. 6-22. Final approximation of the noninferior set by the NISE method for a given maximum allowable error.

consists of the points P_1, P_2, and P_3 and the line segments connecting them. If a smaller value of T had been chosen, then the algorithm would have continued to a more accurate approximation. Notice that even though the line segment from P_2 to P_3 in Fig. 6-22 is part of the actual noninferior set, the analyst cannot know this with the information given by the NISE method. One could get this information by setting $T = 0$, i.e., by forcing the algorithm to find an exact representation of the noninferior set.

6.3.4 The NISE Method for Problems with More than Two Objectives

If the algorithm presented in 6.3.2 were applied to a problem with three or more objectives, parts of the noninferior set could be missed entirely. Consider the objective space in Fig. 6-23 for a problem with three objectives. The points that maximize the three objectives individually are shown in the figure as A, B, and C. The plane in which these three points lie will be called the "primary plane"; it is also drawn in Fig. 6-23.

As the algorithm goes beyond step 2, noninferior points that lie above the primary plane, such as D in Fig. 6-23, will be found. In fact, since only nonnegative weights are allowed in the weighted problem, only those points that lie above the primary plane will be found by the algorithm given in Section 6.2.2. There is no problem with this in two-objective problems, for

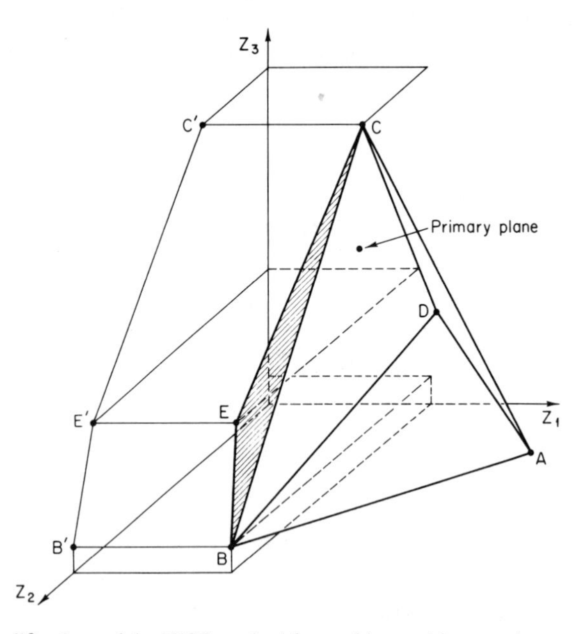

Fig. 6-23. Modifications of the NISE method for problems with more than two objectives.

which noninferior points always lie above the "primary plane," i.e., the line connecting P_1 and P_2 in Fig. 6-21. In higher-dimensional problems, however, there may be parts of the noninferior set that are to the sides of the primary plane. Such an area has been shaded in Fig. 6-23.

The modification that is required to find portions of the noninferior set not above the primary plane employs a familiar technique. One simply performs an analysis along the appropriate side by applying the NISE algorithm for two-objective problems. For the shaded portion of Fig. 6-23, one would apply the NISE method to objectives Z_2 and Z_3 alone. In effect, a projection of the three-dimensional noninferior set onto the Z_2–Z_3 plane is obtained by the two-dimensional NISE method. The projection of the line segments BE and EC onto the Z_2–Z_3 plane, where E is a noninferior extreme point not above the primary plane, is shown in Fig. 6-23. The points B', C', and E' in Fig. 6-23 are the projections of B, C, and E, respectively, onto the Z_2–Z_3 plane. If *all* noninferior extreme points were projected onto the Z_2–Z_3 plane, none would dominate E; otherwise E would be inferior. The two-dimensional NISE method is applied to all three sides of the primary plane.

In four or more dimensions the problem is approached in the same way, although computational complexity is increased. One begins with four individual optima that lead to a primary hyperplane—in this case a three-dimensional surface in a four-dimensional space. The primary hyperplane has four three-dimensional "sides" that must be checked by analyzing (Z_1, Z_2, Z_3), (Z_1, Z_2, Z_4), (Z_1, Z_3, Z_4), and (Z_2, Z_3, Z_4). Of course, each of these "sides" presents its own three-dimensional problem with its own primary plane, etc.

The NISE algorithm for problems with more than two objectives is a topic of current research. The details are being worked out at the time of this writing.

6.3.5 Applicability of the NISE Method

The computations in the demonstration of Section 6.3.3 were rather tedious; one could easily get the impression from a small example that the method is inefficient relative to straightforward approaches such as the weighting and constraint methods. This is not the case, however, since the NISE method is quite powerful when it is applied to real-world problems.

Consider the noninferior set drawn in Fig. 6-24, which appears as a curve with a sharp elbow in it. For noninferior sets shaped like this, the NISE method will produce a good approximation in only three solutions. The three noninferior points and the resulting approximation are shown in Fig. 6-24. Experience with real-world multiobjective problems has indicated

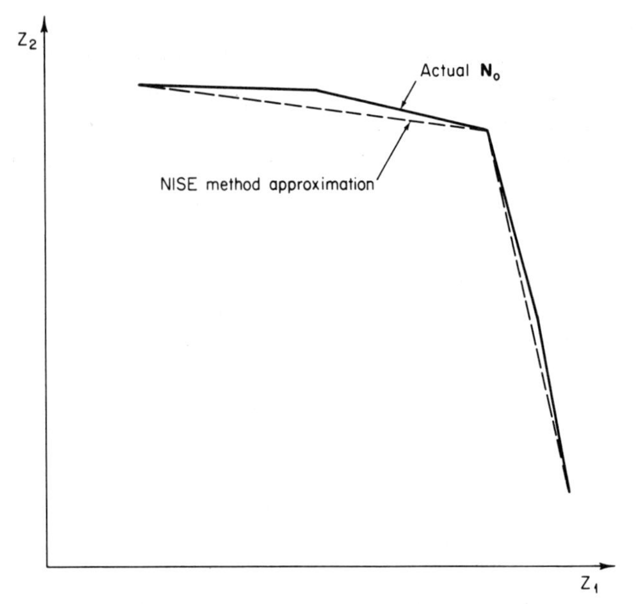

Fig. 6-24. A NISE method approximation for a "kinked" noninferior set.

that the "peaked" shape of the noninferior set in Fig. 6-24 is apparently common. The case study discussed in Chapter 9 demonstrates this.

One last point is that the tedium of the computations for the maximum possible error is avoided, insofar as personal suffering is concerned, by computerizing these operations.

6.4 MULTIOBJECTIVE SIMPLEX METHOD

A very active research area among mathematical programmers is the development of generating techniques that do not depend on the conversion of a multiobjective optimization problem into a single-objective optimization problem. Algorithms have been developed that operate directly on the multiobjective problem. The development of such techniques is a complex and intriguing mathematical problem that is not yet entirely solved. Those approaches that have been suggested are based on the use of the simplex method for linear programming problems, which was discussed in Chapter 3.

Philip (1972) and Ecker and Kouada (1975) presented mathematical characterizations of noninferior solutions and developed the basis for algorithmic approaches for identifying noninferior solutions. Holl (1973), Evans and Steuer (1973), and Zeleny (1974a) have all presented specific simplex-based algorithms for the generation of noninferior solutions. Zeleny's method is the one discussed here.

The presentation that follows omits a substantial amount of mathematical detail that the interested reader can find in Zeleny (1974a, pp. 63–122). Some basic concepts are presented below, followed by a discussion of the algorithm. The method is applied to an example problem at the end of this section.

6.4.1 Mathematical Background

The multiobjective simplex method can be used to generate an *exact* representation of the noninferior set. This is done by moving mathematically from one noninferior extreme point to adjacent noninferior extreme points until all noninferior extreme points have been found. Recall that the simplex method finds the optimal solution for a single-objective linear programming problem by moving from extreme point to adjacent extreme point. These simplex pivots are also used in the multiobjective simplex method for moving from one extreme point to one of its neighbors.

The multiobjective simplex method employs a multiobjective simplex tableau, which is an augmented version of the tableaux presented in Chapter 3. A multiobjective simplex tableau for a problem with $(n + m)$ decision variables (m of them slack variables), m constraints, and p objectives is shown as Table 6-5. This tableau is similar to Table 3-3, but there are some differences. The minor differences are that Table 6-5 is more general in that there are $n + m$ columns (variables) while Table 3-3 was produced for four variables. In addition, the basic-column rows have been left blank in Table 6-5 so that the tableau would not seem too cluttered. If they had been filled in, these rows would appear just as the $\mathbf{a}_3$ and $\mathbf{a}_4$ rows of Table 3-3.

TABLE 6-5

A PARTIAL MULTIOBJECTIVE SIMPLEX TABLEAU FOR A GENERAL PROBLEM

	$\mathbf{a}_1$	$\mathbf{a}_2$	$\cdots$	$\mathbf{a}_n$	$\mathbf{a}_{n+1}$	$\cdots$	$\mathbf{a}_{n+m}$	$\mathbf{a}_0$
c_j^1	c_1^1	c_2^1	$\cdots$	c_n^1	0	$\cdots$	0	0
c_j^2	c_1^2	c_2^2	$\cdots$	c_n^2	0	$\cdots$	0	0
$\vdots$								
c_j^p	c_1^p	c_2^p	$\cdots$	c_n^p	0	$\cdots$	0	0
Basic columns								
f_j^1	f_1^1	f_2^1	$\cdots$	f_n^1	f_{n+1}^1	$\cdots$	f_{n+m}^1	$f_0^1 = Z_1$
f_j^2	f_1^2	f_2^2	$\cdots$	f_n^2	f_{n+1}^2	$\cdots$	f_{n+m}^2	$f_0^2 = Z_2$
$\vdots$								
f_j^p	f_1^p	f_2^p	$\cdots$	f_n^p	f_{n+1}^p	$\cdots$	f_{n+m}^p	$f_0^p = Z_p$

The major differences between Table 6-5 and Table 3-3 lie in the addition of several rows to the top and bottom of the former. The second row of the single-objective tableau shows the coefficients for each decision variable in the objective function. Now since we have p objectives, there are p rows of objective coefficients. The symbol c_j^k stands for the coefficient on the jth decision variable in objective k.

The last row of the single-objective tableau consists of the reduced costs for each variable—the quantity that indicates the impact on the objective function of increasing a nonbasic variable from zero. In the multiobjective simplex tableau of Table 6-5 there are p rows of reduced costs—one for each of the p objectives. Notice also that the notation has been changed here; the reduced cost for objective k, column j is called f_j^k rather than $(z_j^k - c_j^k)$ as the notation of Chapter 3 would have required. This change has been made for notational simplicity.

For each variable we now have a set of reduced costs that will be called $\mathbf{f}_j$; i.e.,

$$\mathbf{f}_j = \begin{bmatrix} f_j^1 \\ f_j^2 \\ \vdots \\ f_j^p \end{bmatrix} \qquad (6\text{-}76)$$

In mathematical parlance, $\mathbf{f}_j$ is a p-dimensional column vector. The relationships of Chapter 3 still hold; i.e., the introduction of a nonbasic variable x_j into the basis at a value of θ changes objective k by

$$Z_k' = Z_k - \theta f_j^k \qquad (6\text{-}77)$$

which is analogous to equation (3-63). The relationship in (6-77) holds for each objective and all nonbasic variables j. It states that if x_j, which is currently equal to zero, is increased to θ, objective k will change by $(-\theta f_j^k)$. x_j will increase k only if $f_j^k < 0$ since $\theta > 0$ for feasibility.

The reduced-cost rows contain some interesting information that can be used to determine the noninferiority of a current basis and its neighbors. Keep in mind that neighboring extreme points or basic feasible solutions have all but one basic column in common with each other. The theorems presented below, from Zeleny (1974a), make use of the reduced-cost rows of the simplex tableau. Proofs are not presented here, but they are straightforward applications of Eq. (6-77). The interested reader can consult the indicated portion of Professor Zeleny's monograph. In the theorems below, $\mathbf{B}$ refers to the collection of basic columns (see Chapter 3).

THEOREM 6-1 [Theorem 3.1.2, Part a, Zeleny (1974a, p. 66)]. Given a current basic feasible solution, if there is a nonbasic column $\mathbf{a}_j$, $j \notin \mathbf{B}$, such that $f_j^k \leq 0$ for $k = 1, 2, \ldots, p$ and $f_j^k < 0$ for at least one $k = 1, 2, \ldots, p$, then the current basic feasible solution is *inferior*.

This theorem says that if a variable can be brought into the basis at a positive value so that all objectives will increase simultaneously (or, at least, some objectives will increase while the others remain the same), then the current solution must be inferior. Putting it another way, if a neighboring extreme point gives more of some objectives and the same amount of all of the others, then the current extreme point (basic feasible) solution is dominated by that neighbor.

THEOREM 6-2 [*Theorem* 3.1.2, *part b*, *Zeleny* (1974*a*, *p*. 66)]. Given a current basic feasible solution, if there is a nonbasic column $\mathbf{a}_j, j \notin \mathbf{B}$, such that $f_j^k \geq 0$ for $k = 1, 2, \ldots, p$ and $f_j^k > 0$ for at least one $k = 1, 2, \ldots, p$, then the introduction of $\mathbf{a}_j$ into the basis will lead to an inferior solution.

Theorem 6-2 states that if the introduction of a nonbasic variable into the basis would decrease all objectives simultaneously ($f_j^k > 0$) or, at least, decrease some without increasing the other objectives, then the new solution would be dominated by the current solution. This is the case of a neighboring extreme point being inferior to the current extreme point.

In the next theorem two nonbasic variables x_j and x_q will be considered. The notation of (6-77) will be altered slightly: θ_j will be the value that x_j would assume if it were introduced into the basis; θ_q is defined similarly. Equation (6-77) becomes

$$Z_k' = Z_k - \theta_j f_j^k \tag{6-78}$$

THEOREM 6-3 [*Theorem* 3.1.4, *Zeleny* (1974*a*, *p*. 67)] Given a current basic feasible solution, if there are two different nonbasic columns $\mathbf{a}_j$ and $\mathbf{a}_q$, $j, q \notin \mathbf{B}$, such that

$$\theta_j f_j^k \leq \theta_q f_q^k, \qquad k = 1, 2, \ldots, p \tag{6-79}$$

where (6-79) is satisfied as a strict inequality for at least one $k = 1, 2, \ldots, p$, then the basic feasible solution resulting from the introduction of $\mathbf{a}_q$ into the current basis is inferior to the basic feasible solution resulting from the introduction of $\mathbf{a}_j$ into the current basis.

Theorems 6-1 and 6-2 addressed the relationship of a current basic feasible solution to its neighbors. Theorem 6-3 concerns the relationship between two neighbors of the current basic feasible solution. It says, as an examination of (6-78) would indicate, that if a neighboring solution gives more of some objectives than another neighbor of the current basic feasible solution, while giving the same amount of the other objectives, then the first neighbor dominates the second.

Theorems 6-1–6-3 will be demonstrated with an example. Assume that, for the problem for which Table 6-5 is a simplex tableau, the first m variables $x_1, x_2, \ldots, x_m$ form the current basis feasible solution. Assume also that $p = 4$. The tableau that would correspond to the basis $\mathbf{B} = \{1, 2, \ldots, m\}$ is shown in Table 6-6, where it has been assumed that $m < n$. The basic-column rows of Table 6-6 have been left blank for simplicity: Ordinarily, the y_{ij}s, which express each column as a function of the basic columns, would be shown there. The last four rows contain the reduced costs for the four objectives. The reduced costs for all columns in the current basic feasible solution, $\mathbf{a}_1, \mathbf{a}_2, \ldots, \mathbf{a}_m$, are equal to zero for all objectives: Z_1, Z_2, Z_3, and Z_4. This familiar result from the single-objective simplex holds for the multiobjective case as well.

The column $\mathbf{a}_{m+1}$ is not in the current basic feasible solution. Its reduced costs are -1, -2, -4, and 0 for objectives Z_1, Z_2, Z_3, and Z_4 respectively. If the variable x_{m+1} were brought into the basis at the level θ_{m+1}, then Z_1 would increase by θ_{m+1}, Z_2 would increase by $2\theta_{m+1}$, Z_3 would increase by $4\theta_{m+1}$, and Z_4 would be unchanged. We can conclude from Theorem 6-1 that the current basic feasible solution is inferior. We can say nothing, however, about the noninferiority of the new basic feasible solution created by bringing x_{m+1} into the basis. Although the new solution would dominate the old one, the new solution may still be inferior.

The column $\mathbf{a}_{m+2}$ is another nonbasic column in Table 6-6. Its reduced costs are all positive so Theorem 6-2 allows us to conclude that the solution reached by the introduction of x_{m+2} into the basis would be inferior. The new solution would *decrease* Z_1 by θ_{m+2}, Z_2 by $2\theta_{m+2}$, and Z_4 by $3\theta_{m+2}$ while Z_3 would be unchanged. Thus, the current solution dominates the new one associated with x_{m+2}.

Two more nonbasic variables of Table 6-6 are x_n and x_{n+1}, to which neither Theorem 6-1 or 6-2 apply since they each have some positive and some negative reduced costs. It is assumed for illustrative purposes that θ_n and θ_{n+1} are such that $f_n^k\theta_n \leq f_{n+1}^k\theta_{n+1}$ for $\forall k = 1, 2, 3, 4$. (Recall that these are determined from appropriate y_{ij}s—see Section 3.8.3.) This will be the case if $\frac{1}{2}\theta_{n+1} \leq \theta_n \leq \frac{3}{4}\theta_{n+1}$, which is derived by using the values of the reduced costs in Table 6-6 to form four simultaneous inequalities for θ_n and θ_{n+1}. If θ_n and θ_{n+1} satisfy the required relationships, then Theorem 6-3 tells us that the new basic feasible solution found by the introduction of $\mathbf{a}_{n+1}$ is dominated by the new basic feasible solution that would result from the introduction of $\mathbf{a}_n$ into the basis. The latter solution would give a larger increase in Z_1 and Z_3 (check the signs of the reduced costs in Table 6-6) and a smaller decrease in Z_2 and Z_4 than would the former ($\mathbf{a}_{n+1}$) solution. Notice that the solution found by the introduction of $\mathbf{a}_n$ may or may not be noninferior.

TABLE 6-6

AN EXAMPLE MULTIOBJECTIVE SIMPLEX THAT DEMONSTRATES THE THEOREMS

	$\mathbf{a}_1$	$\mathbf{a}_2$	$\cdots$	$\mathbf{a}_m$	$\mathbf{a}_{m+1}$	$\mathbf{a}_{m+2}$	$\cdots$	$\mathbf{a}_n$	$\mathbf{a}_{n+1}$	$\cdots$	$\mathbf{a}_{n+m}$	$\mathbf{a}_0$
c^1_j	c^1_1	c^1_2	$\cdots$	c^1_m	c^1_{m+1}	c^1_{m+2}	$\cdots$	c^1_n	0	$\cdots$	0	0
c^2_j	c^2_1	c^2_2	$\cdots$	c^2_m	c^2_{m+1}	c^2_{m+2}	$\cdots$	c^2_n	0	$\cdots$	0	0
c^3_j	c^3_1	c^3_2	$\cdots$	c^3_m	c^3_{m+2}	c^3_{m+2}	$\cdots$	c^3_n	0	$\cdots$	0	0
c^4_j	c^4_1	c^4_2	$\cdots$	c^4_m	c^4_{m+1}	c^4_{m+2}	$\cdots$	c^4_n	0	$\cdots$	0	0
$\mathbf{a}_1$												
$\mathbf{a}_2$												
$\vdots$						The y_{ij}s would go here						
$\mathbf{a}_m$												
f^1_j	0	0	$\cdots$	0	-1	1	$\cdots$	-4	-2	$\cdots$	1	Z_1
f^2_j	0	0	$\cdots$	0	-2	2	$\cdots$	6	8	$\cdots$	7	Z_2
f^3_j	0	0	$\cdots$	0	-4	0	$\cdots$	-1	0	$\cdots$	-3	Z_3
f^4_j	0	0	$\cdots$	0	0	3	$\cdots$	3	4	$\cdots$	2	Z_4

If the reduced costs of all nonbasic variables are covered by Theorems 6-1–6-3, then further action can be determined. That is, Theorem 6-1 tells us that the current basic feasible solution is inferior. If it is determined to be noninferior (by a method discussed below), then Theorem 6-1 is not applicable. If it is noninferior then we know that columns that are covered by Theorem 6-2 can be ignored as candidates for introduction into the basis. By Theorem 6-3 a nonbasic column that dominates all other nonbasic columns should be introduced into the basis in order to move us to another noninferior solution. If Theorem 6-3 does not apply, then we are left with nonbasic columns that have reduced costs of both signs and which are not dominated by any other nonbasic columns. Such a column is $\mathbf{a}_{n+m}$ in Table 6-6. Its introduction will lead to a noninferior solution if the current basic feasible solution is noninferior. Its reduced costs will be termed as not comparable with $\mathbf{0}$, where $\mathbf{0} = (0, 0, 0, 0)$. That is, $\mathbf{f}_{n+m}$ has some negative elements and some positive elements, so one cannot say whether $\mathbf{f}_{n+m}$ is greater than or less than $\mathbf{0}$.

Theorems 6-1–6-3 are directed at the neighbors of the current basic feasible solution. What of the noninferiority of the current solution? There is usually insufficient data in the simplex tableau to answer this question. The only case in which the simplex tableau allows us to establish immediately the noninferiority of a solution is the situation in which that solution is a *unique* maximum for one of the objectives. The simplex tableau indicates a unique maximum when all of the reduced costs for all nonbasic columns for one objective (i.e., in one reduced-cost row) are strictly positive. If some nonbasic columns have positive reduced costs for that objective while the others have zero reduced costs, then the current solution is a maximum for that objective, but it is not unique; the existence of alternative optima means that this solution may be inferior.

The last row of Table 6-6 shows the reduced costs for Z_4. Assuming that the reduced costs for all of those columns represented by the ellipses in the tableau are nonnegative, it can be concluded that the current basic feasible solution is an optimal solution for Z_4; i.e., it maximizes Z_4 since all nonbasic variables have nonnegative reduced costs for this objective. Notice, however, that $f^4_{m+1} = 0$, i.e., x_{m+1}, a nonbasic variable, has a zero reduced cost for Z_4. This means that the current solution is not a unique maximum for Z_4 and may be inferior to another solution which maximizes Z_4. An alternate maximum for Z_4 is the basic feasible solution that would result from the introduction of x_{m+1} into the basis. For this particular case, we have already concluded from Theorem 6-1 that the new solution dominates the current solution. Thus, although the current solution is a maximum for Z_4, it is inferior.

The preceding background forms part of the basis for the algorithm that is presented in the next subsection. One tool that is needed for the algorithm is a method for determining the noninferiority of a given basic feasible solution. Zeleny's algorithm employs a subproblem for this purpose. The subproblem attempts to find a feasible solution that will yield more of at least one objective, but not less of any objective, than is provided by the current basic feasible solution. If such a feasible solution can be found, then the current basic solution is inferior; if not, then it is noninferior.

The subproblem looks very much like the constrained problem of Section 6.2. Define the values of the $n + m$ decision variables at the current basic feasible solution as $(\hat{x}_1, \hat{x}_2, \ldots, \hat{x}_{n+m})$. For the basic feasible solution depicted in Table 6-6, $\hat{x}_j > 0$, $j = 1, 2, \ldots, m$ and $\hat{x}_j = 0$, $j = m + 1, m + 2, \ldots, n$. The collection of the values for the variables is $\hat{\mathbf{x}} = (\hat{x}_1, \hat{x}_2, \ldots, \hat{x}_{n+m})$. The constraints for the subproblem are

$$\mathbf{x} \in \mathbf{F}_d \tag{6-80}$$

$$Z_k(\mathbf{x}) \geq Z_k(\hat{\mathbf{x}}), \qquad k = 1, 2, \ldots, p \tag{6-81}$$

where (6-80) requires that the constraints of the original problem be satisfied and (6-81) requires that the new feasible solution provide at least as much of each objective as that provided by the current solution that is being tested.

The goal of the subproblem is to push each objective as high as possible beyond their current levels. This can be done by converting (6-81) to equality constraints:

$$Z_k(\mathbf{x}) - \delta_k = Z_k(\hat{\mathbf{x}}), \qquad k = 1, 2, \ldots, p \tag{6-82}$$

where δ_k is a surplus variable. The objective function of the subproblem is then

$$\text{maximize} \quad V = \sum_{k=1}^{p} \delta_k \tag{6-83}$$

which is to be maximized subject to (6-80), (6-82), and the nonnegativity restrictions

$$\delta_k \geq 0, \qquad k = 1, 2, \ldots, p \tag{6-84}$$

The basic feasible solution for the original multiobjective problem $\hat{\mathbf{x}}$ is noninferior if $V = 0$ when the subproblem is solved. This result for the subproblem would mean that there are no feasible solutions that give more ($\delta_k > 0$) of any objective without sacrificing some other objectives. If $V > 0$ when the subproblem is solved, then $\hat{\mathbf{x}}$ is inferior. $V > 0$ implies that $\delta_k > 0$ for at least one $k = 1, 2, \ldots, p$ so more of at least one objective can be feasibly obtained without sacrificing other objectives. Note that δ_k cannot be negative by (6-84).

Zeleny (1974a, pp. 68–75, 94–98) presents a very efficient method for solving the subproblem. It will not be presented, although it is an important computational feature of the algorithm.

The last major part of the algorithmic approach is a "bookkeeping" method to account for extreme-point solutions that have been found previously and to generate adjacent extreme points to analyze in subsequent steps of the algorithm. Zeleny (1974a, pp. 80–93) presents two extreme-point labeling and generating schemes. Neither are presented here; it is sufficient to realize that techniques do exist for keeping track of the solutions.

6.4.2 The Algorithm

The multiobjective simplex algorithm makes use of the mathematical background presented in the previous subsection. A flowchart for the algorithm, adapted from Zeleny (1974a, p. 99) is shown in Fig. 6-25.

After an initial basic feasible solution is found in steps 1–3 (if one does not exist, then the problem is infeasible and the algorithm terminates), our knowledge about reduced costs is applied to establish quickly, if possible, its inferiority or noninferiority. The values of the decision variables (including slacks) in the basic feasible solution are referred to as $\mathbf{x}^h$ and the current

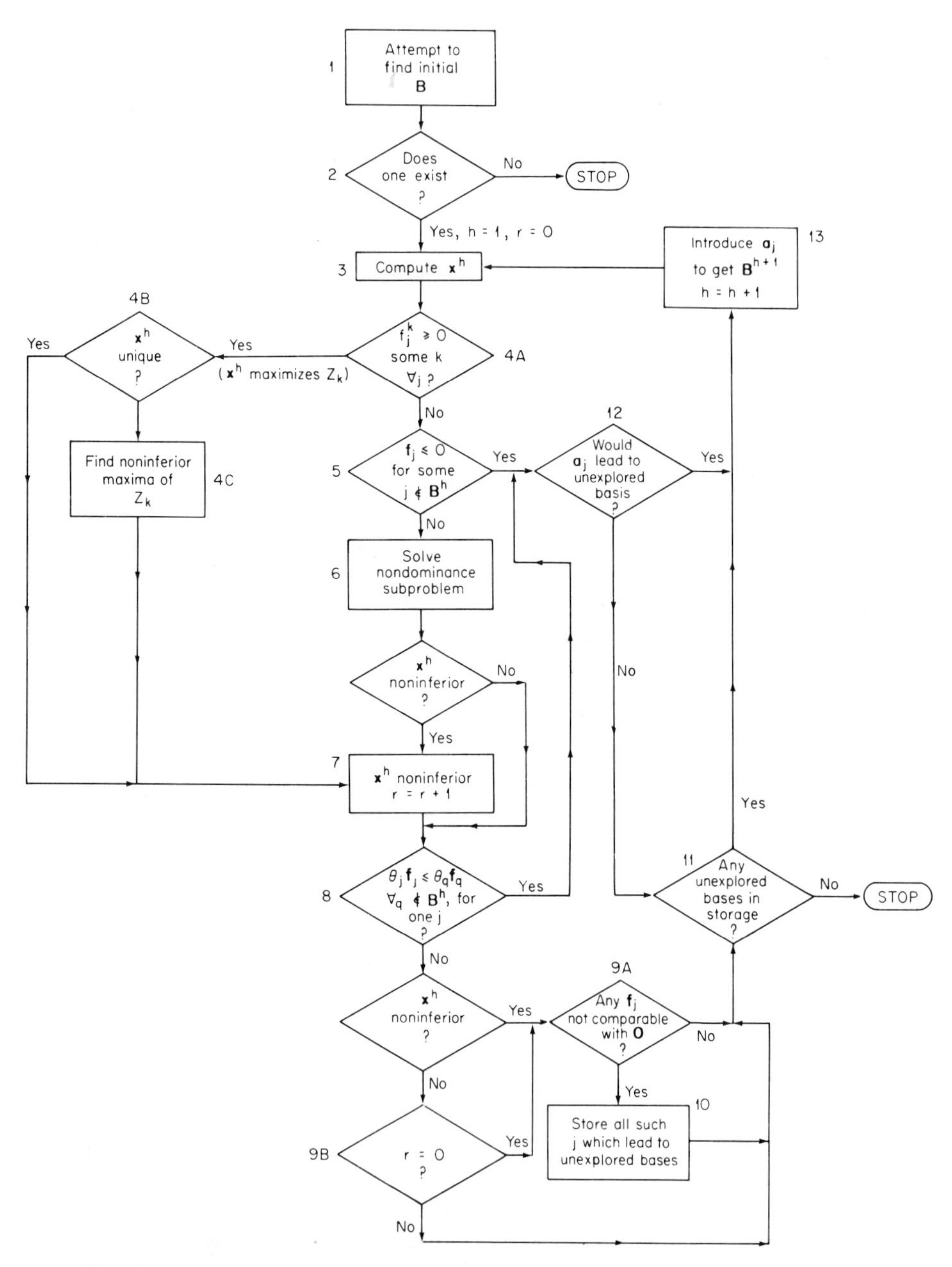

Fig. 6-25. A multiobjective simplex algorithm [adapted from Zeleny (1974a, p. 99, Fig. 3.3.2)].

basis as $\mathbf{B}^h$. Any basic feasible solution can be used to initialize the method, but the use of an individual optimal solution would seem to avoid unnecessary computation that could result from the use of other solutions. In step 4A each objective is checked to see if one of them is maximized by $\mathbf{x}^h$. If this is the case, then in step 4B the uniqueness of the solution is checked by searching for nonbasic variables with $f_j^k = 0$. If $\mathbf{x}^h$ is unique, then the current solution is labeled noninferior in step 7. If it is not unique, then those alternative optima that are also noninferior are found in step 4C. A technique such as that presented in Section 6.2.4 could be used.

If $\mathbf{x}^h$ was found in step 4A not to maximize any of the objectives, then nonbasic columns with all nonpositive reduced costs are sought in step 5. If any are found, then we know that $\mathbf{x}^h$ is inferior (Theorem 6-1). One of those columns with nonpositive reduced costs that would lead to a previously unexplored solution (checked in step 12 with the bookkeeping technique) is introduced into the basis in step 13 to get a new solution, h is incremented, and we return to step 3. If the answer to the question in step 5 is "no," then a quick determination of the noninferiority of $\mathbf{x}^h$ is not possible. In step 6 the non-dominance subproblem is solved to establish the noninferiority of $\mathbf{x}^h$. If $\mathbf{x}^h$ is noninferior, then an index r that counts the number of noninferior solutions is incremented in step 7.

In step 8 we begin to look for promising new directions to pursue. Theorem 6-3 is used to motivate step 8, in which a dominating nonbasic column is sought. If one is found, then that is the most promising new lead whether or not $\mathbf{x}^h$ is noninferior. (If $\mathbf{x}^h$ is inferior, it would be most desirable to find a nonbasic column with nonpositive reduced costs, but that failed in step 5.) The dominant column is brought into the basis if it will lead to an unexplored solution (checked in step 12).

If step 8 fails to identify a dominant nonbasic column *and* if $\mathbf{x}^h$ is non-inferior, then those nonbasic columns that have some positive and some negative reduced costs are identified in step 9A and stored in step 10. If $\mathbf{x}^h$ is inferior, then there is no reason to perform step 9A. However, if no noninferior solutions have been found to this point ($r = 0$ in step 9B), then step 9A is performed even if $\mathbf{x}^h$ is inferior. The algorithm could terminate prematurely without this additional check, which is a modification of Zeleny's original algorithm.

The algorithm continues until all promising leads have been followed, which is checked in step 11. All noninferior extreme-point solutions can be found with this approach because the noninferior set is "connected." This means that any point in the noninferior set can be reached from any other point in the noninferior set through a finite number of simplex pivots. This property of the noninferior set guarantees that the algorithm will locate all noninferior extreme points. Of course, a reasonable question is *when*?

One issue that is not included in the above algorithm is the generation of the entire noninferior set from the noninferior extreme points found with the multiobjective simplex method. The algorithm yields a collection of points in a p-dimensional space, but one must go further to find the faces or hyperplanes that connect these points and which represent the noninferior set. Yu and Zeleny (1975) present a method for doing this.

6.4.3 Application of the Multiobjective Simplex Method to the Sample Problem

The sample problem in (4-7) with slack variables introduced into the constraints is

$$\text{maximize} \quad [Z_1(x_1, x_2), Z_2(x_1, x_2)] \tag{6-85}$$

$$Z_1(x_1, x_2) = 5x_1 - 2x_2 \tag{6-86}$$

$$Z_2(x_1, x_2) = -x_1 + 4x_2 \tag{6-87}$$

$$\text{s.t.} \quad -x_1 + x_2 + x_3 = 3 \tag{6-88}$$

$$x_1 + x_2 + x_4 = 8 \tag{6-89}$$

$$x_1 + x_5 = 6 \tag{6-90}$$

$$x_2 + x_6 = 4 \tag{6-91}$$

$$x_1, \quad x_2, \quad x_3, \quad x_4, \quad x_5, \quad x_6 \geq 0 \tag{6-92}$$

A basis for this problem will have four elements, one for each constraint. A convenient initial basis is $\mathbf{B}^1 = (\mathbf{a}_3, \mathbf{a}_4, \mathbf{a}_5, \mathbf{a}_6)$, i.e., the basic feasible solution consisting of all slack variables. This all-slack solution will be used for illustrative purposes. In practice, it is undoubtedly better to start the algorithm with a solution that maximizes one of the objectives individually. The tableau for $\mathbf{B}^1$ is shown in Table 6-7. We see for step 3 that $\mathbf{x}^1 = (0, 0)$. Notice that the y_{ij}s are just the coefficients in the constraints and that the reduced costs are the negatives of the appropriate objective coefficients.

Going through the algorithm, the answer to the questions in steps 4 and 5 is "no," so the noninferiority of the current basic feasible solution must be established by solving the subproblem

$$\text{maximize} \quad V = \delta_1 + \delta_2 \tag{6-93}$$

$$\text{s.t.} \quad (x_1, x_2, \ldots, x_6) \in \mathbf{F}_d \tag{6-94}$$

$$Z_1(x_1, x_2) = 5x_1 - 2x_2 - \delta_1 = 0 \tag{6-95}$$

$$Z_2(x_1, x_2) = -x_1 + 4x_2 - \delta_2 = 0 \tag{6-96}$$

$$\delta_1, \delta_2 \geq 0 \tag{6-97}$$

TABLE 6-7

APPLICATION OF THE MULTIOBJECTIVE SIMPLEX METHOD TO THE SAMPLE PROBLEM: THE INITIAL BASIC FEASIBLE SOLUTION

		$\mathbf{a}_1$	$\mathbf{a}_2$	$\mathbf{a}_3$	$\mathbf{a}_4$	$\mathbf{a}_5$	$\mathbf{a}_6$	$\mathbf{a}_0$
	c_j^1	5	−2	0	0	0	0	0
	c_j^2	−1	4	0	0	0	0	0
	$\mathbf{a}_3$	−1	(1)	1	0	0	0	3
$\mathbf{B}^1$	$\mathbf{a}_4$	1	1	0	1	0	0	8
	$\mathbf{a}_5$	1	0	0	0	1	0	6
	$\mathbf{a}_6$	0	1	0	0	0	1	4
	f_j^1	−5	2	0	0	0	0	0
	f_j^2	1	−4	0	0	0	0	0

where $\mathbf{F}_d$ represents the set of solutions that satisfy (6-88)–(6-92). The solution to this problem is $V = 28$, $\delta_1 = 26$, $\delta_2 = 2$, $x_1 = 6$, $x_2 = 2$, $x_3 = 7$, $x_4 = x_5 = 0$, $x_6 = 2$. Clearly, $\mathbf{x}^1$ is inferior.

We skip step 7, noting that $r = 0$. In step 8 dominance of a nonbasic column is checked. However, since $\theta_1, \theta_2 > 0$ neither $\mathbf{a}_1$ nor $\mathbf{a}_2$ can dominate. We perform step 9A because $r = 0$. Both columns x_1 and x_2 have reduced costs not comparable with (0, 0). For step 10 we will simply note that $\mathbf{B}^1 = (\mathbf{a}_3, \mathbf{a}_4, \mathbf{a}_5, \mathbf{a}_6)$ and $\mathbf{a}_1$ and $\mathbf{a}_2$ should be considered for entry, since their introduction into the basis would lead to unexplored solutions.

The answer in step 11 is "yes" so we proceed to step 13 where $\mathbf{a}_2$ will be chosen for entry into the basis; $h = h + 1 = 2$. When $\mathbf{a}_2$ enters the basis, $\theta_2 = \min(\frac{3}{1}, \frac{8}{1}, \frac{6}{0}, \frac{4}{1}) = 3$, and the pivot element is the circled entry in Table 6-7; i.e., $\mathbf{a}_3$ will leave the basis. At this point the rules for simplex operations are applied to get a new simplex tableau.

The new simplex tableau, shown in Table 6-8, gives $\mathbf{x}^2 = (0, 3)$. Recall that the last two entries of the reduced-cost rows show the values for the two objectives. In Table 6-8, $Z_1 = -6$ and $Z_2 = 12$. In this pass through step 5 we find that $\mathbf{a}_1$ has all negative reduced costs so that $\mathbf{x}^2$ is inferior. The column $\mathbf{a}_1$ is now brought into the basis to get the new basic feasible solution shown in Table 6-9.

In returning to step 3, $h = 3$ and $\mathbf{x}^3 = (1, 4)$. In step 4 we find that Z_2 is maximized since f_3^2 and f_6^2 (x_3 and x_6 are the nonbasic variables) are both positive. Since this is a unique optimum, $\mathbf{x}^3$ is a noninferior solution and $r = 1$; i.e., one noninferior solution has been found. For step 8 we can compute that

TABLE 6-8

APPLICATION OF THE MULTIOBJECTIVE SIMPLEX METHOD TO THE SAMPLE PROBLEM: SECOND SOLUTION

		$\mathbf{a}_1$	$\mathbf{a}_2$	$\mathbf{a}_3$	$\mathbf{a}_4$	$\mathbf{a}_5$	$\mathbf{a}_6$	$\mathbf{a}_0$
	c_j^1	5	-2	0	0	0	0	0
	c_j^2	-1	4	0	0	0	0	0
	$\mathbf{a}_2$	-1	1	1	0	0	0	3
$\mathbf{B}^2$	$\mathbf{a}_4$	2	0	-1	1	0	0	5
	$\mathbf{a}_5$	1	0	0	0	1	0	6
	$\mathbf{a}_6$	(1)	0	-1	0	0	1	1
	f_j^1	-3	0	-2	0	0	0	-6
	f_j^2	-3	0	4	0	0	0	12

$\theta_3 = 3$ and $\theta_6 = 1$. We find in step 8 that $f_3^1\theta_3 = -15, f_6^1\theta_6 = 3$, and $f_3^2\theta_3 = f_6^2\theta_6 = 3$. Thus, $\mathbf{a}_3$ dominates $\mathbf{a}_6$ and will be brought into the basis, leading to Table 6-10. The computation in step 8 was actually unnecessary since $\mathbf{f}_6 > 0$. It was performed here for illustrative purposes.

The new basic feasible solution is $\mathbf{x}^4 = (4, 4)$. The noninferiority of this solution cannot be established in steps 5 or 6 so that the subproblem in (6-93)–(6-97) must be solved with new right-hand sides of 12 (the current value of Z_1 and Z_2) for equations (6-95) and (6-96). The solution to the subproblem gives $V = 0$, so $\mathbf{x}^4$ is noninferior and $r = 2$ in step 7. A dominat-

TABLE 6-9

APPLICATION OF THE MULTIOBJECTIVE SIMPLEX METHOD TO THE SAMPLE PROBLEM: THIRD SOLUTION

		$\mathbf{a}_1$	$\mathbf{a}_2$	$\mathbf{a}_3$	$\mathbf{a}_4$	$\mathbf{a}_5$	$\mathbf{a}_6$	$\mathbf{a}_0$
	c_j^1	5	-2	0	0	0	0	0
	c_j^2	-1	4	0	0	0	0	0
	$\mathbf{a}_2$	0	1	0	0	0	1	4
$\mathbf{B}^3$	$\mathbf{a}_4$	0	0	(1)	1	0	-2	3
	$\mathbf{a}_5$	0	0	1	0	1	-1	5
	$\mathbf{a}_1$	1	0	-1	0	0	1	1
	f_j^1	0	0	-5	0	0	3	-3
	f_j^2	0	0	1	0	0	3	15

TABLE 6-10

APPLICATION OF THE MULTIOBJECTIVE SIMPLEX METHOD TO THE SAMPLE PROBLEM: FOURTH SOLUTION

		$\mathbf{a}_1$	$\mathbf{a}_2$	$\mathbf{a}_3$	$\mathbf{a}_4$	$\mathbf{a}_5$	$\mathbf{a}_6$	$\mathbf{a}_0$
	c_j^1	5	−2	0	0	0	0	0
	c_j^2	−1	4	0	0	0	0	0
	$\mathbf{a}_2$	0	1	0	0	0	1	4
	$\mathbf{a}_3$	0	0	1	1	0	−2	3
$\mathbf{B}^4$	$\mathbf{a}_5$	0	0	0	−1	1	(1)	2
	$\mathbf{a}_1$	1	0	0	1	0	−1	4
	f_j^1	0	0	0	5	0	−7	12
	f_j^2	0	0	0	−1	0	5	12

ing column is not found in step 8. There are two columns, $\mathbf{a}_4$ and $\mathbf{a}_6$, with $\mathbf{f}_4$ and $\mathbf{f}_6$ found in step 9A not to be comparable with (0, 0). Notice, however, that the introduction of $\mathbf{a}_4$ would lead to $(\mathbf{a}_2, \mathbf{a}_4, \mathbf{a}_5, \mathbf{a}_1)$, which is a previously explored basis, $\mathbf{B}^3$ (Table 6-9). The column $\mathbf{a}_4$ is discarded as a candidate and $\mathbf{a}_6$ is introduced into the basis to get $\mathbf{B}^5 = (\mathbf{a}_2, \mathbf{a}_3, \mathbf{a}_6, \mathbf{a}_1)$.

The new solution, $\mathbf{x}^5 = (6, 2)$, is shown in Table 6-11. Notice that now $Z_1 = 26$, $Z_2 = 2$. The noninferiority of this solution is determined by solving the subproblem with 26 and 2 as the right-hand sides of (6-95) and (6-96), respectively. Step 8 does not find a dominating column, but in step 9

TABLE 6-11

APPLICATION OF THE MULTIOBJECTIVE SIMPLEX METHOD TO THE SAMPLE PROBLEM: FIFTH SOLUTION

		$\mathbf{a}_1$	$\mathbf{a}_2$	$\mathbf{a}_3$	$\mathbf{a}_4$	$\mathbf{a}_5$	$\mathbf{a}_6$	$\mathbf{a}_0$
	c_j^1	5	−2	0	0	0	0	0
	c_j^2	−1	4	0	0	0	0	0
	$\mathbf{a}_2$	0	1	0	(1)	−1	0	2
	$\mathbf{a}_3$	0	0	1	−1	2	0	7
$\mathbf{B}^5$	$\mathbf{a}_6$	0	0	0	−1	1	1	2
	$\mathbf{a}_1$	1	0	0	0	1	0	6
	f_j^1	0	0	0	−2	7	0	26
	f_j^2	0	0	0	4	−5	0	2

TABLE 6-12

APPLICATION OF THE MULTIOBJECTIVE SIMPLEX METHOD TO THE SAMPLE PROBLEM: SIXTH SOLUTION

		$\mathbf{a}_1$	$\mathbf{a}_2$	$\mathbf{a}_3$	$\mathbf{a}_4$	$\mathbf{a}_5$	$\mathbf{a}_6$	$\mathbf{a}_0$
	c_j^1	5	-2	0	0	0	0	0
	c_j^2	-1	4	0	0	0	0	0
$\mathbf{B}^6$	$\mathbf{a}_4$	0	1	0	1	-1	0	2
	$\mathbf{a}_3$	0	1	1	0	1	0	9
	$\mathbf{a}_6$	0	1	0	0	0	1	4
	$\mathbf{a}_1$	1	0	0	0	1	0	6
	f_j^1	0	2	0	0	5	0	30
	f_j^2	0	-4	0	0	-1	0	-6

we see that $\mathbf{f}_4$ and $\mathbf{f}_5$ in Table 6-11 are both not comparable to zero. The introduction of $\mathbf{a}_4$ leads to $(\mathbf{a}_4, \mathbf{a}_3, \mathbf{a}_6, \mathbf{a}_1)$ while $\mathbf{a}_5$ would lead to

$$(\mathbf{a}_2, \mathbf{a}_3, \mathbf{a}_5, \mathbf{a}_1),$$

which is $\mathbf{B}^4$, the basic feasible solution from which we have just come. Thus, only $\mathbf{a}_4$ would lead to an unexplored basis. It is brought into the current basis to get $\mathbf{B}^6$, shown in Table 6-12.

The new simplex tableau indicates that $\mathbf{x}^6 = (6, 0)$ with $Z_1 = 30$ and $Z_2 = -6$. In going through the procedures we find in step 4 that $f_j^1 > 0$ for all nonbasic variables. Therefore this solution is a unique optimum for Z_1 and it is concluded that this solution is also noninferior. In step 8, $\theta_2 = 2$ and $\theta_5 = 6$ so that $f_2^1\theta_2 = 4 < f_5^1\theta_5 = 30$ and $f_2^2\theta_2 = -8 < f_5^2\theta_5 = -6$. The conclusion is that $\mathbf{a}_2$ dominates $\mathbf{a}_5$ as a candidate for a new entry into the basis. In step 12, however, we see that the introduction of $\mathbf{a}_2$ would lead to $(\mathbf{a}_2, \mathbf{a}_3, \mathbf{a}_6, \mathbf{a}_1)$ which is $\mathbf{B}^5$, a previously explored basis.

With an answer of "no" to the question in step 12, the procedure takes us to step 11 where any explored bases in storage are examined. Recall that the basis formed by introducing $\mathbf{a}_1$ into $\mathbf{B}^1 = (\mathbf{a}_3, \mathbf{a}_4, \mathbf{a}_5, \mathbf{a}_6)$ was stored during our first pass through the flowchart. Looking at Table 6-7, if $\mathbf{a}_1$ were introduced into this basis, then $\mathbf{a}_5$ would leave, giving a new basis of $(\mathbf{a}_3, \mathbf{a}_4, \mathbf{a}_1, \mathbf{a}_6)$. But this basis has already been explored: It is $\mathbf{B}^6$ in Table 6-12 (the order of the $\mathbf{a}_j$s does not matter). Therefore there are no unexplored bases in storage; the algorithm terminates.

It is interesting to consider the diagram of the feasible region in decision space in Fig. 6-26. Each extreme point is labeled with the basis number it was given by the algorithm. We began at the origin and worked our way around

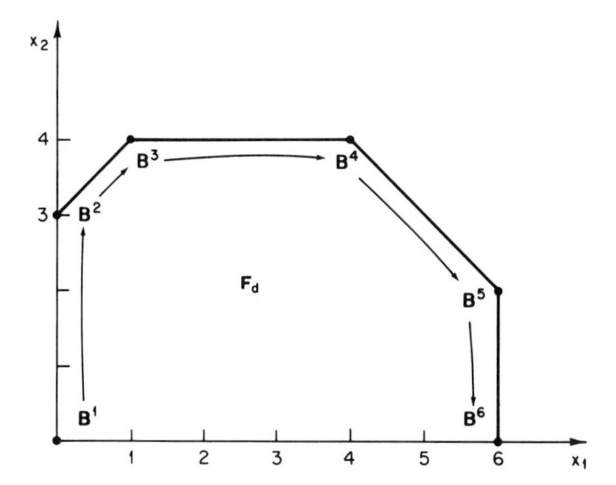

Fig. 6-26. The progression of the multiobjective simplex method around the boundary of the feasible region in decision space.

the boundary of $\mathbf{F}_d$ until all extreme points had been explored. If the initial basic feasible solution had been a maximum for one of the objectives (either $\mathbf{B}^3$ or $\mathbf{B}^6$), then $\mathbf{B}^1$ and $\mathbf{B}^2$ would not have been explored at all since they are inferior extreme points.

6.5 COMPLICATING ISSUES FOR THE USE OF GENERATING TECHNIQUES

Four techniques for the generation of noninferior solutions were presented in this chapter: the weighting method, the constraint method, the noninferior set estimation (NISE) method, and a multiobjective simplex routine. All of these methods are plagued by two practical complexities: computational costs and the display of results. Both issues are aggravated by the size of the problem: As the number of constraints, variables, and objectives goes up, so do computational costs and display difficulty.

In this section these important practical considerations are discussed. First, the appropriate role for each of these methods is considered, after which computational costs are discussed. Finally, various display techniques are presented.

6.5.1 When to Use the Techniques

The fact that a generating technique is to be used at all implies a certain philosophical approach to the planning and decision-making processes on the part of the analyst, as indicated in Chapter 5. The analyst perceives the planning process as one in which the articulation of preferences is postponed until the range of choice is identified. Given this methodological commitment, the question remains as to which generating technique to use.

The choice of a technique depends on the analyst's perception of the required results and, of course, on the analyst's preferences for mathematical procedures. If an exact representation of the noninferior set is sought, then the multiobjective simplex method is far superior to any of the other methods. The weighting, constraint, and NISE methods can be used to find the noninferior set exactly, but they are very inefficient when used in this capacity. Every noninferior solution that is found by a method that converts a multiobjective problem into a single-objective problem requires the solution of a linear programming problem; if there are g noninferior extreme points, then g linear programming problems must be solved. [Actually, through the use of parametric programming (Chapter 3), the solution of many of the problems can be accelerated.] The multiobjective simplex method, on the other hand, identifies these solutions by making simplex pivots; in some cases, a single pivot can lead to a new noninferior solution. Unfortunately, the multiobjective simplex method has yet to be applied to realistically sized problems, as mentioned in the next subsection on computational costs.

When an approximation of the noninferior set is sufficient for planning purposes, the multiobjective simplex method is inferior to the other three generating methods. The simplex method begins at an extreme point and moves to adjacent points; it traverses the noninferior set in a continuous manner, never skipping a noninferior extreme point. An approximation could be obtained with this method by premature termination of the algorithm, but the result would be partial coverage of the noninferior set. A portion of the noninferior set would be known exactly while the remaining part would be totally unknown.

The remaining choice is among the weighting, constraint, and NISE methods. The NISE method guarantees good coverage of the noninferior set in a manner that allows the accuracy of the approximation to be controlled. Furthermore, every noninferior solution that it identifies is chosen so as to reduce the error in the approximation as much as possible. These claims hold regardless of the geometrical character of the noninferior set, but the strength of the NISE method is most dramatic when the noninferior set has sharp "kinks" or elbows in it.

The weighting and constraint methods have certain weaknesses that tend to promote the use of the NISE method. The weighting method can give poor coverage of the noninferior set by getting "stuck" at an extreme point or in a small range of the noninferior set and by skipping over large portions of the noninferior set. The skipping behavior is not necessarily bad since this is an approximation technique; it is the lack of control by the analyst that can be troublesome. One may choose different weight sets with uniform increments, e.g., $(w_1, w_2) = (1, 1), (1, 2), (1, 3)$, and $(1, 4)$, but there is no way to predict or to control the nature of the resulting approximation.

The constraint method provides complete control of the spacing and coverage of the noninferior set. Its major weakness, however, is the rather high occurrence of infeasible formulations. This was demonstrated with Fig. 6-13.

The weighting and constraint methods may be the best techniques in some planning situations. If the weights themselves are considered important results, then some degree of control over their value is a significant attribute of the solution method. For instance, it may be worthwhile to communicate to decision makers that this solution implies that objective Z_1 is equally as important as objective Z_2; this solution results when objective Z_1 is twice as important; etc. In these situations the weighting method is an advantageous approach since the weights can be entirely controlled. The weights, represented as dual variables, can be by-products of all of the generating techniques, but their values are not controllable.

The constraint method is a good approach when, for display purposes, it is desirable to show a "cut" through the noninferior set. That is, one would like to indicate the tradeoffs between, say, Z_1 and Z_2 when Z_3 equals a given number. Tradeoffs between two other objectives with a different objective held constant may also be shown. This display technique is discussed in the subsection on display. It is most easily obtained with the constraint method.

6.5.2 Computational Considerations

There are two issues that will be discussed: computer software requirements and computational burden. Both are important considerations for the implementation of any technique.

All of the generating techniques rely on a computer program that can be used to solve mathematical programming problems. The simplex algorithm for linear programming problems is available in many forms. The Mathematical Programming System (MPS) and the Mathematical Programming System Extended (MPSX) can be rented for use on most International Business Machines (IBM) computers of the 360 or 370 series. References for these programs are IBM (1968, 1971). Other digital computers also have packages for linear programs. A FORTRAN program called PITMFOR was developed at the University of Pittsburgh for use on Digital Equipment Corporation (DEC) 10 computers.

The solution cost of a single linear programming can vary greatly. The cost seems most sensitive to the number of constraints and, of course, to the type of computer being used. On an IBM 360 or 370 with the MPS or MPSX package, a problem with fewer than 100 or 200 constraints is considered small; such a problem can be solved in a few seconds of computer time. The same problem would be considered large on some DEC 10 computers. A linear programming problem with 155 constraints required less than 1 min on an IBM 360/91 computer, but 15 min were not enough to find the optimal solution using a program called SIMPLX on the DEC 10 computer. On any computer, solution times for linear programs increase exponentially with the number of constraints.

The size of a linear program is generally controlled by the scale and complexity of the system under study. The multiobjective character adds a different kind of complexity and expense; it does not really affect the size of the basic model.

The weighting and constraint methods are straightforward applications of parametric programming. In the weighting method, the weighted problem is set up and several noninferior solutions are generated by changing the objective function coefficients. Most linear programming packages include a feature that allows easy parametric variation of the objective function. The use of this feature is more efficient than solving a new linear program with every new set of weights.

The constraint method is used to generate noninferior solutions by changing the right-hand sides of the constraints on the objectives. Most linear programming codes also include a procedure for altering right-hand sides that does not require the solution of the new problem. In some instances, operating with the right-hand sides is more straightforward than altering the objective function coefficients. In this case, the constraint method may be a more favorable approach than the weighting method.

The NISE method requires computation in addition to that which is required for the solution of the weighted problem. At every step of the algorithm, error terms and new weights must be computed. These can be done by hand or performed by a computer program that uses a linear programming code as a subroutine. Both approaches have been taken and neither seems to be a particularly difficult computational task; slightly more effort than that for the weighting or constraint methods is required.

The multiobjective simplex method is the most elaborate generating technique reviewed in this chapter. The extensive computer programming effort that is needed to implement this procedure results in part from the need to identify and store adjacent extreme points. Zeleny (1974a, pp. 202–220) provides a complete listing of a FORTRAN program which he used on an IBM 7040 computer. At the time of this writing, Professor Zeleny reported that the program was limited to problems with eight constraints, 40 variables, and eight objective functions; these are very small problems. The applications discussed in Chapters 9 and 10 required problems with 150–800 constraints, 200–1400 variables, and two–six objectives.

This brings us to the issue of computational burden. The multiobjective simplex method and the three approximating techniques are not really comparable on the basis of costs because they address two different problems. The cost of the multiobjective simplex method is related directly to the number of noninferior extreme points for a given problem. This cannot be predicted beforehand; it can only be stated that there are not more noninferior extreme points than the $n!/[m!(n - m)!]$ number of all extreme points (including infeasible extreme points), where n is the number of decision variables (including slacks) and m is the number of constraints (not including nonnegativity restrictions). Zeleny (1974a, pp. 116–121) discussed two examples with 16 decision variables (including slacks) and eight constraints. Both problems can have as many as $16!/(8!8!) = 12{,}870$ feasible extreme points. One problem with three objectives had only three noninferior extreme points, while the other, with five objectives, had 70 noninferior extreme points! This significant difference is a function of the number of objectives and of the nature of the feasible regions in each problem (they had different constraints and objectives). In its current state of development the multiobjective simplex method cannot be applied to large public decision-making problems with, say, hundreds of decision variables and hundreds of constraints.

The constraint and weighting methods have fairly predictable computational burdens that are directly related to the number of objectives p and the number of weights or right-hand sides used for each objective, which will be called r. The number of different combinations of weights or constraints, which is also the number of linear programming problems that must be solved, is r^{p-1}. The NISE method generally requires fewer than this number of solutions because of its exploitation of the shape of the noninferior set. However, although the number of solutions required cannot be stated exactly for the NISE method, it is easy to see that its computational burden is also exponentially related to the number of objectives.

The exponential relationship between the number of objectives and the computational burden is a major practical problem for the implementation of generating techniques. If there are several objectives and the optimization problem has several hundred constraints and decision variables, the expense of obtaining a sufficiently accurate approximation of the noninferior set may be prohibitive. Computational costs and the complexity of display when there are more than three objectives provide a strong motivation for the use of techniques that avoid the generation of an approximation or an exact representation of the noninferior set. Display mechanisms are taken up in the next section while methods that incorporate preferences, thereby reducing computational burden, are discussed in the next chapter.

6.5.3 Display Mechanisms

In order to be useful, the results from a generating technique must be communicated to decision makers. The results must be communicated in a manner that "tells the full story" without confusing the important issues. Completeness requires that the full range of choice be identified and that the values of the most important decision variables be communicated along with values of the objectives. Along with objective function values an appreciation for the tradeoffs among objectives—necessary sacrifices to achieve desirable gains—must be communicated.

When there are two, or perhaps three, objectives the most effective display mechanism is a graphical representation of the noninferior set in objective space, as in Figs. 6-1 and 6-13. This immediately shows, in a concise and dramatic fashion, the values of the objectives at each alternative and the tradeoffs between objectives. Decision variable values, i.e., the alternatives, can be communicated through tabular displays or other visual methods that are presented in a manner that corresponds to the graphical display of the noninferior set.

The three-objective graphical displays may be considered awkward in many situations, depending on the decision makers' technical backgrounds and views of such diagrams. For more than three objectives, and perhaps for three-objective problems, graphical displays of the entire noninferior set must be abandoned. However, Meisel (1973) has shown how a series of two-dimensional displays can be used to capture higher-dimensional noninferior sets. This amounts to "decomposing" the noninferior set by looking at slices through the noninferior set or at projections of it onto a plane. In the first approach a two-dimensional display, e.g., Fig. 6-1, is shown for two of the p objectives with the other $p - 2$ objectives held fixed. A different curve, on the same diagram, is shown for every different combination of values for the $p - 2$ objectives.

The second decomposition approach suggested by Meisel (1973) uses the projection of the noninferior set onto all planes formed by pairing the objectives. That is, projections onto the Z_1–Z_2, Z_1–$Z_3, \ldots, Z_1$–Z_p, Z_2–$Z_3, \ldots, Z_{p-1}$–Z_p planes of all generated noninferior points are drawn. The intention of this exercise is to give the person viewing these displays a graphical appreciation for the tradeoffs. Consider a three-objective problem and the projection of the noninferior set onto the Z_1–Z_2 plane as in Fig. 6-27. Each known noninferior alternative is shown as a circle with a number next to it. The number is the value that Z_3 assumes at each alternative. Some

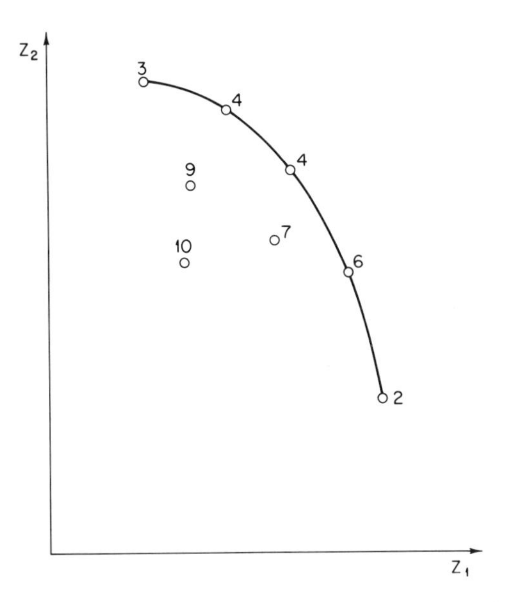

Fig. 6-27. Projection of a hypothetical collection of noninferior points for a three-objective problem onto the Z_1–Z_2 plane. The numbers indicate the value of Z_3.

alternatives, such as those that give 7, 9, and 10 of Z_3, appear inferior in Fig. 6-26; they are not, however, since they give more of Z_3 than do the points on the curve. The distance that a point lies below the curve indicates how much must be sacrificed of Z_1 and Z_2 to achieve the indicated gain in Z_3. Meisel (1973) shows how the projections on the various planes can be interpreted.

The two decomposition approaches give a way to display results, but the complexities of high-dimensional problems remain. In the first approach the more objectives, the more separate curves that must be shown. This can get too messy and intricate for meaningful analysis. The second approach suffers the same limitation in addition to the further problem that each point in a diagram such as Fig. 6-27 has $(p - 2)$ labels. With five objectives, the decision maker must digest the information contained in the curve while mentally computing and comparing three objective values at each point; this must be done with $5!/3!2! = 10$ two-dimensional figures. One technique that may not add much to decision makers' appreciation of tradeoffs, but which has the virtue of insensitivity to dimension, is tabular display. Of course, being confronted with a dense array of numbers, decision makers may simply throw up their hands!

A promising new methodology has been developed by Schilling (1976). The approach makes use of *value paths* shown in Fig. 6-28. A value path

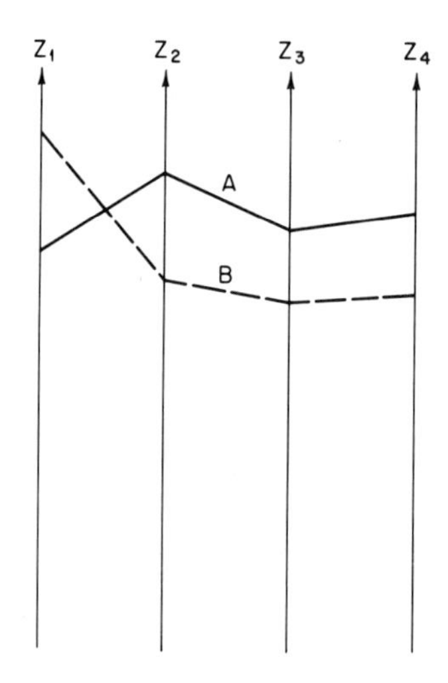

Fig. 6-28. Value-path display for a problem with four objectives.

is a line drawn for each noninferior alternative to indicate the level of achievement of each objective; a line high up on the scale for an objective indicates a high value for that objective. Value paths are shown for two alternatives in Fig. 6-28, where, for example, alternative A gives more of Z_2, Z_3, and Z_4 while B gives more of Z_1. The advantage of the value path methodology is the conciseness of the display regardless of the number of objectives. A possible disadvantage appears when there are many noninferior alternatives to be displayed: the density of the paths may confuse the decision makers. Experience to date, however, has been quite positive.

CHAPTER 7

Solution Techniques That Incorporate Preferences

Generating techniques, discussed in Chapter 6, develop a representation of the noninferior set for consideration by decision makers; preferences are articulated in this process insofar as they are implicitly captured by the decision makers' selection of a best-compromise solution. The techniques presented in this chapter all rely on explicit statements of preferences. Preferences are articulated in various ways: They may be assumed by the analyst; decision makers may be asked to state preferences before any knowledge of the tradeoffs is known; or preferences may be articulated progressively through interaction with a computer and the analyst. The preferences are also represented by various quantities: various weighting schemes, constraints, goals, utility functions, and many other kinds of less popular function and parameters.

Regardless of the approach or quantification schemes, all of the preference-based methods share the common goal of identifying the best-compromise solution. This goal is the unifying thread in the array of methods discussed in this chapter.

The computational burden of the techniques varies substantially from one method to the next. In most cases, preference-oriented methods are less computationally intensive than generating methods. This cost saving is accomplished by avoiding the expense of exploring the entire noninferior set; the specification of preferences allows the bulk of the noninferior solutions to be ignored. However, the computational savings must be traded off against other complexities related to extracting reliable preference statements. The identification of preferences can be a time-consuming and expensive affair.

Keep your notions of the public decision-making process uppermost in your mind while you read this chapter. Consider the demands that a technique places on the analyst and decision makers. Are they excessive? Does one have to assume too much about the decision-making process, such as the identifiability, number, and accessibility of decision makers? Are the types of questions that require asking by the technique answerable in planning contexts with which you are familiar? Are the techniques too elaborate or intricate to be considered meaningful by planners and decision makers? These questions must be considered before any method can be implemented.

The discussion of the techniques begins with multiattribute utility functions since they are the most complete and flexible statement of preferences. Many of the techniques have been developed so as to approximate a multiattribute utility function.

After the discussion of utility functions we shall turn to notions of preference that are more narrow and simple: the prior specification of weights and geometrical concepts of best, including goal programming. This is followed by a presentation of several interactive techniques that seek to draw out relevant preference information from decision makers through iterative person–machine or decision maker–analyst interactions. Some of these methods are based on local approximations of the assumed underlying utility function.

7.1 MULTIATTRIBUTE UTILITY FUNCTIONS

A utility function is a mathematical function that associates a utility with each alternative so that all alternatives may be ordered. Given any two alternatives A and B, the utility function allows us to state that A is preferred to B, B is preferred to A, or A and B are indifferent.

An *ordinal utility function* provides an ordering of alternatives, but does not indicate the degree to which one alternative is preferred to another. The scale by which utility is measured is not important for ordinal utility functions. *Cardinal utility functions* indicate an ordering and a level of preference, e.g., A is preferred to B and is 10 units of utility more desirable than is B. A cardinal utility is a measurable mathematical quantity for which any appropriate unit may be used, but the functions must be such that if a number is added to every utility or if every utility is multiplied by the same positive number, then the relative desirability of the alternatives will not change. Cardinal utility functions are the only ones dealt with in this section. Henceforth, the "cardinal" will be dropped and they will just be called utility functions.

Utility functions are used to capture the preferences of an individual for various *attributes* of alternatives. For instance, an attribute in a fire station location problem is the number of people covered by a location alternative. The argument of the utility function in this case is population coverage; the desirability of population coverage for the person whose preferences are being assessed vary with the level of population coverage. Notice that an attribute and an objective are equivalent terms; the former is used extensively in the utility theory literature.

When there are many attributes or objectives associated with an alternative the utility function has multiple arguments and is referred to as a *multiattribute utility function.* Every alternative implies a value for each objective or attribute, and a multiattribute utility function associates a single number, a utility, with the combination of values for the objectives.

The remainder of this discussion of multiattribute utility functions is divided into three parts: review of the relevant literature, theory for multiattribute utility functions, and an algorithm for finding the best-compromise solution, for which a numerical example is given.

It is important that the reader be aware of the limited scope of the discussion which follows. The literature on utility functions is voluminous, a reliable indication that utility theory is a complex and intricate field of study. The presentation here is very selective with the justification that multiattribute utility theory is worthy of (at least) one book by itself. In fact, such a book has been written by Keeney and Raiffa (1976). It also is worth pointing out that a certain amount of skepticism exists concerning the practical usefulness of utility theory. While this author shares that skepticism, the topic is worth studying because it represents a paradigm for a substantial amount of multiobjective research.

7.1.1 Review of Selected Literature on Utility

The notion of utility as a measurement of preferences for an alternative or outcome is a rather old one, going back at least to the utilitarian philosopher/economist Jeremy Bentham (1948†). Bentham believed in the measurability and comparability of utility: Not only could one measure that a headache is worth −3 and that a drink of cold water from a mountain stream on a hot day

† Originally published in 1789.

after jogging 2 miles is worth +1 to Mr. Jones (this was 200 years ago when the marginal utility of pure cold fresh water was low); one could also conclude that if for Ms. Smith a headache and the drink were worth −4 and +2, respectively, then Ms. Smith would suffer more from the headache and derive more satisfaction from the drink than would Mr. Jones.

The notion that one could establish a cardinal utility scale that would allow such interpersonal comparisons of utility has not been a popular notion since Bentham's time. Utility theory, however, has played an important role in the development of welfare economics. It led Pareto (1971†) to devise ordinal utility functions that supported and still support much of welfare economic theory.

A resurgence of cardinal utility theory began with the landmark work of von Neumann (1928) and, later, von Neumann and Morganstern (1967‡). They developed the basic axiomatic theory for cardinal utility functions for the purpose of individual decision making; there was no claim of interpersonal comparability of utilities. The major part of their book is concerned with the analysis of games, competitive situations involving two or more participants.

The history of utility theory for decision making since World War II is too eventful for review here. The books by Pratt *et al.* (1965), Raiffa (1968), Fishburn (1970), and Keeney and Raiffa (1976) present full treatments of this subject.

The interest in utility functions since von Neumann and Morganstern's treatise has been motivated by their potential applicability to the analysis of decisions in a risky environment. While most of the recent research has been directed at risky decision-making situations, the utility notions that have developed are entirely general.

Our primary area of interest is multiattribute utility theory, which has been a very active area of research since 1970. Aumann (1964) appears to be the first to have considered utility functions for multiobjective problems. His primary concern was the impact of many objectives on single-dimensional utility theory. Briskin (1966) indicated how one could proceed to a best-compromise solution given a utility function defined over two objectives.

Research really seemed to commence, however, with Raiffa's (1969) and Kenney's (1969) development of some of the theory required for multiattribute utility functions. The motivation for much of the research since 1970 is the difficulty of assessing multiattribute utility functions in their general form. Thus, many specialized forms—additive, multiplicative, and more general decomposed forms—have been suggested. Pollak (1967), Keeney

† Originally published in 1909.

‡ Originally published in 1944.

(1971, 1972a, 1972b, 1973a, 1974), Fishburn (1973, 1974), Fishburn and Kenney (1974, 1975), Farquhar (1975, 1976), and Kirkwood (1976) have presented various forms for multiattribute utility functions, developed the underlying theory, and discussed the measurement implications. The interested reader will want to begin with the excellent review by Farquhar (1977) and the more detailed discussion by Keeney and Raiffa (1976).

Multiattribute utility theory has not been applied extensively due to the relative youth of the field and the aforementioned measurement difficulties. Applications have been reported in the areas of water resources (Major, 1974), transportation planning (deNeufville and Keeney, 1973; Keeney, 1973b), and fire systems analysis (Keeney, 1973c). Farquhar (1977) mentions other applications.

7.1.2 Multiattribute Utility Theory

A multiattribute utility function (which frequently will simply be called a utility function below) is a mathematical statement that indicates the utility of all combinations of values for the various attributes or objectives that are under consideration. The general form of a multiattribute utility function is $U[Z_1(\mathbf{x}), Z_2(\mathbf{x}), \ldots, Z_p(\mathbf{x})]$, where the objectives, which are the arguments of the utility functions, are themselves functions of the decision variables $\mathbf{x} = (x_1, x_2, \ldots, x_n)$. For any p numbers, each one representing a value for an objective, the utility function translates them into a single number that indicates the utility or degree of preference which that combination of objectives yields.

There are several axioms that cardinal utility functions must satisfy (von Neumann and Morganstern, 1944; Keeney, 1969). We will not deal with these axioms in a structured manner. Rather, the nature of utility functions, for which the most important axioms hold, is discussed.

A fundamental requirement is that a utility function must provide a complete ordering of alternatives in the sense that, given any two alternatives $\mathbf{x}^1 = (x_1^1, x_2^1, \ldots, x_n^1)$ and $\mathbf{x}^2 = (x_1^2, x_2^2, \ldots, x_n^2)$ with objective values $Z_1(\mathbf{x}^1)$, $Z_2(\mathbf{x}^1), \ldots, Z_p(\mathbf{x}^1)$ and $Z_1(\mathbf{x}^2), \ldots, Z_p(\mathbf{x}^2)$, respectively, either $\mathbf{x}^1$ is preferred to $\mathbf{x}^2$, $\mathbf{x}^2$ is preferred to $\mathbf{x}^1$, or $\mathbf{x}^1$ and $\mathbf{x}^2$ are indifferent (equally good) alternatives. The objective function values alone allow such unambiguous statements only when one alternative dominates the other. When both alternatives are noninferior, the alternatives are not comparable on the basis of the objective values alone. A utility function, on the other hand, allows all alternatives, even noninferior ones, to be compared.

A multiattribute utility function is general in that it can take on any mathematical form that satisfies the axioms. It is usually assumed, however, that utility is defined by a monotonically increasing concave function when the objectives are to be maximized. Mathematically this means that

$$\frac{\partial U(Z_1, Z_2, \ldots, Z_p)}{\partial Z_k} > 0, \qquad k = 1, 2, \ldots, p \tag{7-1}$$

and

$$\frac{\partial^2 U(Z_1, Z_2, \ldots, Z_p)}{\partial Z_k^2} < 0, \qquad k = 1, 2, \ldots, p \tag{7-2}$$

where the notation for the utility function has been simplified by not including the decision variables as arguments.

The above mathematical assumptions are quite reasonable in that they reflect typically observed behavior. The condition in (7-1) states that marginal utility with respect to any objective is positive; i.e., more of any objective, when the others are held constant, is preferred to less. The condition in (7-2) means that marginal utility of any objective decreases as the level of that objective increases. This is intuitively appealing in that an additional unit of something, like money, is worth less when you have a large bank account than it is when you are less well off.

A concave utility function for a *single-objective* problem is shown in Fig. 7-1. Notice that $\Delta U' < \Delta U$ reflecting decreasing marginal utilities as

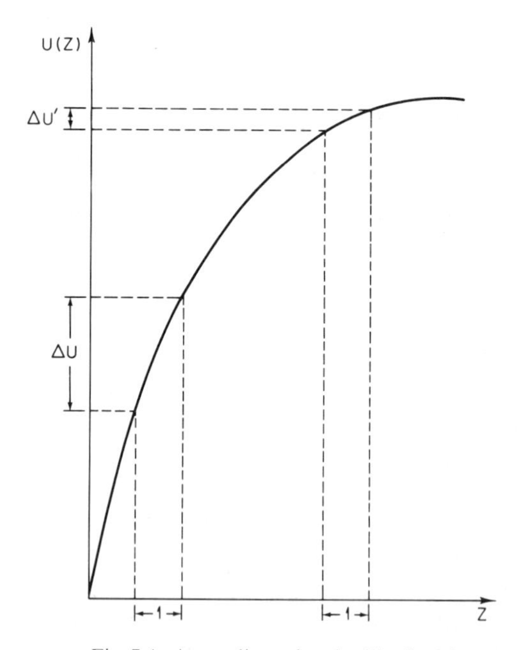

Fig. 7-1. A one-dimensional utility function.

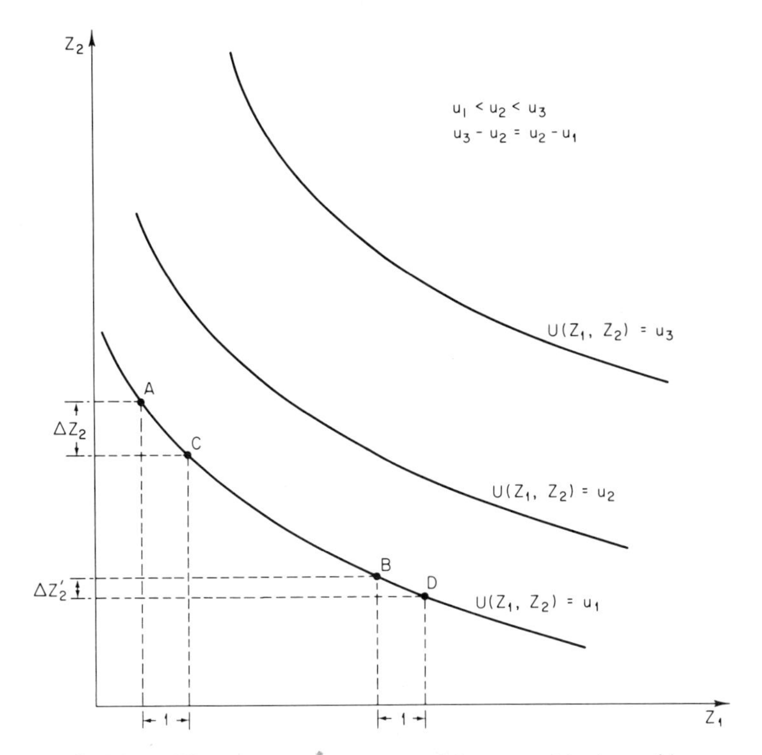

Fig. 7-2. Indifference (isopreference curves) for a two-objective problem.

required by (7-2). A utility function for a two-objective problem would appear as a surface in three dimensions, but a frequently used device for representing multidimensional utility functions is the *isopreference* or indifference curve. If a plane parallel to the Z_1–Z_2 axes is passed through the utility surface, the cross section of the surface is a curve of equal utility called an isopreference or indifference curve. Three indifference (isopreference) curves are shown in Fig. 7-2 for three different levels of utility u_1, u_2, u_3 that have been chosen so that $u_1 < u_2 < u_3$ and $u_2 - u_1 = u_3 - u_2$.

The indifference curves in Fig. 7-2 reflect the concavity of the utility function in two ways. First, in moving from one indifference curve to another in the direction of increasing utility, i.e., to the northeast in Fig. 7-2, we have to move farther to achieve an increase in utility when we are far from the origin than when we are close to the origin. The three indifference curves in Fig. 7-2 are "equidistant" in utility terms, i.e., $u_3 - u_2 = u_2 - u_1 = \Delta u$. However, the distance in terms of Z_1 and Z_2 quantities is greater between the two outer curves. This reflects the fact that we need a larger increase in Z_1 and Z_2 to achieve Δu when $U(Z_1, Z_2) = u_2$ already than we do when $U(Z_1, Z_2) = u_1$.

This is equivalent to the decreasing marginal utility in the one-dimensional case of Fig. 7-1.

Second, in moving along an indifference curve, i.e., letting Z_1 and Z_2 vary with utility fixed, we are less inclined to sacrifice Z_2 to gain more of Z_1 when we have a high level of Z_1 than when we have a low level of Z_1. Consider points A and B, which represent two combinations of Z_1 and Z_2 to which we are indifferent, i.e., they both give u_1 units of utility since they lie on the indifference curve $U(Z_1, Z_2) = u_1$ in Fig. 7-2. At point A we have little Z_1 relative to Z_2. The shape of the indifference curve implies that we are willing to sacrifice ΔZ_2 units of Z_2 to gain one additional unit of Z_1; this trade would take us to point C. At point B, however, we are willing to sacrifice only $\Delta Z'_2, \Delta Z'_2 < \Delta Z_2$, for an additional unit of Z_1; this trade would move us along the indifference curve to point D.

The amount of one objective that we are willing to sacrifice for another objective in the above fashion is called the *marginal rate of substitution* (MRS). This has also been called the tradeoff between objectives. The MRS is defined mathematically by

$$\text{MRS}_{12} = -\left.\frac{dZ_2}{dZ_1}\right|_{U(Z_1, Z_2) = U(Z_1^*, Z_2^*)} \tag{7-3}$$

where Z_1^* and Z_2^* are fixed values of the two objectives. Notice that the MRS is just the negative of the slope of the indifference curve, and since the indifference curve has a constantly changing slope, the MRS varies as we move along the indifference curve. The point (Z_1^*, Z_2^*) must be specified since the MRS varies from point to point. Notice also that MRS_{12} is subscripted to denote the marginal rate of substitution between objectives Z_1 and Z_2. For p objectives the MRS between objectives Z_i and Z_k is

$$\text{MRS}_{ik} = -\left.\frac{\partial Z_k}{\partial Z_i}\right|_{U(Z_1, Z_2, \ldots, Z_p) = U(Z_1^*, Z_2^*, \ldots, Z_p^*)}, \qquad k = 1, 2, \ldots, p \tag{7-4}$$

where objective Z_i is an arbitrarily chosen reference objective or *numeraire*. This means that the units of objective Z_i are chosen as the scale upon which to measure the variability of the other objectives; e.g., if Z_i is net national income (economic efficiency) benefits, then tradeoffs are measured in terms of dollars per unit of environmental quality or dollars per new job (for an employment objective).

The variation in the MRS or the desirable tradeoff between objectives is the characteristic of a utility function that makes it a general representation of preferences. The utility function gives us the ability to capture the behaviorial assumption that the relative desirability of the next unit of an objective goes down as we gain more of that objective.

A utility function is defined over the entire objective space; i.e., all combinations of objectives are ordered by the function. Thus, the indifference curves span the entire objective space. This collection of curves is called an "indifference map," which can be used to identify the best-compromise solution. Recall from Fig. 4-3 that the best-compromise solution is that noninferior point at which an indifference curve is tangent to the noninferior set. This tangency condition is equivalent to finding the feasible solution that is on the indifference curve farthest to the northeast. Consider Fig. 7-3, in which $\mathbf{F}_o$, $\mathbf{N}_o$, and three curves in an indifference map are shown. The indifference curve $U = u_1$ goes through many feasible solutions, but none of these are candidates for the best-compromise solution because we can move to feasible solutions on higher indifference curves. This is true for all feasible solutions except the best-compromise solution, which lies on $U = u_2$ in Fig. 7-3. Higher indifference curves such as $U = u_3$ do not touch any feasible solutions.

The tangency condition can be restated in the following way: The best-compromise solution is that noninferior solution for which the desirable tradeoff between objectives is equal to the feasible tradeoff between objectives. The *desirable tradeoff* is the MRS, the negative of the slope of the indifference

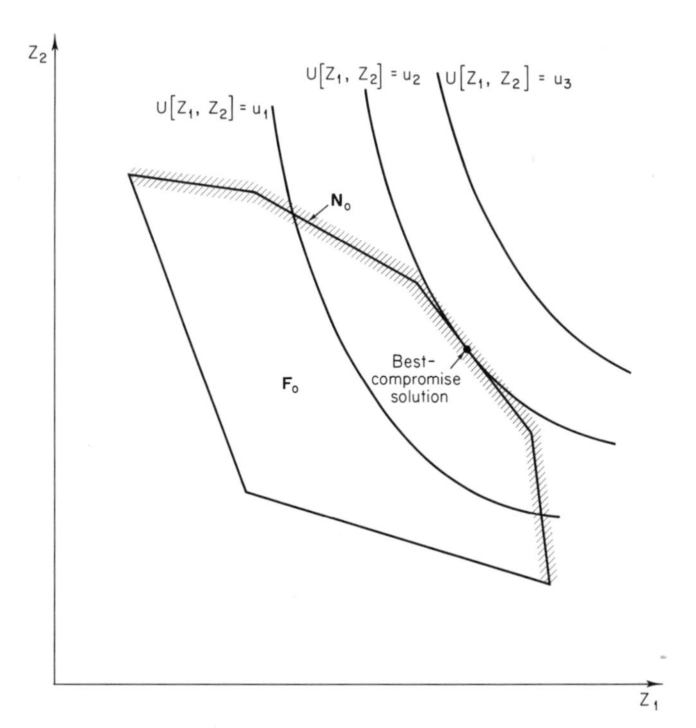

Fig. 7-3. Indifference map and the best-compromise solution.

curve, while the *feasible tradeoff* is often referred to as the marginal rate of transformation (MRT) between Z_1 and Z_2, the negative of the slope of the noninferior set. The equality of the slopes of $\mathbf{N}_0$ and the indifference curve, or equivalently, MRS = MRT, is the tangency condition.

A utility function is an obviously comprehensive statement of preferences. Therefore it is very difficult to measure in a practical situation. Briskin (1966) presented an application of a multiattribute utility function to a routing problem in which the two objectives were the minimization of travel time on the route Z_1 and the minimization of the cost of the route Z_2. A functional form of the utility function was assumed and then "fit" to the decision makers' preferences by getting them to describe tradeoff situations. The assumed equation for an indifference curve was

$$Z_2 = \rho(1 - e^{\sigma Z_1}) \tag{7-5}$$

where ρ and σ are constants to be determined. This function is actually convex (the opposite of concave) because the objectives are being minimized. If we were to convert the objectives to maximizations, we would multiply them and the utility function by (-1), which would convert the utility function to a concave function.

The MRS along the prescribed indifference curve is

$$\text{MRS}_{12} = -(dZ_2/dZ_1) = \rho\sigma e^{\sigma Z_1} \tag{7-6}$$

Briskin suggested that ρ and σ could be found by getting decision makers to describe hypothetical situations, e.g., for routes that cost a dollars and take b hours to travel, cost and time should be traded off at the rate c. This gives $Z_2 = a, Z_1 = b$, and $(dZ_2/dZ_1) = c$, which are used in (7-5) and (7-6) to find ρ and σ. The indifference curves found in this way are then used to identify the best-compromise solution. Notice that the information may not be easy to provide; hypothetical situations are difficult to assess.

As mentioned in the review in the previous sections, many "decomposition" methods have been proposed for multiattribute utility functions. A decomposition method requires additional assumptions, stronger than the axioms of utility theory, to be made. But the increased restrictions buy greater ease of measurement. One of the most restrictive approaches is the additive function, in which it is assumed that the utility function is the sum of separable utility functions for each objective;

$$U(Z_1, Z_2, \ldots, Z_p) = U_1(Z_1) + U_2(Z_2) + \cdots + U_p(Z_p) \tag{7-7}$$

This form allows the decision maker to concentrate on each objective individually, ignoring the other objectives. The simplification and the underlying assumptions are apparent: The preference interactions among objectives are not captured by this function.

Other decomposition schemes are more realistic and more complicated. Regardless of the approach, however, multiattribute utility functions are very difficult to construct in a practical situation. This complexity has motivated the remainder of the techniques discussed in this chapter in that the approaches represent attempts to capture decision makers' preferences by working with value judgments that are applicable to relevant ranges of alternatives without the comprehensiveness or generality of the utility function.

7.1.3 An Algorithm for Proceeding More Directly to the Best-Compromise Solution

Given a utility function and a representation of the noninferior set, the best-compromise solution can be found by graphical or numerical methods. Geoffrion (1967) developed algorithms that proceed more directly to the best-compromise solution in that they avoid the development of a representation of the entire noninferior set by focusing on only as much of the noninferior set as is needed.

The problem that Geoffrion considers is

$$\text{maximize} \quad U[Z_1(\mathbf{x}), Z_2(\mathbf{x})] \tag{7-8}$$

$$\text{s.t.} \quad \mathbf{x} \in \mathbf{F}_d \tag{7-9}$$

where $U[\quad]$ is a two-dimensional utility function defined over the two objectives. Geoffrion assumed, as we have throughout this book, that $\mathbf{F}_d$ is a convex set and that $U[\quad]$ gives indifference curves that have the shape of those in Fig. 7-2. Actually, Geoffrion made the weaker assumption that the utility function is "quasi-concave," i.e., that utility is monotonically nondecreasing (instead of strictly increasing) in each objective. All concave functions are quasi-concave, but the inverse is not true. A quasi-concave single-dimensional utility function that is not concave is shown in Fig. 7-4. Notice that dips and flat spots in the function are allowed; the important characteristic is that utility never drops as Z increases. We will make the stronger assumption of a concave utility function as in Fig. 7-1. The essential features of the algorithm are not altered by this assumption.

The algorithm discussed here is based on two major ideas. First, that the best-compromise solution is a noninferior solution and noninferior solutions can be found by solving a weighted problem. We know both of these observations to hold. The particular weighting scheme that Geoffrion used is the solution of

$$\text{maximize} \quad Z(\mathbf{x}, \alpha) = \alpha Z_1(\mathbf{x}) + (1 - \alpha) Z_2(\mathbf{x}) \tag{7-10}$$

$$\text{s.t.} \quad \mathbf{x} \in \mathbf{F}_d \tag{7-11}$$

in which the weights on Z_1 and Z_2 have been chosen to sum to one.

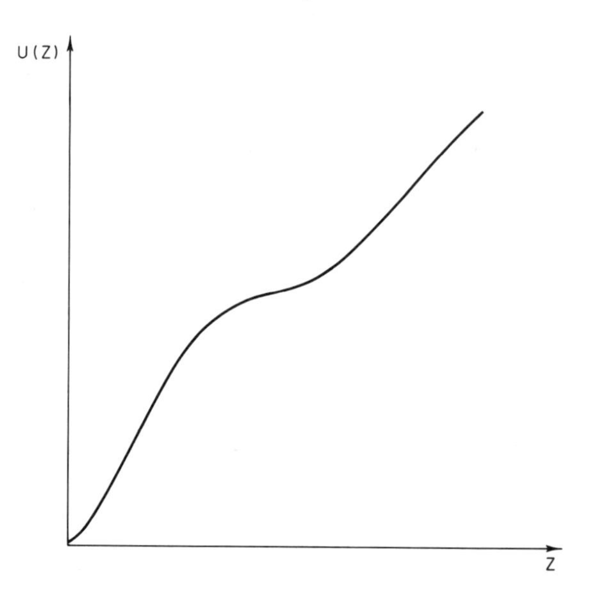

Fig. 7-4. A quasi-concave one-dimensional utility function.

The second important idea is that the utility function is *unimodal* over the noninferior set. This means that if one began at one end of the noninferior set and traversed it to the other end, $U[Z_1, Z_2]$ would increase to some maximum, and then decrease thereafter. Utility would not go up, then down, then up again. To see that this is the case consider Fig. 7-5. Every point in $\mathbf{N}_o$ to the right of point A and to the left of C is on an increasingly higher indifference curve; compare B on $U = u_3$ to A, which gives u_1 units of utility. From point C, the best-compromise solution, to point E utility decreases; compare C to D and D to E.

One way unimodality could be violated is if moving away from the origin in a northeasterly direction were to result in lower utility, but that would violate the concavity of the utility function. Another way is if feasible moves along the noninferior set were to take us to lower utility, but this would violate the convexity of $\mathbf{F}_o$. Thus, the unimodality of the utility function over $\mathbf{N}_o$ is a result of the convexity of $\mathbf{F}_o$ (and therefore of $\mathbf{F}_d$) and the quasi-concavity (concavity) of $U[\quad]$.

The algorithm discussed below [adapted from Algorithm 2 of Geoffrion (1967)†] combines the two observations by starting at the end of the noninferior set and continually generating noninferior extreme point solutions until the utility function is observed to decrease. When this occurs we know that the

† Adapted by permission from Arthur Geoffrion, "Solving Bicriterion Mathematical Programs," OPERATIONS RESEARCH Vol. 15, p. 39, copyright 1967, ORSA.

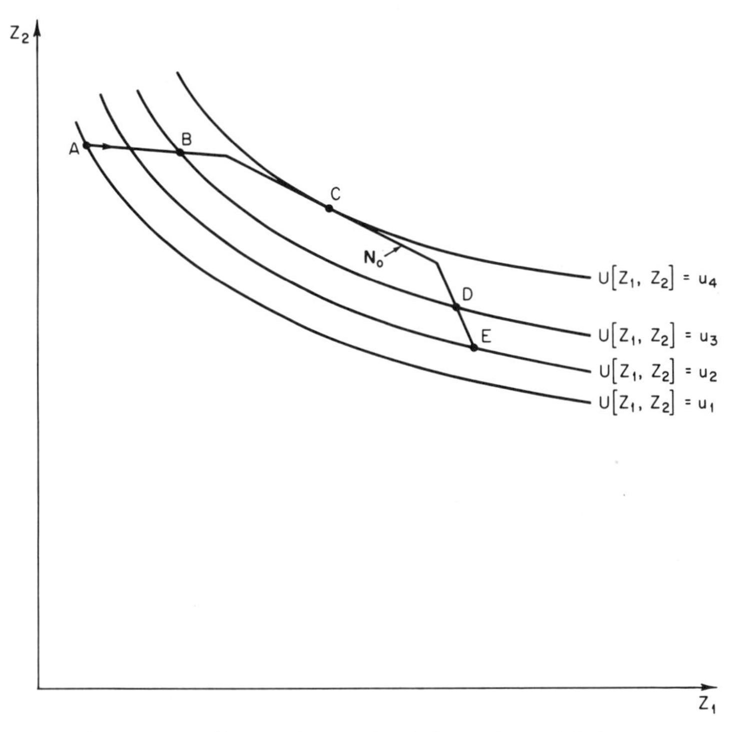

Fig. 7-5. The utility function is unimodal over the noninferior set.

best-compromise solution has recently been passed. A subproblem is then solved over a small range within which the best-compromise solution is known to lie. In the steps listed below the indices i and j are used to define the range over which the subproblem is to be solved. The quantity α, the weight on objective Z_1, is increased from zero so as to move along the noninferior set.

Step 1 Set $i = 0$, $j = 1$, and $\alpha = 0$. Solve (7-10) and (7-11) to find $\mathbf{x}^0$.

Step 2 Increase α parametrically until a new solution $\mathbf{x}^{i+1}$ to (7-10) and (7-11) is found or $\alpha = 1$. In the first case go to step 3; in the second case go to step 4.

Step 3 Compare $U[Z_1(\mathbf{x}^i), Z_2(\mathbf{x}^i)]$ with $U[Z_1(\mathbf{x}^{i+1}), Z_2(\mathbf{x}^{i+1})]$:

(a) If $U[Z_1(\mathbf{x}^i), Z_2(\mathbf{x}^i)] < U[Z_1(\mathbf{x}^{i+1}), Z_2(\mathbf{x}^{i+1})]$, put $j = i$, increase i by 1, and go to step 2.

(b) If $U[Z_1(\mathbf{x}^i), Z_2(\mathbf{x}^i)] = U[Z_1(\mathbf{x}^{i+1}), Z_2(\mathbf{x}^{i+1})]$, increase i by 1 and go to step 4. (Go to step 2 if the utility function is quasi-concave.)

(c) If $U[Z_1(\mathbf{x}^i), Z_2(\mathbf{x}^i)] > U[Z_1(\mathbf{x}^{i+1}), Z_2(\mathbf{x}^{i+1})]$, increase i by 1 and go to step 4.

Step 4 If $i = 0$, STOP; $\mathbf{x}^0$ is the best-compromise solution. If $i \geq 1$, then the best-compromise solution occurs somewhere in between $\mathbf{x}^i$ and $\mathbf{x}^j$; go to step 5.

Step 5 Solve for $q = j, \ldots, i - 1$

$$\begin{aligned} \text{maximize} \quad & U\{t[Z_1(\mathbf{x}^q), Z_2(\mathbf{x}^q)] + (1 - t)[Z_1(\mathbf{x}^{q+1}), Z_2(\mathbf{x}^{q+1})]\} \\ \text{s.t.} \quad & 0 \leq t \leq 1 \end{aligned} \tag{7-12}$$

The solution that maximizes $U[\quad]$ over the interval is the best-compromise solution. STOP.

The subproblem in (7-12) is usually an easy problem to solve. The parameter t is used to form a *convex combination* of the solutions $\mathbf{x}^q$, $\mathbf{x}^{q+1}$. A convex combination of two points lies along the line segment connecting the two points. Thus, the subproblem in (7-12) is a search along the line segment for that point that maximizes the utility.

Step 3(b) is stated in a manner that allows for quasi-concave utility functions. When the utility function is concave, two successive solutions that yield equal utility imply that the best-compromise solution has been passed. Consider, for example, points B and D in Fig. 7-5. For concave functions, then, we go to step 4. If the utility function is quasi-concave, successive equal utility solutions can occur prior to passing the best-compromise solution. In this case we return to step 2 and continue searching.

The algorithm will be demonstrated with the sample problem in (4-7). We will assume that a utility function has been specified for us;

$$\text{maximize} \quad U[Z_1(\mathbf{x}), Z_2(\mathbf{x})] = Z_1(\mathbf{x})Z_2(\mathbf{x}) \tag{7-13}$$

As an aside, one may want to "normalize" the objectives so as to avoid negative utilities; e.g., at $Z_1 = 15$, $Z_2 = -3$, $U = -45$.

It is interesting to find the best-compromise solution graphically before we use the algorithm. The noninferior set is shown in Fig. 7-6 with three indifference curves for $U = 50, 100, 150$. It appears that the best-compromise solution is between (12, 12) and (26, 2). The algorithm will tell us exactly. Notice in Fig. 7-6 that the noninferior extreme points are numbered from 0 to 3, corresponding to the values of i and j in the algorithm; this is the sequence in which they will be generated by the algorithm until it terminates.

Step 1 $\alpha = 0$, $i = 0$, $j = 1$, and we solve

$$\text{maximize} \quad Z_2(x_1, x_2) = -x_1 + 4x_2 \tag{7-14}$$

$$\text{s.t.} \quad (\mathbf{x}_1, \mathbf{x}_2) \in \mathbf{F}_d \tag{7-15}$$

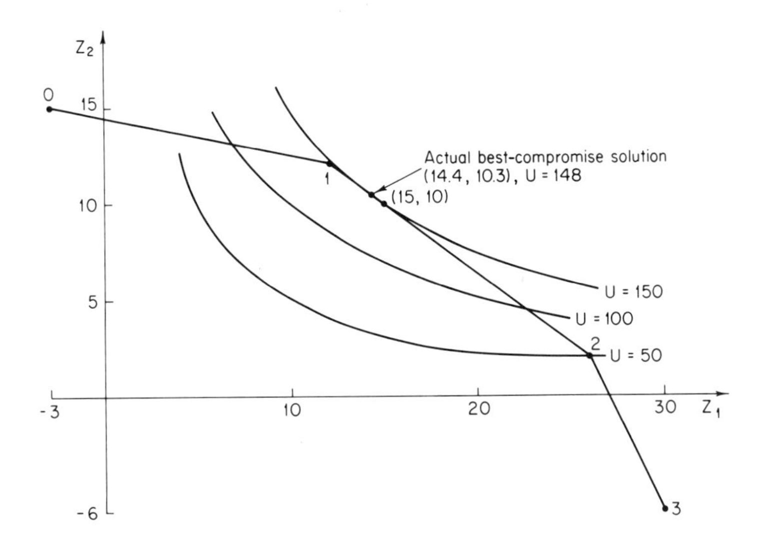

Fig. 7-6. Application of Geoffrion's algorithm to the sample problem.

where (7-15) requires the solution to satisfy the constraints of (4-7). The solution is $\mathbf{x}^0 = (1, 4)$ at which $Z_1(\mathbf{x}^0) = -3$, $Z_2(\mathbf{x}^0) = 15$, and $U[Z_1\mathbf{x}^0)$, $Z_2(\mathbf{x}^0)] = -45$.

Step 2 Increase α until a new extreme point is found. For this problem $0 \leq \alpha \leq \frac{1}{6}$ gives $\mathbf{x}^0$ and $\frac{1}{6} \leq \alpha \leq \frac{5}{12}$ gives a new noninferior extreme point $\mathbf{x}^1 = (4, 4)$, at which $Z_1(\mathbf{x}^1) = 12$, $Z_2(\mathbf{x}^1) = 12$, and $U[Z_1(\mathbf{x}^1), Z_2(\mathbf{x}^1)] = 144$. We go to Step 3.

Step 3 We see that $U[Z_1(\mathbf{x}^0), Z_2(\mathbf{x}^0)] < U[Z_1(\mathbf{x}^1), Z_2(\mathbf{x}^1)]$, so $j = i = 0$, $i = i + 1 = 1$, and we return to step 2.

Step 2′ α is increased beyond $\frac{5}{12}$ to get the new solution $\mathbf{x}^2 = (6, 2)$ at which $Z_1(\mathbf{x}^2) = 26$, $Z_2(\mathbf{x}^2) = 2$, and $U[Z_1(\mathbf{x}^2), Z_2(\mathbf{x}^2)] = 52$.

Step 3′ Now $U[Z_1(\mathbf{x}^1), Z_2(\mathbf{x}^1)] = 144 > U[Z_1(\mathbf{x}^2), Z_2(\mathbf{x}^2)]$, so $i = i + 1 = 2$ and we go to step 4.

Step 4 Since $i > 1$, we know the best-compromise solution must be between $\mathbf{x}^j$ and $\mathbf{x}^i$, i.e., it is $\mathbf{x}^1$, on the line segment between $\mathbf{x}^0$ and $\mathbf{x}^1$ or on the line segment between $\mathbf{x}^1$ and $\mathbf{x}^2$ (see Fig. 7-6). We go to step 5.

Step 5 For $q = j = 0$ we solve

$$\begin{aligned} \text{maximize} \quad & U\{t[Z_1(\mathbf{x}^0), Z_2(\mathbf{x}^0)] + (1 - t)[Z_1(\mathbf{x}^1), Z_2(\mathbf{x}^1)]\} \\ \text{s.t.} \quad & 0 \leq t \leq 1 \end{aligned} \tag{7-16}$$

or

$$\begin{aligned} \text{maximize} \quad & U[t(-3, 15) + (1 - t)(12, 12)] \\ \text{s.t.} \quad & 0 \leq t \leq 1 \end{aligned} \tag{7-17}$$

or

$$\begin{aligned} \text{maximize} \quad & U[(-3t, 15t) + (12 - 12t, 12 - 12t)] \\ \text{s.t.} \quad & 0 \leq t \leq 1 \end{aligned} \tag{7-18}$$

or

$$\begin{aligned} \text{maximize} \quad & U[(-15t + 12), (3t + 12)] \\ \text{s.t.} \quad & 0 \leq t \leq 1 \end{aligned} \tag{7-19}$$

Using the form of the utility function gives

$$\begin{aligned} \text{maximize} \quad & [(-15t + 12)(3t + 12)] \\ \text{s.t.} \quad & 0 \leq t \leq 1 \end{aligned} \tag{7-20}$$

or

$$\begin{aligned} \text{maximize} \quad & (-45t^2 - 114t + 144) \\ \text{s.t.} \quad & 0 \leq t \leq 1 \end{aligned} \tag{7-21}$$

which is maximized at $t = 0$. This result was expected from the graphical presentation in Fig. 7-6 since $t = 0$ means that on the segment from $\mathbf{x}^0$ to $\mathbf{x}^1$ the utility function is maximized at $\{0[Z_1(\mathbf{x}^0), Z_2(\mathbf{x}^0)] + [Z_1(\mathbf{x}^1), Z_2(\mathbf{x}^1)]\} = [Z_1(\mathbf{x}^1), Z_2(\mathbf{x}^1)]$, i.e., at the solution $\mathbf{x}^1$.

For $q = 1 = i - 1$ we solve

$$\begin{aligned} \text{maximize} \quad & U\{t[Z_1(\mathbf{x}^1), Z_2(\mathbf{x}^1)] + (1 - t)[Z_1(\mathbf{x}^2), Z_2(\mathbf{x}^2)]\} \\ \text{s.t.} \quad & 0 \leq t \leq 1 \end{aligned} \tag{7-22}$$

which becomes, by following the steps in (7-16)–(7-21) above,

$$\begin{aligned} \text{maximize} \quad & U(-140t^2 + 232t + 52) \\ \text{s.t.} \quad & 0 \leq t \leq 1 \end{aligned} \tag{7-23}$$

This is maximized at $t = 0.83$, where $U = 148$ and the best-compromise solution in objective space is computed from

$$(0.83)(12, 12) + (0.17)(26, 2) \tag{7-24}$$

which gives (14.4, 10.3); i.e., $Z_1(\mathbf{x}^*) = 14.4$ and $Z_2(\mathbf{x}^*) = 10.3$, where $\mathbf{x}^*$ is the best-compromise solution. The point in decision space is found from

$$\mathbf{x}^* = 0.83(4, 4) + 0.17(6, 2) \tag{7-25}$$

which gives $\mathbf{x}^* = (4.34, 3.68)$. The algorithm ends here.

Geoffrion (1967) points out that the above algorithm is applicable to p-dimensional problems, $p \geq 3$, but the computational burden of the method goes up rapidly with the number of objectives. One caution is that the "ends" of the noninferior set may not be found with Geoffrion's weighting scheme. The discussion of the noninferior set estimation method for higher-dimensional problems (Section 6.3.4) is applicable here.

7.2 PRIOR ASSESSMENTS OF WEIGHTS

Recall the hypothetical fire station location problem that was used to motivate the weighting method in Chapter 6. It was stated then that *if* decision makers were willing to state that population coverage is w times more important than property value coverage, then the multiobjective problem could be reduced to a single-objective problem and solved directly.

Weights on objectives represent the most simple form in which preferences can be stated. In Section 4.3 the use of weights was shown to be equivalent to the construction of linear indifference curves with the slope of a curve equal to the negative of the ratio of the weights. In other words, weights are equivalent to a utility function that is linear in each objective and that exhibits constant marginal rates of substitution (MRS) between objectives. Therefore the use of weights is based on strong behavioral assumptions. The linearity implies that the marginal utility, which is a constant equal to the weight, does not decrease with the level of an objective. The constant MRS means that the willingness to tradeoff one objective for another is independent of the level of objectives.

The simplicity is perhaps unrealistic, but it does promote a straightforward and computationally efficient planning process. Economists such as Marglin (1967) [and in UNIDO (1972)] and Major (1969) have advocated the prior assessment of weights for public decision-making problems. They have suggested a planning process in which decision makers ponder the value question of the relative importance of objectives before specific design or policy issues are raised. UNIDO (1972) proposed a separate national policy committee that would set the values for "national parameters," such as the weight on environmental quality indices or employment in terms of net national income. The policy committee would have no specific design or policy responsibilities other than the setting of the parameters. The weights articulated by the national committee would be used in all design and policy formulation activities in all federal agencies.

The notion of a national policy committee is raised periodically by observers of the federal bureaucracy. The motivation for a separate policy body stems from the belief that bureaucratic objectives often obscure social objectives.

Personnel in an agency that owes its existence to the construction of large-scale public facilities are less apt to reject projects on, say, environmental grounds than are national policy makers who are divorced from that bureaucratic setting. These proposals to strip design and construction agencies of their policy responsibilities have yet to be implemented.

One potential difficulty with preset national parameters is the unique nature of every project. The objectives differ from project to project considered by the federal bureaucracy and by a single agency. Furthermore, while the overall objectives may not differ, the measures frequently will. In two reservoir design problems for example, Miller and Byers (1973) measured environmental quality as the sediment load in the water while Major (1974) measured it as the number of acres inundated by the reservoir impoundment.

Prior specification of weights can be used at any level of decision making; it is not confined to national policy making. It may be a favorable approach *because* of its simplicity, but the assumptions of constant marginal utility and MRS should be kept in mind. Whenever one follows this approach, extensive sensitivity analysis should be pursued whenever possible. Varying the weights to see the implications for the results is an obvious approach to take. The results from sensitivity analyses should prove useful to decision makers. Zeleny (1974b, p. 485) presents arguments against the direct assessment of weights. He criticizes the approach on behavioral grounds. His claims that human ability for overall evaluation is limited and that the precision of weighting does not reflect the fuzziness of multiobjective decision making are valid criticisms that should qualify the use of prior specification of weights.

7.3 METHODS BASED ON GEOMETRICAL DEFINITIONS OF BEST

Some methodologies for proceeding to the best-compromise solution are based on a geometrical notion of best. They proceed by first defining an ideal solution, i.e., a generally infeasible alternative that one would like to achieve if only one could. Computational procedures are then applied to find the feasible solution that is closest to the ideal solution on the basis of some distance measure. Two methods are considered here: minimum distance from an ideal solution and goal programming. The approaches differ in their definition of what is ideal and in the distance metric.

Before the discussion of the two methods, the notion of a distance metric and its various forms will be presented. The distance between two points, say $\mathbf{x}$ and $\mathbf{y}$, with coordinates $(x_1, x_2, \ldots, x_p)$ and $(y_1, y_2, \ldots, y_p)$ is, in general,

$$d_\alpha = \left\{ \sum_{k=1}^{p} |x_k - y_k|^\alpha \right\}^{1/\alpha} \tag{7-26}$$

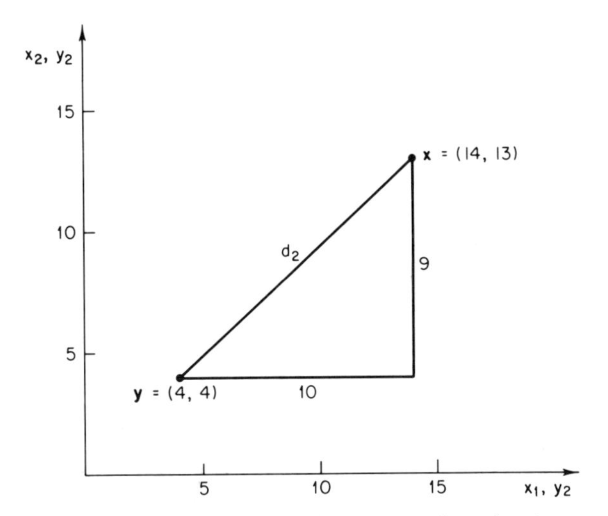

Fig. 7-7. Distance between two points in a two-dimensional space.

where $\alpha \geq 1$. This definition of distance corresponds to our usual notion when $\alpha = 2$. However, α may take on any value from 1 to $+\infty$.

The metrics d_1 and d_∞ represent bounds on the distance between any two points; i.e.,

$$d_\infty \leq d_\alpha \leq d_1 \tag{7-27}$$

To see this consider a two-dimensional example with $\mathbf{x} = (14, 13)$ and $\mathbf{y} = (4, 4)$. These two points are drawn in Fig. 7-7 and the distance between them for several different metrics is shown in Table 7-1.

When $\alpha = 1$, (7-26) reduces to

$$d_1 = |x_1 - y_1| + |x_2 - y_2| \tag{7-28}$$

which states that the absolute differences on a component-by-component basis should be summed. These differences appear as the sides of the triangle of length 9 and 10 in Fig. 7-7. Therefore Table 7-1 shows that $d_1 = 19$.

TABLE 7-1

DISTANCE BETWEEN (4, 4) AND (14, 13) UNDER SEVERAL DISTANCE METRICS

α	$d_\alpha = [(14 - 4)^\alpha + (13 - 4)^\alpha]^{1/\alpha}$	α	$d_\alpha = [(14 - 4)^\alpha + (13 - 4)^\alpha]^{1/\alpha}$
1	19	6	10.74
2	13.45	⋮	⋮
3	11.97	10	10.30
4	11.34	⋮	⋮
5	10.97	∞	10

When $\alpha = 2$, (7-26) reduces to our usual notion of distance—Euclidean distance.

$$d_2 = [(x_1 - y_1)^2 + (x_2 - y_2)^2]^{1/2} \tag{7-29}$$

where the absolute value signs have been dropped since the component distances are squared. The distance d_2 is shown in Fig. 7-7 as just the straight-line distance from **x** to **y**. The value of d_2 is shown in Table 7-1.

As α increases beyond two, more weight is attached to large component differences through the operation of raising all $(x_k - y_k)$ to the exponent α. Thus, as α increases the distance d_α decreases, gradually approaching the largest component difference as in Table 7-1. At d_∞, an infinite weight is put on the largest component difference, so the distance just equals this largest difference. Therefore Table 7-1 shows that $d_\infty = x_1 - y_1 = 10$.

Notice in Table 7-1 that d_1 is the largest distance and that d_∞ is the smallest. This is generally true as it results from the form of the distance metric: As α increases more weight is associated with the largest $(x_k - y_k)$, forcing d_α down toward that component difference.

More insight into the distance metric and how it varies with the value of α can be gained through a graphical interpretation. We will consider the loci of points r units from the origin of a two-dimensional space for three different values of α: $\alpha = 1, 2, \infty$. Since we have chosen the origin as one of our points, we can set $\mathbf{y} = (0, 0)$ and (7-26) becomes for a two-dimensional case,

$$d_\alpha = [|x_1|^\alpha + |x_2|^\alpha]^{1/\alpha} \tag{7-30}$$

For $\alpha = 1$, the locus of points r units from the origin, using the definition of d_1 in (7-30), is

$$|x_1| + |x_2| = r \tag{7-31}$$

The locus of the points that satisfy (7-31) is drawn as the square with solid lines in Fig. 7-8. To understand the figure, consider the positive quadrant where $x_1, x_2 \geq 0$. In this quadrant (7-31) is just the equation of a straight line since the absolute value signs may be dropped.

As α increases beyond one, more weight is put on large values of $(x_k - y_k)$. When $\alpha = 2$, (7-30) becomes, dropping the absolute value signs,

$$d_2 = [x_1^2 + x_2^2]^{1/2} \tag{7-32}$$

Setting (7-32) equal to r and squaring both sides gives

$$x_1^2 + x_2^2 = r^2 \tag{7-33}$$

which is just the equation of a circle of radius r, centered at the origin. This circle is shown in Fig. 7-8.

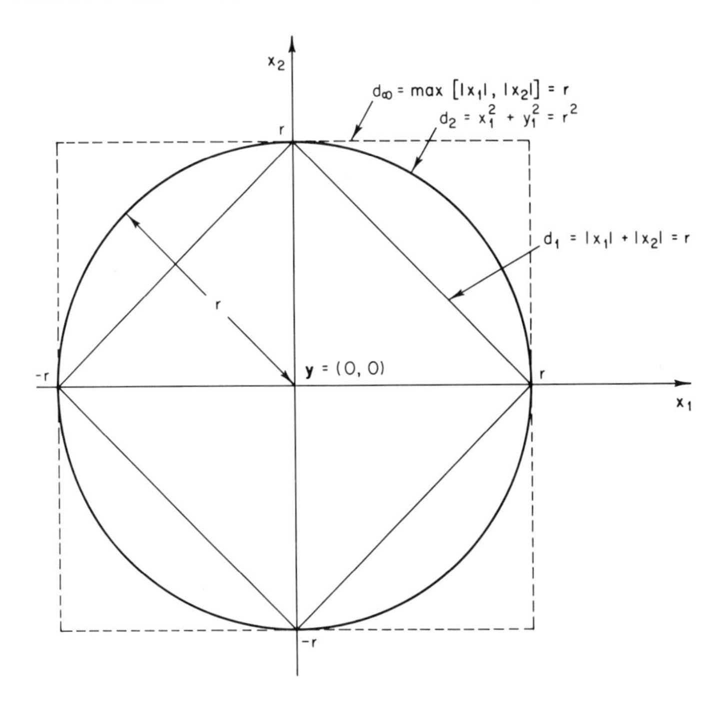

Fig. 7-8. The locus of points r units from the origin for distance metrics d_α with $\alpha = 1, 2, \infty$.

The circle in Fig. 7-8, which shows the locus of points $d_2 = r$ from the origin, circumscribes the square that is the locus of points $d_1 = r$ from the origin. This is consistent with what we discovered in Table 7-1. That is, we would expect d_2 to give us points that appear farther out from the origin since $d_2 < d_1$.

As α increases beyond two, the locus of points $d_\alpha = r$ from the origin shifts farther out. At the extreme of $\alpha = \infty$, (7-30) becomes

$$d_\infty = \max[|x_1|, |x_2|] \tag{7-34}$$

and the locus of points $d_\infty = r$ from the origin is the square with dashed lines in Fig. 7-8 described by the equation

$$\max[|x_1|, |x_2|] = r \tag{7-35}$$

Thus, the inner square of Fig. 7-8, corresponding to $d_1 = r$, and the outer square of Fig. 7-8, corresponding to $d_\infty = r$, represent bounds on the loci of points $d_\alpha = r$ from the origin for all values of $\alpha \geq 1$. As α increases, the locus shifts out since d_α goes down; i.e., a larger α requires points that appear farther from the origin to maintain $d_\alpha = r$.

One last remark pertains to points where d_α is the same for all values of α. It should be obvious from (7-26) that for points, and only such points, for which all but one component difference equal zero, d_α is identical for all α. Thus, for example, the points (5, 0) and (3, 0) are $d_\alpha = 2$ apart for any value of α. Figure 7-8 shows this in that $(r, 0)$, $(0, r)$, $(-r, 0)$, and $(0, -r)$ are points that are on the loci for $\alpha = 1, 2, \infty$.

7.3.1 Minimum Distance from the Ideal Solution

Yu (1973) and Zeleny (1974a, pp. 171–176, 1974b, pp. 486–489) define the ideal solution (or "utopia point," in Yu's terminology) as that solution that would simultaneously optimize each objective individually. This ideal solution is generally infeasible; it it were not, then there would be no conflict among objectives. The coordinates of the ideal solution in objective space are the optima for each objective, $[Z_1^*, Z_2^*, \ldots, Z_p^*]$. An ideal solution is shown in Fig. 7-9 for an arbitrary, linear two-objective problem.

The best-compromise solution is defined by Yu (1973) and Zeleny (1974a, 1974b) as the solution that is the minimum distance from the ideal solution.

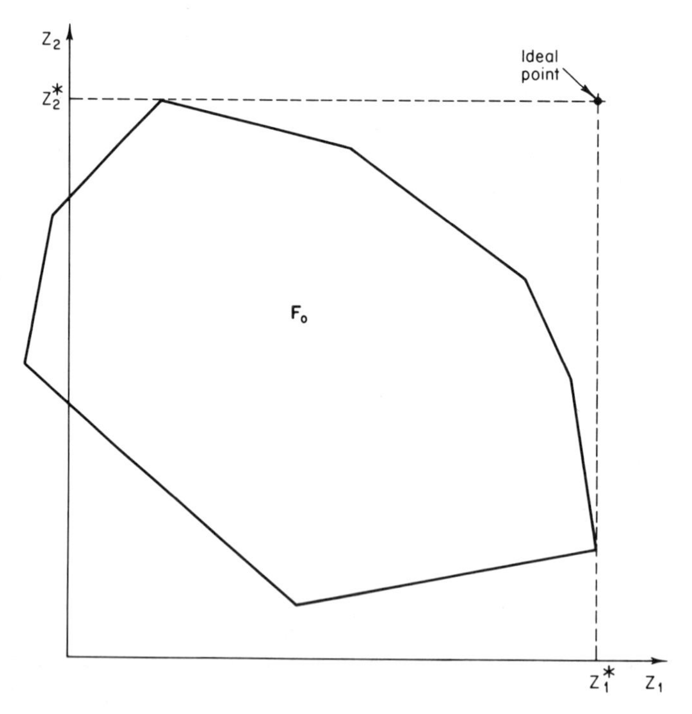

Fig. 7-9. The ideal point for a two-objective problem.

Rewriting (7-26) in terms of objectives and the ideal solution, we get

$$\text{minimize} \quad d_\alpha = \left\{ \sum_{k=1}^{p} |Z_k^* - Z_k(x)|^\alpha \right\}^{1/\alpha} \tag{7-36}$$

The choice of a metric, i.e., a value for α, is problematical. Let us consider the extremes for α, $\alpha = 1$ and $\alpha = \infty$.

When $\alpha = 1$, (7-36) gives the minimization of the absolute deviations from the ideal. For this particular definition of the ideal solution, however, we can drop the absolute value sign since, because all objectives are to be maximized, $Z_k(\mathbf{x}) \leq Z_k^* \; \forall k = 1, 2, \ldots, p$ and $\forall$ feasible $\mathbf{x}$.

For the d_1 metric, the best-compromise solution is found by solving

$$\text{minimize} \quad d_1 = \sum_{k=1}^{p} [Z_k^* - Z_k(\mathbf{x})] \tag{7-37}$$

$$\text{s.t.} \quad \mathbf{x} \in \mathbf{F}_\text{d} \tag{7-38}$$

Notice in (7-37) that the Z_k^*, $k = 1, 2, \ldots, p$, are constants that are not subject to minimization so they can be removed from the objective function to get

$$\text{minimize} \quad d_1 = -\sum_{k=1}^{p} Z_k(\mathbf{x}) \tag{7-39}$$

which is equivalent to

$$\text{maximize} \quad \sum_{k=1}^{p} Z_k(\mathbf{x}) \tag{7-40}$$

The d_1 metric for this definition of the ideal, for which the absolute value can be ignored, is equivalent to an equal weighting of objectives. That is, minimizing the shortfall from the maximum of each objective, (7-37), is identical to maximizing the equally weighted sum of the objectives.

When $\alpha = \infty$, only the largest deviation counts so that the metric in (7-36), again dropping the absolute value sign, becomes

$$d_\infty = \max_{k=1,2,\ldots,p} [Z_k^* - Z_k(\mathbf{x})] \tag{7-41}$$

and the formulation for the best-compromise solution is

$$\text{minimize} \quad d_\infty = \max_{k=1,2,\ldots,p} [Z_k^* - Z_k(\mathbf{x})] \tag{7-42}$$

$$\text{s.t.} \quad \mathbf{x} \in \mathbf{F}_\text{d}$$

which is equivalent to

$$\text{minimize} \quad d_\infty \tag{7-43}$$

$$\text{s.t.} \quad \mathbf{x} \in \mathbf{F}_\mathrm{d} \tag{7-44}$$

$$Z_k^* - Z_k(\mathbf{x}) \leq d_\infty, \qquad k = 1, 2, \ldots, p \tag{7-45}$$

The fact that the selection of a metric determines the best-compromise solution will be demonstrated with the example problem from (4-7) in Chapter 4. The noninferior set in objective space and the ideal solution $(Z_1^*, Z_2^*) = (30, 15)$ are shown in Fig. 7-10. For $\alpha = 1$ we can solve the equal weight problem from (7-40):

$$\text{maximize} \quad [Z_1(x_1, x_2) + Z_2(x_1, x_2)] = 4x_1 + 2x_2 \tag{7-46}$$

$$\text{s.t.} \quad (x_1, x_2) \in \mathbf{F}_\mathrm{d} \tag{7-47}$$

where (7-47) requires that the constraints in (4-7) be satisfied. The solution to this problem, $x_1 = 6$, $x_2 = 2$, $Z_1 = 26$, and $Z_2 = 2$, is shown in Fig. 7-10.

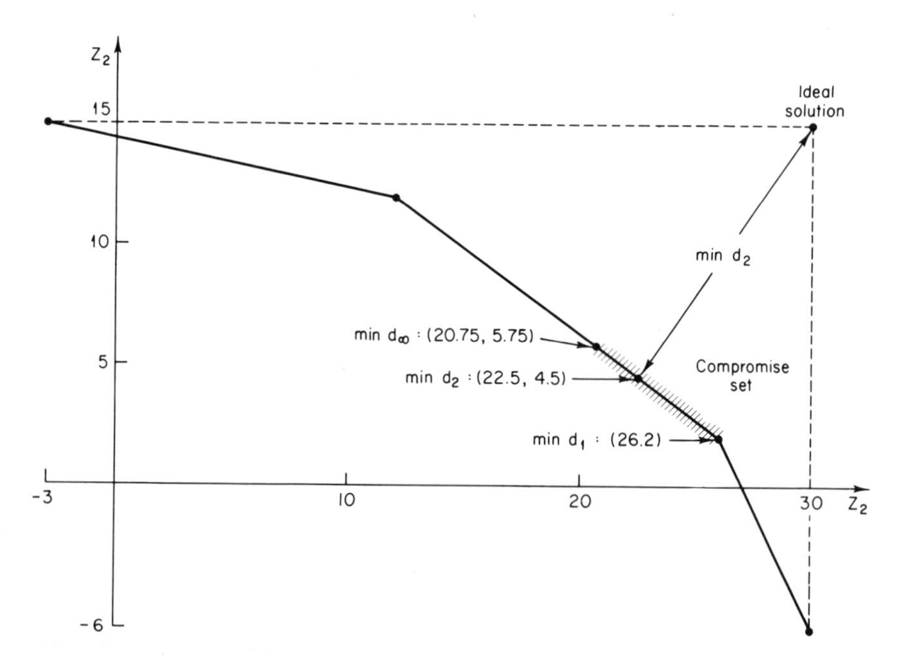

Fig. 7-10. Minimizing the distance from the ideal solution of the sample problem, using all distance metrics d_α, $\alpha \geq 1$.

For $\alpha = \infty$ we solve (7-43)–(7-45) for the sample problem

$$\text{minimize} \quad d_\infty \tag{7-48}$$

$$\text{s.t.} \quad (x_1, x_2) \in \mathbf{F}_d \tag{7-49}$$

$$30 - 5x_1 + 2x_2 \leq d_\infty \tag{7-50}$$

$$75 + x_1 - 4x_2 \leq d_\infty \tag{7-51}$$

where d_∞, x_1, and x_2 are the decision variables. The optimal solution to this problem, which is the best-compromise solution under the d_∞ metric, is $x_1 = 2.25$, $x_2 = 2.75$, $Z_1 = 20.75$, and $Z_2 = 5.75$. This solution is also shown in Fig. 7-10. Yu (1973) has shown that, in general, the best-compromise solutions define a subset of the noninferior set that ranges from the minimum d^1 solution to the minimum d^∞ solution. Zeleny (1974b, p. 488) calls this set the "compromise set." All other best-compromise solutions for d_α, $1 < \alpha < \infty$, fall between the two solutions. Consider, for example the minimum d_2 solution shown in Fig. 7-10. This solution, $Z_1 = 22.5$ and $Z_2 = 4.5$, can be found by computing the distance from the ideal solution to the line connecting (12, 12) and (26, 2) (since we know $\mathbf{N}_0$) or by solving the minimization problem with d^2 as the objective function.

Since there is no *a priori* justification for choosing any solution in the compromise set over any of the other solutions in the compromise set, it is suggested that the selection be performed by decision makers. Therefore the approach is a type of generating technique. The compromise set, rather than the noninferior set, is generated. Of course, if one metric is considered more appropriate than any other, then the best-compromise solution can be considered directly.

The generation of the compromise set avoids the specification of preferences, but is not devoid of value judgments. The minimization of distance, regardless of the metric, from the ideal solution as a representation of best constitutes an important value judgment by the analyst. The appropriateness of this assumption must be given careful consideration before implementing the technique. Perhaps most important is whether the value judgments inherent in the method are meaningful. Should we expect that decision makers will understand the distinction among metrics? The answer to that question must await application of the technique.

7.3.2 Goal Programming

Goal programming is undoubtedly the most well-known multiobjective method. For many, goal programming *is* multiobjective programming. While it certainly is not, the misunderstanding is easily understandable in light of the method's apparent ubiquity.

The methodology known as goal programming was proposed by Charnes and Cooper (1961, pp. 299–310). Other texts on goal programming have been prepared by Ijiri (1965), Lee (1972), and Ignizio (1976). The method has been applied primarily to private sector problems by Charnes *et al.* (1968, 1969), Spivey and Tamura (1970), Lee and Clayton (1972), Price and Piskor (1972), and others, although Lee and Moore (1977) used goal programming to analyze school busing to achieve racial desegregation. Werczberger (1976) used goal programming for land-use planning.

Goal programming employs a minimum-distance notion of best. The metric in (7-26) with $\alpha = 1$ is usually used, although other metrics may be employed. The feature that distinguishes goal programming from the method discussed in the last section, however, is the specification of goals by decision makers. An ideal solution from which deviations are minimized is defined as the alternative that would yield the goals (target values) for all objectives simultaneously. The goal for the kth objective will be called G_k .The goal programming problem is to minimize the d_1 distance from the goals:

$$\text{minimize} \quad d_1 = \sum_{k=1}^{p} |G_k - Z_k(\mathbf{x})| \tag{7-52}$$

$$\text{s.t.} \quad \mathbf{x} \in \mathbf{F}_\text{d} \tag{7-53}$$

where the objective function in (7-52) is just (7-36) with $\alpha = 1$ and G_k in place of Z_k^*. In the previous section, the absolute value in (7-42) was dropped because $Z_k(\mathbf{x})$ was always less than the kth component of the ideal solution. We cannot guarantee that $G_k \geq Z_k(\mathbf{x})$ for all feasible $\mathbf{x}$ since the goals are set by decision makers. Therefore we must use absolute values in the distance metric.

The absolute value in (7-52) is nonlinear, but a linear programming formulation that is equivalent to (7-52) and (7-53) is

$$\text{minimize} \quad \sum_{k=1}^{p} (d_k^+ + d_k^-) \tag{7-54}$$

$$\text{s.t.} \quad \mathbf{x} \in \mathbf{F}_\text{d} \tag{7-55}$$

$$G_k - Z_k(\mathbf{x}) = d_k^- - d_k^+, \qquad k = 1, 2, \ldots, p \tag{7-56}$$

$$d_k^+, d_k^- \geq 0, \qquad k = 1, 2, \ldots, p \tag{7-57}$$

where d_k^+ and d_k^- are positive and negative deviations, respectively, of objective k from its goal. If $Z_k(\mathbf{x}) < G_k$, then we have less than the desired amount of objective k; this is a negative deviation, so $d_k^- = G_k - Z_k(\mathbf{x})$ and $d_k^+ = 0$. If $Z_k(\mathbf{x}) > G_k$, then we have more than a desirable amount so that from (7-57), $d_k^+ = -[G_k - Z_k(\mathbf{x})] = Z_k(\mathbf{x}) - G_k$ and $d_k^- = 0$. The two deviations d_k^+ and d_k^- will never both be nonzero for the same goal since total deviations are being minimized in (7-54).

A simple numerical example will convince the skeptical reader that the formulation in (7-54)–(7-57) is equivalent to the absolute value formulation in (7-52) and (7-53). Suppose that $G_k = 5$ and that for a particular solution $\mathbf{x}$, $Z_k(\mathbf{x}) = 4$. We have then in (7-56) that

$$5 - 4 = 1 = d_k^- - d_k^+ \tag{7-58}$$

This constraint can be satisfied by infinite combinations of (d_k^-, d_k^+), e.g., (1, 0), (2, 1), (3, 2), (5, 4), (5.4, 4.4). But since this is an optimization problem, only the combination that minimizes (7-54) will be selected so $d_k^- = 1$, $d_k^+ = 0$. Any feasible combination of d_k^- and d_k^+, other than that combination with only one nonzero deviation, would give a higher value for the sum of deviations. Work through this line of reasoning for $G_k = 5$ and $Z_k(\mathbf{x}) = 6$.

There are a number of minor modifications that can be incorporated into the basic formulation of (7-54)–(7-57). Decision makers may wish to weight the deviations in (7-54) to reflect the relative importance of the objectives and the relative significance of positive or negative deviations. If we define w_k^+ and w_k^- as the weights on the positive and negative deviations, respectively, of objective k from its goal, then (7-54) becomes

$$\text{minimize} \quad \sum_{k=1}^{p} (w_k^+ d_k^+ + w_k^- d_k^-) \tag{7-59}$$

The constraints in (7-55)–(7-57) are unchanged.

The weights in (7-59) can assume any nonnegative values. Two special cases are when w_k^+ or w_k^- is set to zero and when the weights on one objective are such that w_k^+ and w_k^- are "much greater" ($\gg$) than the weights on any other objectives.

The first case corresponds to "one-sided" goal programming. When $w_k^+ = 0$, for example, then we are minimizing negative deviations only in (7-59), which means that positive deviations do not matter. This approach is consistent with some environmental quality objectives. In water quality problems, an important environmental constituent and quality indicator is dissolved oxygen (DO) concentration. When DO falls between 3 and 4 milligrams per liter (mg/liter), some of the lifeforms in the water will die. When DO falls below 1 mg/liter, most of the lifeforms cannot be sustained, and the water is visually unpleasant and foul smelling. Thus, in a water quality planning problem a typical goal might be 5 mg/liter of DO. Higher concentrations are certainly acceptable, but it is a lower concentration that is of the greatest concern. For this objective a one-sided approach with $w_k^+ = 0$ would be appropriate.

The second case has $w_k^+ \gg w_q^+$ and $w_k^- \gg w_q^-$ for all $q \neq k$. The deviations from the kth objective's goal receive so much weight in (7-59) that the linear program will first meet that goal (if it is feasible) before considering the other goals. The use of weights such as these allows the analyst to represent *lexicographic* orderings, which some decision makers may find desirable. In effect, the decision maker's preferences order alternatives the way that a lexicon (dictionary) orders words. The applicable situation is one in which decision makers consider one of the p objectives to be dominant over the others to the extent that the sacrifice of the goal on that objective would not be considered.

Goal programming has been shown to be a powerful tool for the analysis of private sector problems. Its applicability to public decision-making problems has yet to be established, due perhaps to the inconvenience of specifying target values for some social objectives. For example, highways and reservoirs are not designed or planned to generate a certain quantity of economic efficiency benefits; these benefits are merely an indicator of the project's impact on a single dimension of economic welfare. Other objectives, such as certain formulations of environmental quality or employment, may be more amenable to a goal programming approach.

A warning on the use of goal programming (for public or private problems) will complete this section. Because the technique relies heavily on decision makers' perceptions of the range of choice and feasibility, a set of goals may lead to an *inferior* solution. Consider Fig. 7-11, which shows a noninferior set for a problem with two objectives. The noninferior set is unknown to the decision makers when they set G_1 and G_2 as the goals for Z_1 and Z_2, respectively. Since (G_1, G_2) is feasible in $\mathbf{F}_o$ in Fig. 7-11, the solution to the goal program will give total deviations of zero; i.e., both goals are attained. Unfortunately, (G_1, G_2) is an inferior solution. If the decision makers stick with their original goals, then they will settle for less than they should.

In general, a set of goals will lead to a noninferior goal programming solution only if the ideal solution is in the noninferior set or if it is infeasible (to the northeast in Fig. 7-11). If the ideal solution is interior to $\mathbf{F}_o$, then it can be obtained with zero deviations. Thus, whenever a goal programming solution yields zero deviations, the analyst should suspect that the set of goals has led to an inferior solution. Noninferiority is guaranteed only when strictly positive deviations are obtained. Of course, the analyst could use the constrained problem (see Section 6.2) to check the noninferiority of a solution.

It is the analyst's responsibility to check the noninferiority of the decision makers' goals. Sensitivity analyses should be conducted on the goals and the weights to examine the change in the solution as the decision parameters change and to push the decision makers toward noninferiority.

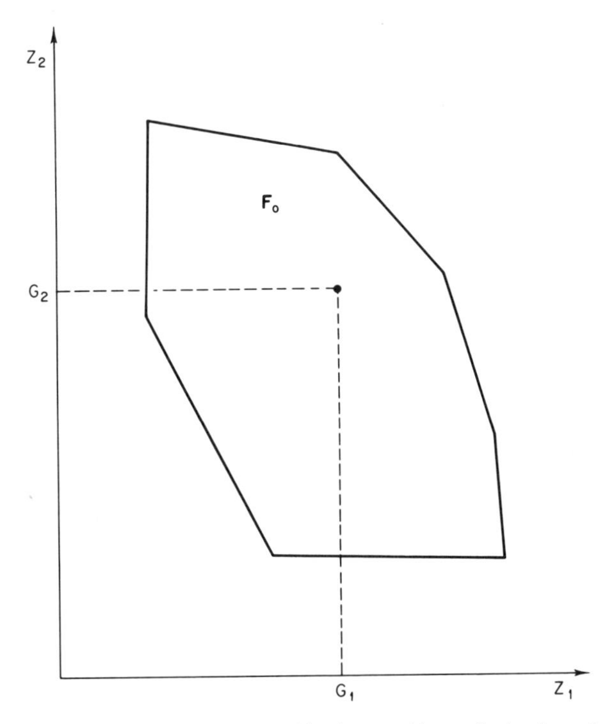

Fig. 7-11. Goals for an arbitrary two-objective problem: Optimal goal programming solutions may be inferior.

7.4 THE SURROGATE WORTH TRADEOFF METHOD

The surrogate worth tradeoff (SWT) method was developed by Haimes and Hall (1974) and Haimes *et al.* (1975) and applied to water resource planning by Haimes (1977). This unique approach to multiobjective problems is based on local approximations of a decision maker's assumed underlying multiattribute utility function. It is based on levels of objectives and marginal rates of substitution among objectives. The SWT method is different from Briskin's (1966) approach in Section 7.1.2 and Geoffrion's (1967) algorithm, however, because the full utility function is not estimated. The emphasis of the SWT method is on the indifference curves that are tangent to selected "slices" of the noninferior set. The method is described in detail and applied to our numerical example below.

7.4.1 The Theoretical Procedure

In the SWT method preference information is obtained from a decision maker by getting the person's reactions to a subset of the noninferior set.

The solutions are generated with a modified form of the constraint method (see Section 6.2). Given a problem with p objectives, $p - 2$ of them are set at predetermined values and one of the remaining two objectives is maximized with the other objective constrained at varying levels. Assume that Z_2 is to be maximized, Z_1 is varied over some range of values, and $Z_3, Z_4, \ldots, Z_p$ are fixed at the values $L_3, L_4, \ldots, L_p$. The problem to be solved is

$$\text{maximize} \quad Z_2(\mathbf{x}) \tag{7-60}$$

$$\text{s.t.} \quad \mathbf{x} \in \mathbf{F}_\mathrm{d} \tag{7-61}$$

$$Z_1(\mathbf{x}) \geq L_1 \tag{7-62}$$

$$Z_k(\mathbf{x}) = L_k, \qquad k = 3, 4, \ldots, p \tag{7-63}$$

If effect, we have for the moment reduced the original p-objective problem to a two-objective problem. The parametric variation of L_1 in (7-62) will trace out an approximation of a portion of the noninferior set as shown in Fig. 7-12. It is very important to realize that the portion of $\mathbf{N}_\mathrm{o}$ that is found

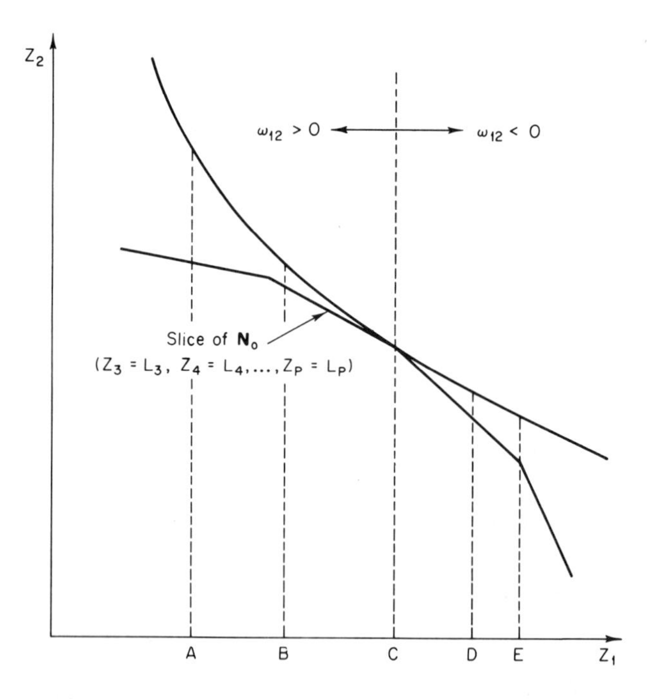

Fig. 7-12. The surrogate worth tradeoff method: Points of indifference are identified on slices of the noninferior set.

with (7-60)–(7-63) depends on the values of L_k, $k = 3, 4, \ldots, p$. If this is unclear, return to the discussion of decomposition approaches to the display of the noninferior set in Section 6.5.3.

A portion of the noninferior set, as in Fig. 7-12, is generated to discover the tradeoff between Z_1 and Z_2. It was shown in Section 6.2.2 that the dual variable associated with (7-62) is the tradeoff t_{12} between Z_1 and Z_2; it tells us the amount by which Z_2 would decrease (increase) by increasing (decreasing) Z_1 by one unit. Notice that the value of t_{12} depends on the value of L_k, $k = 3, 4, \ldots, p$, in (7-63).

Several values of t_{12} can be generated by using several different values of L_1 in (7-62). This information is then presented to the decision maker in order to get an estimate of the decision maker's utility function at the particular location in $\mathbf{N}_o$ defined by the L_k, $k = 3, 4, \ldots, p$. The decision maker is asked to assign a value to each t_{12} generated with the constrained problem. The decision maker may or may not be willing to sacrifice more than t_{12} or less than t_{12} units of Z_2 to gain one unit of Z_1. If more than t_{12} units of Z_2 should be sacrificed, then the decision maker is asked to assign a value greater than zero and less than 10 to t_{12}; if the decision maker does not wish to sacrifice as much as t_{12}, then a value less than zero and greater than -10 is assigned to t_{12}. By eliciting values for t_{12} over the entire range of the slice of the noninferior set, a point of indifference can be found if it can be assumed that the decision maker has a typical underlying utility function.

The value that is assigned to a particular t_{12} is ω_{12}, the *surrogate worth*. Since t_{12} varies with the value of Z_1, ω_{12} is also a function of Z_1. The function $\omega_{12}(Z_1)$ is called the *surrogate worth function*. It will generally have high values at low Z_1, decreasing to negative values at high Z_1 as in Fig. 7-13. Other than the monotonic decrease in ω_{12}, it is difficult to generalize about the shape of $\omega_{12}(Z_1)$.

To understand better the nature of the surrogate worth function we must consider the decision maker's underlying utility function. An indifference curve that is tangent to the slice of $\mathbf{N}_o$ under consideration is shown in Fig. 7-12. The SWT method is based on the premise that given a noninferior solution for evaluation, a decision maker will compare the magnitude of t_{12}, the tradeoff or slope of the noninferior set, with the marginal rate of substitution (MRS) between Z_1 and Z_2 or the slope of the decision maker's indifference curve. Consider $Z_1 = A$ and $Z_1 = B$ in Fig. 7-12; $t_{12} < \text{MRS}_{12}$ at both of these levels of Z_1 so that the decision maker is willing to sacrifice more than t_{12} units of Z_2 to gain one additional unit of Z_1. Therefore when asked to attach a surrogate worth to the corresponding noninferior solutions, the decision maker should give $\omega_{12} > 0$ in both cases. Furthermore, $\omega_{12}(A)$ should be greater than $\omega_{12}(B)$ since $[\text{MRS}_{12}(A) - |t_{12}(A)|] > [\text{MRS}_{12}(B) - |t_{12}(B)|]$.

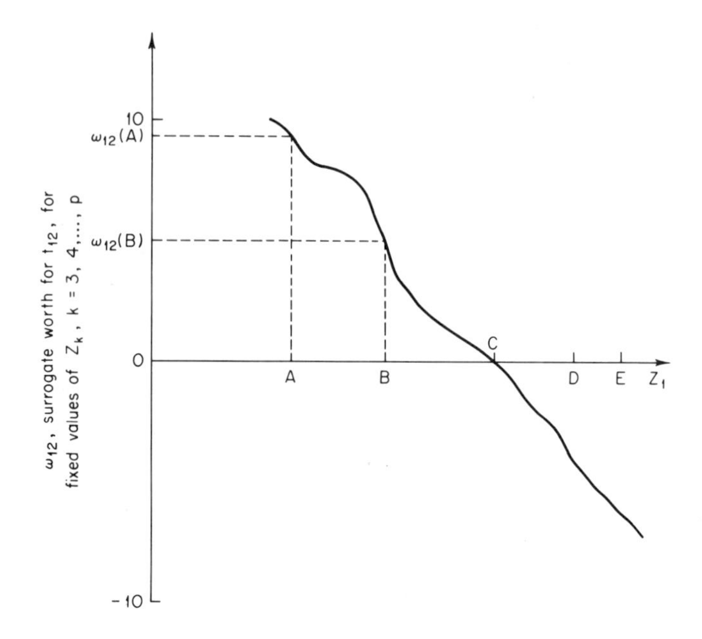

Fig. 7-13. A possible surrogate worth function.

When $Z_1 = C$ in Fig. 7-12, the slopes of the indifference curve and the noninferior set are equal so the decision maker should evaluate $t_{12}(C)$ with a surrogate worth of zero. For $Z_1 > C$, $\text{MRS}_{12} < |t_{12}|$ and the decision maker should respond with $\omega_{12}(Z_1) < 0$. These relationships give rise to the surrogate worth function in Fig. 7-13. The curve is drawn with a rather irregular shape because the decision maker cannot be expected to articulate the underlying utility function exactly.

The goal of the SWT method is the identification of an indifference point defined by $\omega_{12}(Z_1) = 0$, e.g., at $Z_1 = C$ in Figs. 7-12 and 7-13. The point of indifference, where $|t_{12}| = \text{MRS}_{12}$, is by definition the best-compromise solution. By exploring the entire range of choice over a given slice of the noninferior set, the analyst can be confident that both positive and negative values for the surrogate worth function will be obtained. Even if the decision maker does not respond with a zero value, a function that is fitted to the decision maker's responses will pass through zero, defining an approximate point of indifference.

To this point, we have only dealt with two of the p objectives. Tradeoffs exist between all of the p objectives so the characterization of a best-compromise solution requires all pairwise combinations of objectives to be analyzed.

A general form for the formulation in (7-60)–(7-63) is

$$\text{maximize} \quad Z_r(\mathbf{x}) \tag{7-64}$$

$$\text{s.t.} \quad \mathbf{x} \in \mathbf{F}_\mathrm{d} \tag{7-65}$$

$$Z_q(\mathbf{x}) \geq L_q \tag{7-66}$$

$$Z_k(\mathbf{x}) = L_k \qquad \forall k \neq q, r \tag{7-67}$$

This formulation is used to generate approximate slices of the noninferior set between r (an arbitrarily chosen numeraire objective) and $q = 1, 2, \ldots, r-1, r+1, \ldots, p$. That is, we use (7-64)–(7-67) to generate tradeoffs $t_{qr}(Z_q)$ and surrogate worth functions $\omega_{qr}(Z_q)$ $\forall q \neq r$.

In total, $(p-1)$ surrogate worth functions are developed through interaction with the decision makers. Each function is used to identify an indifference point where $\omega_{qr}(Z_q) = 0$. The value of Z_q at which the surrogate worth function crosses the Z_q axis will be called Z_q^*. These preferred values for all but the rth objective are then used in the following problem:

$$\text{maximize} \quad Z_r(\mathbf{x}) \tag{7-68}$$

$$\text{s.t.} \quad \mathbf{x} \in \mathbf{F}_\mathrm{d} \tag{7-69}$$

$$Z_q(\mathbf{x}) \geq Z_q^*, \qquad q = 1, 2, \ldots, r-1, r+1, \ldots, p \tag{7-70}$$

The solution to this problem is the best-compromise solution.

One of the uncertainties of the SWT method is the value of the L_k to be used in (7-67); i.e., how should a slice of the noninferior set be chosen? This is an important issue since one would expect decision maker's evaluations of tradeoffs to be tempered by the values of Z_k, $k \neq q, r$. The developers of the method offer no guidance here. Presumably, in an actual decision-making context the values of Z_k should be such that the decision maker is willing to trade off Z_q and Z_r; i.e., Z_k should be sufficiently large for $k \neq q, r$ to allow decision makers to consider the tradeoff currently under consideration. This is not a precise guideline, but we shall attribute this remaining uncertainty to the difficulties of higher-dimensional problems.

It is important to realize that the SWT method, although it employs a form of the constraint method, does not generate as much information as the constraint method. The SWT approach is not a generating technique. The intent of the SWT method is to gain an articulation of preferences by generating portions of the noninferior set. The method would appear to be reliable when the generated slices of $\mathbf{N}_\mathrm{o}$ are in the vicinity of the actual (unknown) best-compromise solution. It is impossible to guarantee this, so sensitivity analyses should be performed to gain an appreciation of the accuracy of the approximation. The sensitivity analyses could take the form of an entirely new pass through the method with different values for the L_k in (7-67). This can get to be a very expensive proposition if the number of objectives is large.

Another potential difficulty, which is not a unique problem of the SWT method, is the potential for inconsistency on the part of the decision maker. It would not be surprising to find contradictory evaluations in a problem with many dimensions since the decision maker does not really have a utility function in mind when the tradeoffs are evaluated. One way to improve consistency, or at least to check for it, is to use $t_{rq}(Z_r)$ in addition to $t_{qr}(Z_q)$ to construct surrogate worth functions; and since $t_{rq} = t_{rq}$ at a given noninferior point, as Haimes and Hall (1974) have demonstrated, the decision maker's responses should be numerically close, if not identical. This inverted version of the tradeoffs should lead to the same, or nearly the same, best-compromise solution. If it does not, then the inconsistencies must be pointed out to the decision maker in the hope of understanding and resolving them.

7.4.2 Numerical Example of the Surrogate Worth Tradeoff Method

In this section the surrogate worth tradeoff method (SWT) is applied to the two-objective problem in (4-7). The difficulty inherent in the choice of a slice of the noninferior set is avoided here because $\mathbf{N}_o$ is two dimensional, so decomposition is unnecessary. The constraint method was previously applied to the sample problem in Section 6.2.5. Note that the SWT method employs the constrained problem with no modifications only when there are only two objectives.

The constrained problem in (6-42)–(6-44) was used to generate the four solutions shown in decision space in Fig. 6-15 and in objective space in Fig. 6-16. These solutions are listed in Table 7-2 along with the tradeoff

TABLE 7-2

NONINFERIOR POINTS AND TRADEOFF VALUES FOR THE APPLICATION OF THE SURROGATE WORTH TRADEOFF METHOD TO THE SAMPLE PROBLEM

x_1	x_2	Z_1	Z_2	$\lvert t_{12}(Z_1)\rvert$
1	4	-3	15	
				0.34
4.8	3.2	17.6	8	
				0.79
6	1.75	26.5	1	
				2.0
6	0	30	-6	

$t_{12}(Z_1)$. The tradeoff is computed as the slope of the line segment connecting adjacent noninferior points in objective space. Thus, for example, the first value of t_{12} shown in Table 7-2 is computed from $(15 - 8)/(-3 - 17.6) = -0.34$ so that $|t_{12}(Z_1)| = 0.34$. Only the magnitude of t_{12} is shown, but we shall remember that Z_2 must be *decreased* by t_{12} to gain one additional unit of Z_1; i.e., t_{12} is negative. Notice that t_{12} is constant over ranges of Z_1, reflecting the piecewise linearity of $\mathbf{N}_o$ (see Fig. 6-16). For example, Table 7-2 indicates that $|t_{12}(Z_1)| = 0.34$ for $-3 \leq Z_1 \leq 17.6$. It is also worth noting that the four solutions listed in Table 7-2 and the line segments connecting them represent an approximation of $\mathbf{N}_o$ as shown in Fig. 6-16. The computed tradeoffs are also approximations of the slopes of the real noninferior set.

The piecewise linearity of $\mathbf{N}_o$ presents a problem for the SWT method since there is no unique value of Z_1 for a given tradeoff. In eliciting responses from the decision maker which value should be used? One approach is to take the average value in an interval and then to construct a *tradeoff function* that shows $t_{12}(Z_1)$. For example, the results in Table 7-2 are shown in Fig. 7-14. The actual $t_{12}(Z_1)$ is a step function, but it has been approximated by a piecewise linear tradeoff function that was constructed by connecting the midpoints (average values in the intervals) with straight lines. The approximated function can now be used to give information to the decision maker.

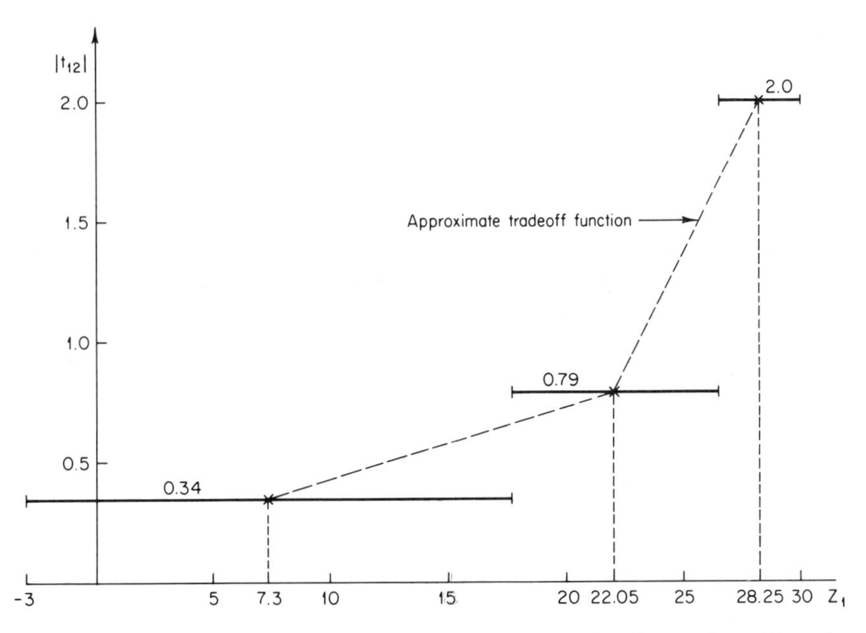

Fig. 7-14. Application of the surrogate worth tradeoff method to the sample problem: the tradeoff function.

For the purpose of continuing with the example, we shall assume that our decision maker has an underlying utility function

$$U = Z_1 Z_2 \tag{7-71}$$

which was used previously in Section 7.1.3 to demonstrate Geoffrion's (1967) algorithm. Three indifference curves that correspond to this utility function are shown in Fig. 7-6. In order to check the SWT method, it is useful to recall from Section 7.1.3 that utility for this decision maker is maximized at $(x_1, x_2) = (4.34, 3.68)$, at which $(Z_1, Z_2) = (14.4, 10.3)$ and $U = 148$ when the *exact* $\mathbf{N}_o$ is known.

The tradeoff $|t_{12}|$ is the quantity of Z_2 that must be given up for a gain of one unit of Z_1. The marginal rate of substitution MRS_{12} is the quantity of Z_2 that the decision maker is willing to give up for a one unit gain of Z_1. The MRS is computed from Eq. (7-3) as

$$\mathrm{MRS}_{12} = -\left.\frac{dZ_2}{dZ_1}\right|_{\text{fixed } U} = -\frac{d}{dZ_1}\left(\frac{U}{Z_1}\right) = \frac{U}{(Z_1)^2} = \frac{Z_2}{Z_1} \tag{7-72}$$

We now query our decision maker by presenting values of $|t_{12}|$ and the corresponding values of Z_1 and Z_2. We ask, "How do you feel about sacrificing 0.34 units of Z_2 to gain one unit of Z_1 when you currently have 7.3 units of Z_1 (the midpoint of the first interval in Fig. 7-14)?" An actual decision maker would contemplate the offered situation and respond subjectively. Our "decision maker" computes MRS_{12} using (7-72) with

$$(Z_1, Z_2) = \tfrac{1}{2}(-3, 15) + \tfrac{1}{2}(17.6, 7) = (7.3, 11.5).$$

This gives $\mathrm{MRS}_{12} = 11.5/7.3 = 1.58$. That is, the decision maker is willing to give up much more than 0.34 units of Z_2 to gain one unit of Z_1; in fact, the decision maker is prepared to sacrifice 1.58 units of Z_2.

The SWT method requires that the decision maker attach a surrogate worth value to $|t_{12}|$ to indicate the desirability of that tradeoff. This is an obviously subjective quantity (as is the MRS, in reality), but we shall compute ω_{12}—the surrogate worth—from

$$\omega_{12} = \frac{\mathrm{MRS}_{12}}{|t_{12}|} - 1 \tag{7-73}$$

This function was chosen so that $\omega_{12} = 0$ when $\mathrm{MRS}_{12} = |t_{12}|$. We could also multiply the entire quantity by a positive number in order to scale it, but this is not necessary.

For $|t_{12}| = 0.34$, the decision maker would respond that $\omega_{12} = 1.58/0.34 - 1 = 3.65$. We record this response, $\omega_{12}(7.3) = 3.65$, and offer a new situation to the decision maker. "What value do you assign to the situation in which

0.79 units of Z_2 would be sacrificed to gain one unit of Z_1 when you have 22 units of Z_1 (the midpoint of the second interval of Fig. 7-14)?" In this case, MRS_{12} is computed from (7-72) with $(Z_1, Z_2) = \frac{1}{2}(17.6, 8) + \frac{1}{2}(26.5, 1) = (22.0, 4.5)$. This gives $\mathrm{MRS}_{12} = 4.5/22.0 = 0.20$, and from (7-73), $\omega_{12} = (0.2/0.79) - 1 = -0.75$.

We have crossed the indifference line; i.e., the surrogate worth value has gone from a positive to a negative value. The surrogate worth function is next constructed as a line segment between the two responses we have obtained, as in Fig. 7-15. The point at which the surrogate worth function crosses the Z_1 axis—the point of indifference—is $Z_1^* = 19.6$. The following problem is then solved.

$$\text{maximize} \quad Z_2(x_1, x_2) \tag{7-74}$$

$$\text{s.t.} \quad (x_1, x_2) \in \mathbf{F}_\mathrm{d} \tag{7-75}$$

$$Z_1(x_1, x_2) \geq 19.6 \tag{7-76}$$

where (7-75) requires that the constraint set of (4-7) be satisfied. Solving this problem gives $(x_1, x_2) = (5.1, 2.9)$ and $(Z_1, Z_2) = (19.6, 6.5)$ as the best-compromise solution at which $U = 127$. Compare this to the actual best-compromise solution of (4.3, 3.7), (14.4, 10.3), and $U = 148$ that was found with Geoffrion's (1967) algorithm in Section 7.1.3. This discrepancy should

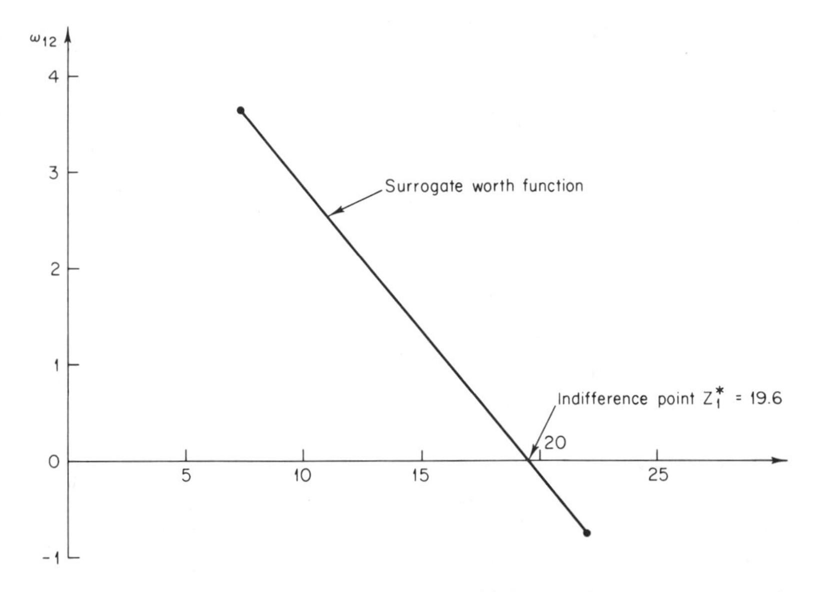

Fig. 7-15. Application of the surrogate worth tradeoff method to the sample problem: surrogate worth function and the indifference point.

not be surprising, given our method of approximating the tradeoff function and the relatively long ranges over which Z_1 is constant for a given t_{12}.

Before leaving the SWT method, it is important to point out that the evaluation of a tradeoff is closely dependent on the values of the objectives; i.e., a decision maker's preferences are defined by the context within which a choice is offered. We chose, for example, the midpoints on the ranges in Fig. 7-14 to define values of the objectives for a given tradeoff value. There was no real justification for this, however, since the particular value of the tradeoff holds at all points in the indicated range. In a real decision-making situation sensitivity analyses should be performed, particularly in linear cases, for which alternative approximations of the tradeoff function should be investigated.

7.5 ITERATIVE TECHNIQUES

The development of iterative techniques for the solution of multiobjective problems has been one of the most active areas of research in recent years. Several techniques have been developed; among them are the step method (Benayoun *et al.*, 1971), interactive goal programming (Dyer, 1972), the method developed by Geoffrion *et al.* (1972), the sequential multiobjective programming system (Monarchi *et al.*, 1973), Belenson and Kapur's (1973) method, and the approach of Zionts and Wallenius (1976). Wallenius (1975) reviewed these methods and reported on experiments with some of the iterative techniques. All of the iterative approaches are based on a formal mechanism by which the decision maker interacts either with the computer directly or with the analyst as intermediary. The methods operate in an iterative fashion by moving from one noninferior solution to another in directions defined by the decision maker. Termination occurs when the decision maker is satisfied or when further iterations cannot be handled by the solution procedure.

In this section the step method of Benayoun *et al.* (1971), which was applied by Loucks (1977) to a water resource problem, and the approach developed by Geoffrion *et al.* (1972) are discussed in detail below. An understanding of these two methods is sufficient for an appreciation of the range of iterative techniques that are available. Of those methods that are not discussed further, Dyer (1972) and Monarchi *et al.* (1973) presented similar approaches in which the decision maker is led to a satisfactory solution by eliciting goals and weights, the same information required in goal programming (see Section 7.3.2). Belenson and Kapur (1973) viewed the resolution of conflict between two objectives as a problem in two-person game theory.

Zionts and Wallenius (1976) presented a method that is similar to the Geoffrion *et al.* (1972) method. While their method seems to provide more information to the decision maker at each iteration, the nature of the information and the value judgments made by the decision maker are not very different from the method of Geoffrion *et al.* (1972).

7.5.1 The Step Method

Benayoun *et al.* (1971) developed the step method as an iterative technique that should converge to the best-compromise solution in no more than p iterations, where p is the number of objectives. The method is based on a geometric notion of best, i.e., the minimum distance from an ideal solution, with modifications of this criterion derived from a decision maker's reactions to a generated solution.

The method begins with the construction of a payoff table (see Table 6-2 in Section 6.2.3). The table is found by optimizing each of the p objectives individually, where the solution to the kth such individual optimization, called $\mathbf{x}^k$, gives by definition the maximum value for the kth objective, which is called M_k; i.e., $Z_k(\mathbf{x}^k) = M_k$. The values of the other $p - 1$ objectives implied by $\mathbf{x}^k$ are shown in the kth row of the payoff table. The payoff table is used to develop weights on the distance of a solution from the ideal solution.

The step method employs the ideal solution, which has components M_k (or Z_k^* as in Section 7.3) for $k = 1, 2, \ldots, p$. Recall from Section 7.3 that the ideal solution is generally infeasible. The d_∞ metric is used to measure distance from the ideal solution, which yields a formulation similar to (7-43)–(7-45). Two modifications in those equations are made: the distance in (7-45) is scaled by a weight based on the range of objective Z_k and the feasible region in (7-44) is allowed to change at each iteration of the algorithm. The basic problem in the step method is

$$\text{minimize} \quad d'_\infty \tag{7-77}$$

$$\text{s.t.} \quad \pi_k[M_k - Z_k(\mathbf{x})] - d'_\infty \leq 0, \qquad k = 1, 2, \ldots, p \tag{7-78}$$

$$\mathbf{x} \in \mathbf{F}_\mathrm{d}^i, \qquad d'_\infty \geq 0 \tag{7-79}$$

where $\mathbf{F}_\mathrm{d}^i$ is the feasible region at the ith iteration and d'_∞ is used to indicate that the original metric has been modified. Initially, $\mathbf{F}_\mathrm{d}^0 = \mathbf{F}_\mathrm{d}$; i.e., at the start of the algorithm the original feasible region is used in (7-79).

The weights π_k in (7-78) are defined as

$$\pi_k = \frac{\alpha_k}{\Sigma_k \alpha_k} \tag{7-80}$$

where

$$\alpha_k = \frac{M_k - n_k}{M_k}\left[\sum_{j=1}^{n}(c_j^k)^2\right]^{-1/2} \tag{7-81}$$

where n_k is the minimum value for the kth objective; i.e., it is the smallest number in the kth column of the payoff table. The c_j^k are objective function coefficients, where it is assumed that each objective is linear.

$$Z_k(\mathbf{x}) = c_1^k x_1 + c_2^k x_2 + \cdots + c_n^k x_n, \qquad k = 1, 2, \ldots, p \tag{7-82}$$

This rather elaborate procedure for the calculation of the weights in (7-78) is supposed to avoid certain value judgments. The weights are used to capture the relative variation in the value of the objectives and to suppress the inordinate weight one objective may receive by virtue of its scale. Benayoun *et al.* (1971, p. 370) argue that if the range $M_k - n_k$ is small, then "the corresponding objective is not sensitive to a variation in the weighting values; so a small weight $[\pi_k]$ can be assigned to this objective function. As the variation gets larger, the weight $[\pi_k]$ will become correspondingly bigger." The portion of (7-81) in brackets is for normalization, while the expression in (7-80) is used to ensure that the π_ks will sum to one. This latter aspect is considered important so that "different weighting strategies can be easily compared" (Benayoun *et al.*, 1971, p. 370.).

The solution of (7-77)–(7-79) with $\mathbf{F}_\mathrm{d}$ in (7-79) yields a noninferior solution $\mathbf{x}(0)$, which is closest, given the modified metric in (7-78), to the ideal solution. The decision maker is asked to evaluate this solution. If it is satisfactory, the method terminates; if it is unsatisfactory, then the decision maker specifies an amount ΔZ_k^* by which objective k^* may be decreased in order to improve the level of unsatisfactory objectives, where objective k^* is at a more than satisfactory level. A problem with a new feasible region in decision space is then solved. A solution is feasible to the new problem, $\mathbf{x} \in \mathbf{F}_\mathrm{d}^{i+1}$, if and only if the following three conditions are satisfied.

$$\mathbf{x} \in \mathbf{F}_\mathrm{d}^i \tag{7-83}$$

$$Z_k(\mathbf{x}) \geq Z_k(\mathbf{x}^i) \qquad \forall k \neq k^* \tag{7-84}$$

$$Z_{k^*}(\mathbf{x}) \geq Z_{k^*}(\mathbf{x}^i) - \Delta Z_{k^*} \tag{7-85}$$

For the new problem $\alpha_{k^*} = 0$, $\pi_{k^*} = 0$, and the other π_ks are recomputed from (7-80) for $k \neq k^*$. The problem in (7-77)–(7-79) is then resolved with $i = i + 1$, and since $\pi_{k^*} = 0$, (7-78) includes constraints for $k \neq k^*$ only. The solution to the new problem yields a new noninferior solution, which the decision maker evaluates. The method continues until the decision maker is satisfied, which the authors claim occurs in fewer than p iterations.

An Algorithm for the Step Method

To provide a more formal procedure an algorithm is presented.

Step 1 Construct a payoff table by optimizing the p objectives individually. Find M_k and n_k, $k = 1, 2, \ldots, p$.

Step 2 Compute α_k from (7-81) for $k = 1, 2, \ldots, p$. Set $i = 0$.

Step 3 Compute π_k from (7-80) and solve the problem in (7-77)–(7-79). Call the solution obtained $\mathbf{x}(i)$. $[\mathbf{F}_d^0 = \mathbf{F}_d.]$

Step 4 Show the decision maker $Z_k[\mathbf{x}(i)]$ $\forall k$:

(a) If satisfied, STOP; the best-compromise solution is $\mathbf{x}(i)$.
(b) If not satisfied and $i < (p - 1)$, go to step 5.
(c) If not satisfied and $i = (p - 1)$, STOP; some other procedure is required.

Step 5 The decision maker defines k^* and ΔZ_{k^*}. If the decision maker cannot do this, STOP; some other procedure is required. Otherwise go to step 6.

Step 6 Define a new feasible region $\mathbf{F}_d^{i+1}$ that satisfies (7-83)–(7-85). Set $\alpha_{k^*} = 0$ and go to step 3. Increment i by 1.

In step 4(c) and in step 5, the situation in which the decision maker has not been satisfied after $(p - 1)$ iterations and that in which the decision maker cannot be satisfied are considered. In either case, the step method cannot help us. In the situation of step 4(c), the problem solved at the $(p - 1)$th iteration is equivalent to the maximization of one objective subject to lower-bound constraints on the other $(p - 1)$ objectives. Check this by setting $(p - 1)$ of the π_ks to zero in (7-78) and by constructing $\mathbf{F}_d^i$ for $i = p - 1$. At the next (pth) iteration there would be no more objectives to relax—the usefulness of the method has been exhausted. Perhaps one could ask the decision maker to backtrack and further relax one or more of the objectives; i.e., some of the ΔZs used in constraints of the form of (7-85) could be increased.

In the second situation, the provision of step 5, the decision maker is not willing to sacrifice *any* of the objectives. Actually Benayoun *et al.* (1971, p. 373) suggest that "there is no solution" to the multiobjective problem in this case. Such a conclusion to the planning process conveys a mental picture of sulking analysts walking off into the sunset while the decision maker gazes after them with a stunned look. The conclusion that a solution does not exist when some course of action will be taken in any event is a silly and totally useless observation. The aim of multiobjective analysis is to aid and support decision making; this aim is not served by telling a decision maker that "there is no solution." If the algorithm comes down to this, then another procedure must be used.

Application of the Step Method to the Example Problem

The problem in (4-7) will be used to demonstrate the step method. A payoff table for this problem was previously presented as Table 6-4 in Section 6.2.5. This table is used to compute the αs in step 2. For objective Z_1 we get

$$\alpha_1 = \frac{M_1 - n_1}{M_1}\left[\sum_{j=1}(c_j^1)^2\right]^{-1/2} \tag{7-86}$$

From Table 6-4, $M_1 = 30$ and $n_1 = -3$, while from the form of $Z_1(x_1, x_2)$ in (4-7), $c_1^1 = 5$ and $c_2^1 = -2$; using these values in (7-86) gives

$$\alpha_1 = \frac{30 - (-3)}{30}[(5)^2 + (-2)^2]^{-1/2} = 0.20 \tag{7-87}$$

A similar computation gives $\alpha_2 = 0.34$. The index i is set to zero. In step 3 of the algorithm π_1 and π_2 are computed.

$$\pi_1 = 0.20/0.54 = 0.37, \qquad \pi_2 = 0.34/0.54 = 0.63 \tag{7-88}$$

Next, the following problem is solved.

$$\text{minimize} \quad d'_\infty \tag{7-89}$$

$$\text{s.t.} \quad 0.37[30 - 5x_1 + 2x_2] - d'_\infty \leq 0 \tag{7-90}$$

$$0.63[15 + x_1 - 4x_2] - d'_\infty \leq 0 \tag{7-91}$$

$$(x_1, x_2) \in \mathbf{F}_\mathrm{d}^0 \quad \text{and} \quad d'_\infty \geq 0 \tag{7-92}$$

where $\mathbf{F}_\mathrm{d}^0 = \mathbf{F}_\mathrm{d}$, which is the feasible region defined by the constraints in (4-7) and shown in Fig. 4-2. The optimal solution to this problem is

$$[x_1(0), x_2(0)] = (4.83, 3.17) \tag{7-93}$$

at which $Z_1[x_1(0), x_2(0)] = 17.9$ and $Z_2[x_1(0), x_2(0)] = 7.9$. As an aside, compare this solution to the point (20.75, 5.75) in Fig. 7-10. That point was found as the solution to (7-48)–(7-51), which is an *unweighted* version of the problem to be solved. Furthermore, it was stated in Section 7.3 that the solution to the minimum d_∞ distance problem was at one end of the compromise set, the collection of solutions to minimum-distance problems with all metrices d_α, $1 \leq \alpha \leq \infty$. The solution that was just found, however, lies *outside* of the compromise set in Fig. 7-10. This has occurred because of the weights used in (7-90) and (7-91). We are, in fact, using a different metric by assigning unequal weights to the deviations from the maxima of the individual objectives. Specifically, we have put even more weight on the largest deviation than does the original d_∞.

At this stage of the algorithm, the solution $\mathbf{x}(0)$ is presented to the decision maker who, we shall assume, states that the solution is unsatisfactory because it yields insufficient Z_2. The decision maker indicates further that three units of Z_1 may be sacrificed to increase Z_2, i.e., $k^* = 1$ and $\Delta Z_1 = 3$. In fact, we could have computed ΔZ_1 exactly if we had used the utility function of previous sections. This is not necessary, however, to demonstrate the algorithm.

We continue to step 6 of the algorithm; $i = 1$, $\alpha_1 = 0$, and $\mathbf{F}_\mathrm{d}^1$ is defined such that

$$(x_1, x_2) \in \mathbf{F}_\mathrm{d} \tag{7-94}$$

$$Z_2(x_1, x_2) = -x_1 + 4x_2 \geq 7.9 \tag{7-95}$$

$$Z_1(x_1, x_2) = 5x_1 - 2x_2 \geq 17.9 - 3 = 14.9 \tag{7-96}$$

and we return to step 3.

The new values for the weights are $\pi_1 = 0$ and $\pi_2 = 1$, so the problem to be solved is

$$\text{minimize} \quad d'_\infty \tag{7-97}$$

$$\text{s.t.} \quad 15 + x_1 - 4x_2 - d'_\infty \leq 0 \tag{7-98}$$

$$(x_1, x_2) \in \mathbf{F}_\mathrm{d}^1 \tag{7-99}$$

The expressions in (7-97) and (7-98) are equivalent to

$$\text{minimize} \quad (15 + x_1 - 4x_2) \tag{7-100}$$

which is equivalent to maximizing Z_2. The original problem can be further reduced by noting that (7-95) is not needed since we are maximizing Z_2. Removing (7-95) and incorporating the previous simplifications gives

$$\text{maximize} \quad Z_2(x_1, x_2) = -x_1 + 4x_2 \tag{7-101}$$

$$\text{s.t.} \quad (x_1, x_2) \in \mathbf{F}_\mathrm{d} \tag{7-102}$$

$$Z_1(x_1, x_2) = 5x_1 - 2x_2 \geq 14.9 \tag{7-103}$$

which can be solved by imposing the new constraint in (7-103) on the original feasible region in Fig. 4-2.

The new solution is $\mathbf{x}(1) = [x_1(1), x_2(1)] = (4.41, 3.59)$, at which $Z_1 = 14.9$ and $Z_2 = 10.0$. At this stage the decision maker would evaluate the new solution. The behavioral assumptions that underlie the step method require that the decision maker should now be satisfied; if this is not the case, then a satisfactory solution must not exist since the decision maker was willing to sacrifice only three units of Z_1 in order to gain Z_2. In a real decision problem, if satisfaction has not been attained, we would have to abandon the step method in order to search for other noninferior solutions that would increase the decision maker's understanding of the feasible range of choice. It seems reasonable that an individual's degree of satisfaction would depend on that person's expectations. Therefore the definition of the noninferior set, or portions of it, would perhaps lead to a redefinition of the decision maker's expectations and an increase in the likelihood that the decision maker will be able to find a satisfactory solution.

7.5.2 An Iterative Algorithm That Employs Local Approximations of an Underlying Utility Function

Geoffrion *et al.* (1972) presented an iterative technique that assumes an underlying preference (utility) function that is approximated locally as the algorithm proceeds. In the words of Geoffrion *et al.* (1972, p. 357), "the algorithm calls only for such local information about the preference function as is actually needed to carry out the optimizing calculations." Barber (1976) applied this method to a land-use planning problem.

The "optimizing calculations" referred to above are the procedures of the Frank–Wolfe algorithm, a method for solving optimization problems that moves from an initial feasible solution towards the optimal solution by following directions of steepest ascent, i.e., directions that provide the maximum rate of increase in the objective function.

There are actually two major parts of the algorithm: the determination of the best direction and the step size along that direction. Both subproblems require the involvement of the decision maker. Geoffrion *et al.* (1972) showed that if the objective function is a multiattribute utility function defined over the p objectives as in Section 7.1.2, i.e.,

$$\text{maximize} \quad U[Z_1(\mathbf{x}), Z_2(\mathbf{x}), \ldots, Z_p(\mathbf{x})] \tag{7-104}$$

then the direction problem may be solved if the decision maker can specify the marginal rates of substitution (MRS) [see Section 7.1.2 and Eq. (7-3)] between a "reference objective," i.e., a numeraire, and each of the other objectives. The MRSs provide enough local information about the utility function to find a direction that will lead to an improvement over the current solution.

The step size problem employs a simple one-dimensional optimization subproblem that can be solved parametrically to indicate several alternative solutions that lie in the preferred direction. The decision maker chooses one of the solutions generated in the subproblem to be used as the new solution in the direction subproblem. The procedure continues until subsequent solutions of the direction and step size problems converge, i.e., the "new" solution referred to above is the same as the previous solution.

The above procedure has been completely computerized (Dyer, 1973); the decision maker interacts directly with the algorithm through a computer terminal. The programmed version queries the decision maker for the necessary information (MRSs and step sizes). The decision maker can control the fineness of the step size problem parametrization.

The mathematical details of the algorithm are presented below for those readers with an interest in and a background for such things. Our numerical example is subsequently used to demonstrate the algorithm.

**Mathematical Details*

Geoffrion *et al.* (1972, p. 358) show that the direction subproblem of the Frank–Wolfe algorithm for the objective function in (7-104) is to find a direction $\mathbf{y}^i - \mathbf{x}^i$, where $\mathbf{y}^i$ is the optimal solution to

$$\text{maximize} \quad \{\nabla_{\mathbf{x}} U[Z_1(\mathbf{x}^i), Z_2(\mathbf{x}^i), \ldots, Z_p(\mathbf{x}^i)]\, \mathbf{y}\} \tag{7-105}$$

$$\text{s.t.} \quad \mathbf{y} \in \mathbf{F}_{\mathrm{d}} \tag{7-106}$$

where $\mathbf{x}^i$ is a given feasible solution, $\mathbf{x}^i \in \mathbf{F}_{\mathrm{d}}$, $\mathbf{F}_{\mathrm{d}}$ is defined over an original feasible region in decision space, and $\nabla_{\mathbf{x}}$ denotes the "del" operator, where

$$\nabla_{\mathbf{x}} = \left[\frac{\partial}{\partial x_1}, \frac{\partial}{\partial x_2}, \ldots, \frac{\partial}{\partial x_n}\right] \tag{7-107}$$

Geoffrion *et al.* (1972, p. 359) show that the gradient of the utility function $\nabla_{\mathbf{x}} U[\ ]$ is equivalent to, upon application of the chain rule for differentiation,

$$\nabla_{\mathbf{x}} U[Z_1(\mathbf{x}^i), Z_2(\mathbf{x}^i), \ldots, Z_p(\mathbf{x}^i)] = \sum_{k=1}^{p} \left(\left.\frac{\partial U}{\partial Z_k}\right|_{\mathbf{Z}(\mathbf{x}^i)}\right) \nabla_{\mathbf{x}} Z_k(\mathbf{x}^i) \tag{7-108}$$

Substituting (7-108) into (7-105) gives

$$\text{maximize} \quad \sum_{k=1}^{p} \left(\left.\frac{\partial U}{\partial Z_k}\right|_{\mathbf{Z}(\mathbf{x}^i)}\right) \nabla_{\mathbf{x}} Z_k(\mathbf{x}^i)\, \mathbf{y} \tag{7-109}$$

Since the utility function increases with each objective, the first partial derivative of U with respect to any Z_k must be positive. Therefore dividing through by $(\partial U/\partial Z_r)$ evaluated at $\mathbf{Z}(\mathbf{x}^i)$ in (7-109) will not alter the solution to the problem. We shall get in each term, using the definition of the marginal rate of substitution in (7-4),

$$\left.\frac{\partial U/\partial Z_k}{\partial U/\partial Z_r}\right|_{\mathbf{Z}(\mathbf{x}^i)} = -\left.\frac{\partial Z_r}{\partial Z_k}\right|_{\mathbf{Z}(\mathbf{x}^i)} = \mathrm{MRS}_{kr}(\mathbf{x}^i) \tag{7-110}$$

i.e., the marginal rate of substitution between objectives Z_k and Z_r at the point in objective space $\mathbf{Z}(\mathbf{x}^i)$. Objective Z_r is the reference objective or numeraire referred to in the previous section.

Using (7-108)–(7-110) in (7-105) and (7-106) gives the direction subproblem as

$$\text{maximize} \quad \sum_{k=1}^{p} \mathrm{MRS}_{kr}(\mathbf{x}^i)\, \nabla_{\mathbf{x}} Z_k(\mathbf{x}^i)\, \mathbf{y} \tag{7-111}$$

$$\text{s.t.} \quad \mathbf{y} \in \mathbf{F}_\mathrm{d} \tag{7-112}$$

The solution $\mathbf{y}^i$, which maximizes (7-111) and (7-112), defines the direction $\mathbf{d}^i = \mathbf{y}^i - \mathbf{x}^i$. The step size problem is

$$\text{maximize} \quad U[Z_1(\mathbf{x}^i + t\mathbf{d}^i), Z_2(\mathbf{x}^i + t\mathbf{d}^i), \ldots, Z_p(\mathbf{x}^i + t\mathbf{d}^i)] \tag{7-113}$$

$$\text{s.t.} \quad 0 \leq t \leq 1 \tag{7-114}$$

Since the utility function is not specified, the authors suggest plotting each objective over the range of interest: $Z_k(\mathbf{x}^i + t\mathbf{d}^i)$, $0 \leq t \leq 1$. The decision maker then chooses a solution based on the objective function values. This implies a value $t = t^*$, which is used to define the new solution $\mathbf{x}^{i+1} = \mathbf{x}^i + t^*\mathbf{d}^i$. Rather than plot the objectives, Dyer (1973) indicates that several alternative solutions at various values of t, $0 \leq t \leq 1$, can be presented to the decision maker instead.

Numerical Example

We shall use the numerical example in (4-7) and the utility function in (7-13), in which the objectives are multiplied, to demonstrate the above procedure.

To begin, we need an initial noninferior solution. In order to be expeditious one could use weights that the analyst would guess to be appropriate for the decision maker in the weighted problem. We shall start with equal weights $w_1 = w_2 = 1$ and solve

$$\text{maximize} \quad Z_1(x_1, x_2) + Z_2(x_1, x_2) = 4x_1 + 2x_2 \tag{7-115}$$

$$\text{s.t.} \quad (x_1, x_2) \in \mathbf{F}_\mathrm{d} \tag{7-116}$$

where $\mathbf{F}_\mathrm{d}$ is defined so that the constraints in (4-7) are satisfied. The optimal solution to this problem is $\mathbf{x}^0 = (x_1^0, x_2^0) = (6, 2)$, at which $Z_1(\mathbf{x}^0) = 26$, $Z_2(\mathbf{x}^0) = 2$.

The next step is to solve the direction problem, for which we need the decision maker's preferred tradeoff or MRS between objectives Z_1 and Z_2.

Since our decision maker cannot articulate the MRS we shall compute it from the assumed utility function. From (7-4),

$$\mathrm{MRS}(\mathbf{x}^0)_{21} = -\left.\frac{dZ_1}{dZ_2}\right|_{\mathbf{Z}(\mathbf{x}^0)} = \frac{Z_1(\mathbf{x}^0)}{Z_2(\mathbf{x}^0)} = \frac{26}{2} = 13 \qquad \text{(7-117)}$$

We also need the gradients of each objective function. For the case of linear objectives the gradients assume a particularly simple form in that they are not functions of the point $\mathbf{x}^0$: $\nabla Z_1(\mathbf{x}) = (5, -2)$ and $\nabla Z_2(\mathbf{x}) = (-1, 4)$. Using these values and (7-117) and noting that $\mathrm{MRS}_{11} = 1$ [Z_1 is the numeraire in (7-111)] gives

$$\text{maximize} \quad (1)(5, -2)(y_1, y_2)^{\mathrm{T}} + (13)(-1, 4)(y_1, y_2)^{\mathrm{T}} \qquad \text{(7-118)}$$

or

$$\text{maximize} \quad -8y_1 + 50y_2 \qquad \text{(7-119)}$$

$$\text{s.t.} \quad (y_1, y_2) \in \mathbf{F}_{\mathrm{d}} \qquad \text{(7-120)}$$

The optimal solution to this problem is $\mathbf{y}^0 = (y_1^0, y_2^0) = (1, 4)$, which gives as the preferred direction

$$\mathbf{d}^0 = (1, 4) - (6, 2) = (-5, 2) \qquad \text{(7-121)}$$

The preferred direction is shown in the objective space of Fig. 7-16 by noting that $\mathbf{Z}(\mathbf{x}^0) = (26, 2)$ and $\mathbf{Z}(\mathbf{y}^0) = (-3, 15)$.

At this point, Geoffrion *et al.* (1972) would compute several values of the objectives for points that lie along the line between (26, 2) and (−3, 15) in Fig. 7-16. The decision maker's choice would then define t^*, the optimal step size. In our case we have presumed to know the decision maker's utility function so that we can compute t^* by solving

$$\begin{aligned} &\text{maximize} \quad U[Z_1(\mathbf{x}^0 + t\mathbf{d}^0), Z_2(\mathbf{x}^0 + t\mathbf{d}^0)] \\ &\text{s.t.} \quad 0 \le t \le 1 \end{aligned} \qquad \text{(7-122)}$$

$$\begin{aligned} &\text{maximize} \quad \{[5(6 - 5t) - 2(2 + 2t)][-(6 - 5t) + 4(2 + 2t)]\} \\ &\text{s.t.} \quad 0 \le t \le 1 \end{aligned} \qquad \text{(7-123)}$$

which is equivalent to

$$\begin{aligned} &\text{maximize} \quad (-277t^2 + 280t + 52) \\ &\text{s.t.} \quad 0 \le t \le 1 \end{aligned} \qquad \text{(7-124)}$$

which is maximized at $t^* = 0.37$. This gives $\mathbf{x}^1 = \mathbf{x}^0 + t^*\mathbf{d}^0 = (4.2, 2.7)$, at which $Z_1(\mathbf{x}^1) = 15.6$ and $Z_2(\mathbf{x}^1) = 6.6$. The new point is shown in Fig. 7-16; notice that it is inferior.

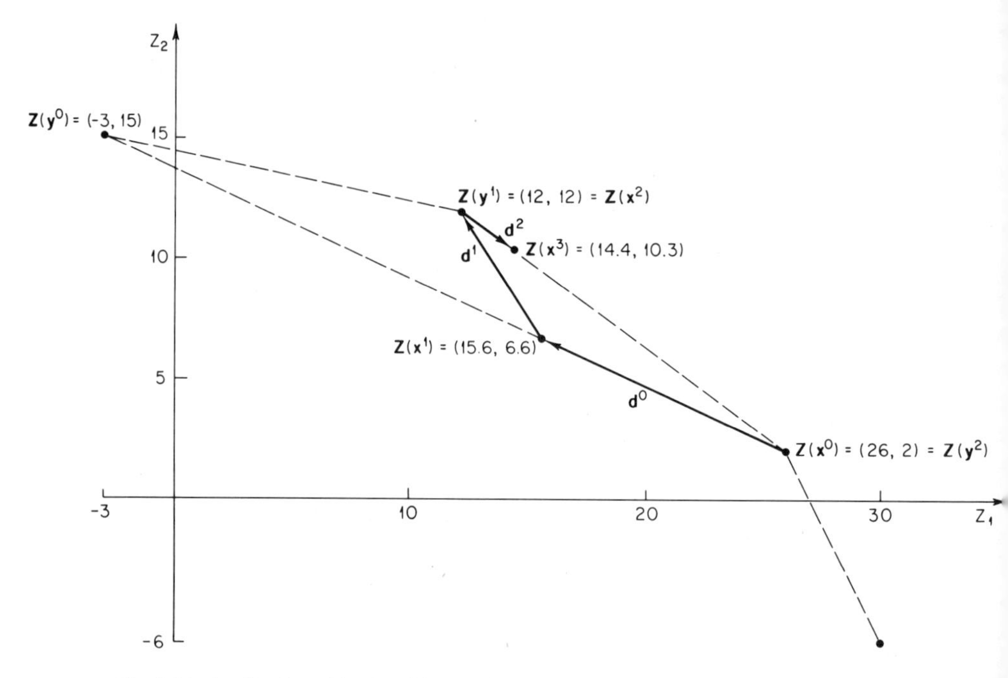

Fig. 7-16. Application of the Geoffrion *et al.* (1972) iterative algorithm to the sample problem.

The algorithm continues by solving a new direction problem with $\mathbf{x}^1$ as the new point. A recomputation gives $\mathrm{MRS}_{21}(\mathbf{x}^1) = 15.6/6.6 = 2.4$. Using this in (7-111) gives the objective function for the new direction problem,

$$\text{maximize} \quad 2.6y_1 + 7.6y_2 \tag{7-125}$$

which is maximized over $\mathbf{F}_\mathrm{d}$ at $\mathbf{y}^1 = (y_1^1, y_2^1) = (4, 4)$, defining the new direction

$$\mathbf{d}^1 = \mathbf{y}^1 - \mathbf{x}^1 = (-0.2, 1.3) \tag{7-126}$$

the image of which in objective space is shown in Fig. 7-16. The step size problem is

$$\begin{aligned} &\text{maximize} \quad (-19.4t^2 + 60.5t + 103) \\ &\text{s.t.} \quad 0 \leq t \leq 1 \end{aligned} \tag{7-127}$$

which is solved optimally at $t^* = 1$. This gives a new point $\mathbf{x}^2 = \mathbf{x}^1 + t^*\mathbf{d}^1 = (4, 4)$, at which $\mathbf{Z}(\mathbf{x}^2) = (12, 12)$.

We will continue with the algorithm since $\mathbf{x}^2 \neq \mathbf{x}^1$. $\mathrm{MRS}_{21}(\mathbf{x}^2) = 1$ so the objective function for the new direction corresponds to an equally weighted sum of the objectives. We solved this problem to begin the algorithm so we know that $\mathbf{y}^2$, the solution to the new direction problem, is (6, 2) with $\mathbf{Z}(\mathbf{y}^2) = (26, 2)$. This defines a new direction $\mathbf{d}^2 = (2, -2)$. A corresponding objective space direction is shown in Fig. 7-16.

The new step size problem is identical to the problem previously stated in Eq. (7-23) in Section 7.1.3. The solution of that problem is $\mathbf{x}^3 = (4.3, 3.7)$, at which $\mathbf{Z}(\mathbf{x}^3) = (14.4, 10.3)$. The correspondence of this new step size problem with the problem solved in Geoffrion's (1967) other algorithm of Section 7.1.3 lends support to the rationale for the iterative algorithm. Of course, one must keep in mind that we have assumed a decision maker who can articulate preferences exactly. Therefore it is not surprising that eventually the iterative approach will reduce to the previous algorithm in which the utility function is prespecified.

The iterations continue since $\mathbf{x}^3 \neq \mathbf{x}^2$. We find for our direction problem that $\text{MRS}_{21}(\mathbf{x}^3) = 1.4$ so our new objective function is

$$\text{maximize} \quad 3.6y_1 + 3.6y_2 \tag{7-128}$$

which is maximized along the line segment between (4, 4) and (6, 2). Our current solution $\mathbf{x}^3$ lies along this segment so $\mathbf{d}^3 = (0, 0)$ and $\mathbf{x}^3 = \mathbf{x}^4$. The algorithm stops with the best-compromise solution (4.3, 3.7), at which $(Z_1, Z_2) = (14.4, 10.3)$. This is, of course, the same best-compromise solution found in Section 7.1.3.

7.5.3 Some Empirical Results on the Utility of Iterative Methods

The philosophy of iterative approaches to multiobjective problems is an appealing one: involve the decision maker directly in the solution process in a manner that will allow that person's best-compromise solution to be discovered. We have seen that the various methods attempt to accomplish this goal in various ways. The step method (Section 7.5.1) involves the decision maker in a limited way by eliciting the amount by which an objective at a currently satisfactory level can be reduced. The method of the previous section was much more explicit and direct about preferences by requiring the decision maker to articulate marginal rates of substitution and to pick one solution from among many that lie in a preferred direction. For either method, it is not clear that the required value judgments can be made reliably or that the decision maker receives enough information to make intelligent responses.

Doubts about the usefulness of these two iterative procedures led Wallenius (1975) to conduct experiments in which subjects (business students and managers) were presented with a multiobjective decision problem. Each subject was asked to find a best-compromise solution for the problem by using three different approaches that were previously programmed for direct use on a computer terminal by the subjects. The three methods were the step method, the technique of Geoffrion *et al.* (1972) discussed previously, and a trial-and-error procedure in which the only information that the subject received was whether or not a solution picked by the subject was feasible.

After using each of the three methods, each subject was asked to rank the three methods in terms of such criteria as ease of use and understanding, the subject's confidence in the solution, and the subject's overall preference for the methods. The results were, not surprisingly, somewhat inconclusive since the evaluation process is itself a multiobjective problem. However, the results tended to indicate quite strongly that the two formal interactive procedures were found to be no better, and in many cases worse, than the unstructured trial-and-error methods.

Wallenius' (1975) results are fascinating since they would seem to indicate, to this author at least, that generating techniques (see Chapters 5 and 6) are perhaps more meaningful to decision makers than are formal procedures for drawing out their preferences. This conclusion is based on the similarity of the trial-and-error method to generating techniques. In both cases, the decision maker is in the role of probing the noninferior set with the goal of gaining information about what is possible and what must be traded off.

Wallenius (1975) interpreted his results more narrowly. He saw them as relevant to the particular characteristics of the two formal procedures that were tested. The results motivated the method presented in Zionts and Wallenius (1976), which is intended to overcome the observed deficiencies.

CHAPTER 8

Multiple-Decision-Maker Methods

In the preceding two chapters it was assumed that there was a single decision maker who articulated preferences or at least was the recipient of the information from a generating technique; i.e., the analyst had a well-defined client. It may be more reasonable to assume, however, that the public sector analyst will simultaneously serve several clients with conflicting viewpoints or, in many situations, serve a single decision maker who is one of many decision makers in a political process. In either case, the analyst must be sensitive to the multiplicity of interests that characterize public decision-making problems.

The analysis of multiple-decision-maker problems is simultaneously the most complex and least understood area of analysis. In this chapter some of the techniques that have been developed are reviewed in a manner that will hopefully give the reader a feel for the complexity and the analytical richness of multiple-party decision making.

People from at least three disciplines—economics, political science, and operations research—have been concerned with multiple-decision-maker problems. Rather than keep these disciplinary lines, however, we shall attempt to categorize the various analytical methods by their implied analytical goal and planning context. We shall talk about three types of method: techniques for the aggregation of multiple preference orderings into a single ordering; methods that are used to counsel a single decision maker; and techniques for the prediction of outcomes from a decision-making process.

Aggregation and counseling techniques represent two very different philosophies about the fundamental but elusive concept of the "public interest": what it is and how to characterize it. [Steiner (1969) presents an excellent discussion of the public interest and its implications for the evaluation of public investments.] Aggregation rests on the definition of public interest in a democratic society as the combination of each individual's interest. Many welfare economists have perceived resource allocation in just this way: A social welfare function (an expression of the way in which society should order public alternatives) is derived from individuals' utilities. A discussion of welfare economics, its basic resource allocation problem, and the ways in which individuals' utilities are aggregated will be presented.

Those who define their analytical role as counselor to a single public decision maker reject the aggregationists' mechanistic view of the public interest. In effect, the public interest is defined by public decision makers so it is meaningful to talk about a social welfare function only to the extent that it expresses a decision maker's preferences (or that person's perception of society's preferences—the public interest) for alternative public actions. The theories of some modern welfare economists and multiobjective generating methods are consistent to varying degrees with this view of public decision making.

The prediction of political outcomes implies an entirely different analytical goal from the first two types of method. Aggregation and counseling are normative in nature, i.e., they attempt to identify what *should* be done, while prediction is a positive analysis with normative consequences. The analyst's choice to predict rather than prescribe a solution implies a rather cynical (but perhaps realistic) view of public decision making: It is useless to identify what should be done if the socially optimal alternative will not be implemented anyhow. What really seems to count are politically powerful interest groups that are assumed to control public decision making. Prediction methods concentrate on those participants with the most clout since they are assumed to matter most in the political process. The methods which fall into this category are Paretian analysis, game theory, and various models of voting procedures.

We shall examine the aggregation, counseling, and prediction categories of method in that order.

8.1 AGGREGATION OF INDIVIDUAL PREFERENCES

The combination of individual preferences into an overall ordering of social choices has been a focus for much of the theoretical work in welfare economics. The basic problem is given a set of possible social actions (policies and projects) that impact various people both positively and negatively to varying degrees, which action should be chosen? An excellent mathematical survey of social orderings—their nature, rules for their derivation, and their implications—is contained in Sen (1970).

In order to understand fully the contribution that welfare economists have made to the analysis of this basic problem, we shall delve into the history of welfare economics. We shall see how current views and controversies over what is best and how it should be defined have evolved.

8.1.1 A History of Welfare Economics

The aggregation of individual preferences into an overall social welfare function has preoccupied welfare economists since the 18th century. Procedures and points of view that have evolved since that time are relevant for our consideration of multiple-decision-maker problems. The reader is referred to Graaf (1971) for a detailed comprehensive review of theoretical welfare economics.

Welfare economists find it difficult to distinguish their field from economics at large. Borrowing from Boulding (1952, pp. 1–3), welfare economics will be characterized as that part of economics that attempts to incorporate human values into its analysis of public policy. Thus, as Professor Boulding has pointed out, welfare economists do not have the luxury of confining their analysis to the ideal world of commodities in which value is determined by production and exchange. It is the lot of welfare economists that they must also worry about the human evaluation of those commodities: What are riches and how do you compare one person's riches with another's? It is the way in which individual riches (or utility) and interpersonal comparisons of them have been quantified (or avoided) that distinguishes one school of thought from another.

The history of welfare economics can be segmented into three major periods, each of which represents a particular methodological movement. The first is the old welfare economics, which is usually associated with the period that ended around 1930 and that began with the utilitarian philosophers/economists Bentham and Mill. The major work of this period, however, was that of Pigou (1920), which for many signaled the beginning of welfare economics as a legitimate discipline.

The second major period in the history of welfare economics found its roots in the work of Pareto (1971†), who was not a contemporary of the period itself. The methodological movement that is traced to Pareto has become known as the new welfare economics. Kaldor and Hicks with their "compensation tests" were the major figures in that period.

The last and current period covers the past 40 years, in which economists have gradually moved away from the principles of the new welfare economics in two very different directions. The modern welfare economists, which include Bergson (1938, 1966‡) and Samuelson (1965§), have developed a theory that fits most conveniently into the counseling category. These modern welfare economists are frequently included in the new school (Boulding, 1952), but it makes sense to distinguish the Bergson–Samuelson point of view from the methodology of Kaldor and Hicks since the two schools deal with the basic problem in very different ways.

The second direction of modern analysis was defined by Arrow (1963‖), who brought mathematical formalism to the aggregation problem. His analysis has affected, and in many cases led directly to, all of the work in welfare economics over the past 25 years.

The ideas about welfare economics in each of these historical periods are discussed below. Before we begin the discussion, however, the general problem of welfare economics will be stated. Old, new, and modern welfare economists all grappled with the same fundamental problem; only their approaches to evaluating potential solutions to the general problem have differed.

8.1.2 The General Problem of Welfare Economics

The problem of welfare economics is How should resources be allocated for the production and ultimate consumption of commodities so as to maximize social welfare? All of the terms in this problem statement are self-explanatory except "social welfare," which is the term of most interest to us. Indeed, consumption, production, and resource allocation are the topics with which all of economics is concerned. It is the welfare economist's concern with the impact of economic activities on social welfare that distinguishes the discipline from the rest of economic theory.

A mathematical statement of the general problem was provided by Samuelson (1947, p. 229). Beginning with constraints on consumption, we shall

† Originally published in 1909.

‡ Originally published in 1954.

§ Originally published in 1947.

‖ Originally published in 1952.

define the total amount of a commodity j produced as Q_j and the amount of j consumed by individual i as q_{ij}. There are a total of n commodities and p individuals. Individual i consumes q_{ij} of each commodity j so individual is total consumption can be represented by an n-dimensional commodity consumption vector $\mathbf{q}_i = (q_{i1}, q_{i2}, \ldots, q_{in})$. An obvious set of constraints on consumption is that the total amount of each commodity consumed by all individuals cannot exceed the availability of that commodity. Mathematically, these constraints may be written as

$$\sum_{i=1}^{p} q_{ij} - Q_j \leq 0, \qquad j = 1, 2, \ldots, n \tag{8-1}$$

Each individual in our society of p persons is also endowed with resources and productive services that may be supplied to production processes. We shall call v_{ik}^* the initial endowment of resource k for individual i. The amount of resource k that is supplied (not just owned) by individual i is v_{ik}. There are m resources in all, and the total amount of resource k available in society is V_k. Analogous to the commodity consumption vector, each individual has an r-dimensional resource supply vector $\mathbf{v}_i = (v_{i1}, v_{i2}, \ldots, v_{ir})$. Another set of constraints requires that the supply of each resource by an individual cannot exceed that individual's endowment. Thus,

$$v_{ik} \leq v_{ik}^*, \qquad i = 1, 2, \ldots, p; \quad k = 1, 2, \ldots, r \tag{8-2}$$

(and $\sum_{i=1}^{p} v_{ik}^* = V_k$ by definition).

The availability of commodities must be related to the supply of resources and services. This relationship is the production function, which will be represented as

$$T(Q_1, Q_2, \ldots, Q_n; \mathbf{v}_1, \mathbf{v}_2, \ldots, \mathbf{v}_p) = 0 \tag{8-3}$$

which is a purely technical (and, in this case, general) statement of how resources can be transformed into commodities for consumption.

The final constraints require nonnegative consumption, production, and resource supply, i.e.,

$$q_{ij}, v_{ik}, Q_j \geq 0, \qquad \forall i, j, k \tag{8-4}$$

The relationships stated above define a set of feasible solutions for the problem. Each feasible solution defines a potential *social state*, i.e., resources supplied and commodities consumed by each individual. Social welfare, which we are trying to maximize, is the quantity that distinguishes one social state from another. The basic element of social welfare is the utility U_i of each member of society. The utility or satisfaction that person i realizes is a result of the commodities that i consumes. Thus, U_i is a function of the commodity consumption vector $\mathbf{q}_i$ so utility will be denoted by $U_i(\mathbf{q}_i)$.

An objective for the general problem can be stated in terms of the individual utilities. That is, social welfare can be related to the utilities of all of the individuals in society. This relationship will form the objective function of the problem. To keep the statement of the general problem as general as possible, we must be careful about the assumptions that we build into the objective function. The objective function will be stated as

$$\text{maximize} \quad \mathbf{U} = [U_1(\mathbf{q}_1), U_2(\mathbf{q}_2), \ldots, U_p(\mathbf{q}_p)] \tag{8-5}$$

where $\mathbf{U}$ is a p-dimensional vector of utilities, which implies that the utilities of all individuals in society are to be maximized simultaneously. The only major assumption inherent in this statement of the objective is that individual utilities are functions only of the individual's own consumption; i.e., there are no consumption externalities: I am no less pleased with my bundle of commodities when you are starving than when you have plenty for yourself.

We know from multiobjective programming theory that the multidimensionality of the objective function in (8-5) implies that the general problem of maximize $\mathbf{U}$, subject to the constraints in (8-1)–(8-4), can be solved only in a limited sense unless we are willing to make some further assumptions. More precisely, we can identify an optimal social state, i.e., optimal values for $\mathbf{q}_i$ and $\mathbf{v}_i$ $\forall i$ and Q_j $\forall j$, only if we make value judgments regarding the relative importance of the individual utilities $U_1(\mathbf{q}_1), \ldots, U_p(\mathbf{q}_p)$. If such value judgments are made, then it must be assumed that utility is comparable on an interpersonal level.

The multiobjective programming problem above provides a convenient way for characterizing the schools of thought in welfare economics. The constraint set (8-1)–(8-4) does not vary significantly throughout the history of welfare economics. The objective function (8-5) and its transformation into a single-dimensional aggregate measure of welfare have attracted much of the interest of the welfare economist. We shall focus our following discussion on that relationship.

8.1.3 Old Welfare Economics

The ethical content of old welfare economics can be traced back to utilitarian philosophy, which had its beginnings with Bentham (1948†). For Bentham, social welfare was a summation of the utility ("pleasure," in

† Originally published in 1789.

Bentham's phraseology) of each individual. The statement of the old welfare economist's objective function is

$$\text{maximize} \quad W = \sum_{i=1}^{p} U_i(\mathbf{q}_i) \tag{8-6}$$

where W is used to denote welfare, the "quantity" that is to be maximized.

The ethically neutral vector objective function of (8-5) is quite different from the relationship in (8-6). The old welfare economist implicitly assumed two major characteristics of individual utility: (1) utility is measurable and (2) the utilities of different individuals can be compared, i.e., summed. The first assumption is clearly documented in the first six chapters of Bentham (1948). The latter condition, however, remained implicit until the new welfare economists reacted to it in the 1930s.

In order to understand fully the ethical content of (8-6), it will be rewritten as

$$\text{maximize} \quad W(\boldsymbol{\lambda}) = \sum_{i=1}^{p} \lambda_i U_i(\mathbf{q}_i) \tag{8-7}$$

in which λ_i is a strictly positive weight on the utility of each individual and $\boldsymbol{\lambda}$ is a vector of weights, $\boldsymbol{\lambda} = [\lambda_1, \lambda_2, \ldots, \lambda_p]$. Clearly, the set of weights $\boldsymbol{\lambda}$ that is used in (8-7) will, in general, alter the solution of that problem, subject to (8-1)–(8-4). We know this to be true from the weighting method presented in Chapter 6. In this case, the weights selected for (8-7) represent a social welfare function, an aggregation of individual utilities, with all of its ethical content relative to the distribution of economic production. For old welfare economists, however, one particular set of weights, with all $\lambda_i = \lambda_j$, out of the infinite number of possible sets was employed. Of course, there was probably no conscious selection of weights by the old welfare economists. The construction in (8-6) and (8-7) is simply the mathematical consequence of the implied value judgment that the utilities of individuals in society should be weighed equally.

8.1.4 New Welfare Economics

The new welfare economics represented a wave of reaction to the value-laden utilitarian approach of the older school. The reaction is evident in the work of Robbins (1935), in which the desire to make economics a science was so forcefully stated. The methodology of the new welfare economist of the 1930s was based on the work of Pareto (1971), in which cardinal utility was replaced by the indifference curve and Pareto's "optimality" condition was formulated.

Pareto optimality, identical mathematically to noninferiority (see Chapter 4), is a central concept to the new welfare economics. A social state is Pareto optimal if no individual can be made better off without making at least one other individual worse off. The identification of Pareto-optimal social states does not require the articulation of value judgments relative to distribution. Thus, the new welfare economist kept the objective function in (8-5) intact. To find solutions to this vector optimization problem, one can use the weighted function, as in (8-7), by choosing a sufficient number of arbitrary sets of positive weights. It has been shown in the discussion of the weighting method of Chapter 6 (in particular, Section 6.1.4) that the solution to the problem

$$\begin{aligned} \text{maximize} \quad & W(\boldsymbol{\lambda}) = \sum_{i=1}^{p} \lambda_i U_i(\mathbf{q}_i) \qquad (8\text{-}8) \\ \text{s.t.} \quad & (8\text{-}1)\text{–}(8\text{-}4) \end{aligned}$$

will be Pareto optimal if $\lambda_i > 0 \; \forall i$.

An alternative mathematical representation of the new welfare economist's problem is obtained by using the constraint method (Section 6.2):

$$\begin{aligned} \text{maximize} \quad & U_r(\mathbf{q}_r) && (8\text{-}9) \\ \text{s.t.} \quad & (8\text{-}1)\text{–}(8\text{-}4) && \\ & U_i(\mathbf{q}_i) \geq L_i \qquad \forall i \neq r && (8\text{-}10) \end{aligned}$$

in which the utility of one individual (say r) is maximized subject to the previously stated constraints (8-1)–(8-4) and to lower-bound constraints on the utilities of all individuals other than r.

Either formulation can be used to generate the "welfare frontier" [in Graaff's (1971) terminology] or Pareto-optimal set, analogous to our noninferior set. A welfare frontier for a two-individual society is shown in Fig. 8-1. Notice that it resembles the noninferior set for a two-objective problem, as in Fig. 6-3. All of the points on the welfare frontier are Pareto optimal since more of individual 1's utility U_1 can be gained only by sacrificing some of U_2 as you move along the frontier, e.g., from C to B in Fig. 8-1. Feasible social states that lie below the frontier, such as A, are not Pareto optimal since both individuals can be made better off by moving to state B.

The interpretation in Chapter 6 of the slopes of lines tangent to the noninferior set and their relationship to weights on the objectives applies to the welfare frontier and the weights on individual utilities. The slope of the line tangent to point C in Fig. 8-1 is $-(\lambda_1^C/\lambda_2^C)$, where the weights $\boldsymbol{\lambda}^C = (\lambda_1^C, \lambda_2^C)$, if used in the weighted objective function of (8-9), would lead to point C as the optimum of the weighted problem. Notice that as we move to point B, the tangent becomes steeper, indicating that the weight on U_1 is increasing.

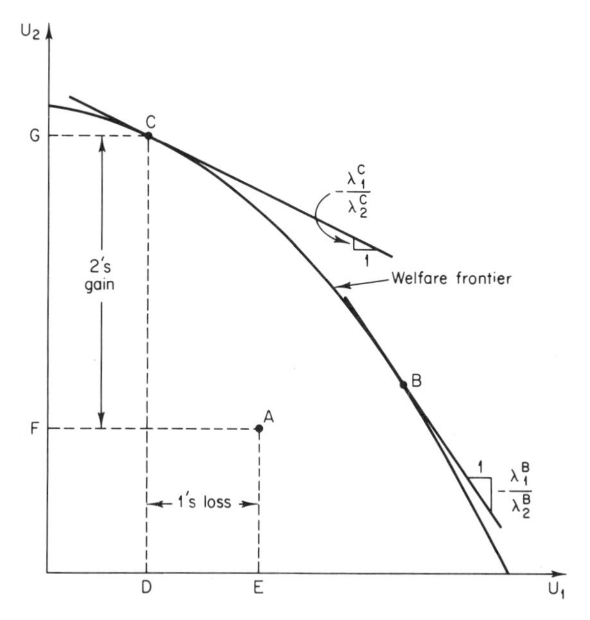

Fig. 8-1. A welfare frontier for a two-individual society.

The welfare frontier, since it can be generated with no value judgments relative to distribution, is consistent with the value-free "scientific" economics espoused by the new welfare economists. By itself, however, the welfare frontier provides insufficient guidance for the analysis of economic policies, assuming that the analyst perceives the analytical goal to be the identification of the optimal social state. There are, in general, an infinite number of Pareto-optimal (noninferior) social states, no single one of which can be considered better than any other in the absence of further value judgments.

The only practical use for the concept of Pareto optimality was for the evaluation of a movement from a present inferior social state to a new Pareto optimal one. But, even in this case, the usefulness of the idea is limited. In Fig. 8-1, state B is clearly preferable to A by simple domination: Everybody can be made better off by moving from A to B. That is, B is Pareto optimal since it lies on society's welfare frontier, while A does not. A movement from B to C cannot be evaluated since they are both Pareto-optimal states. Even a movement from A, which is not Pareto optimal, to C, which is, cannot be evaluated in a strict sense since the utilities of the two individuals are not both increased.

Economists who hoped to do something useful while maintaining their scientific objectivity found their champions in Kaldor (1939) and Hicks (1940), who proposed "compensation tests" to allow the evaluation of movements from an existing to a new social state, such as from A to C in

Fig. 8-1. The general idea was that those who benefited by such a movement would have gained enough to compensate the losers and still have positive gains left over. For example, compare *FG*, individual 2's gain due to a shift from *A* to *C*, to *DE*, individual 1's loss. Everybody would be better off *after* redistribution takes place since person 2 can compensate person 1 for the loss *DE*. A major objection to the compensation tests, in addition to the technical flaws discussed in Graaf (1971, pp. 84–90), is that their use was proposed even in cases where compensation, though feasible, would not necessarily actually occur. The use of the tests when compensation is not actually paid implies the interpersonal comparison of utilities. If individual 2 has not actually compensated individual 1 for the loss *DE*, then we can only surmise that $FG > DE$ since an individual's utility scale is unique. Thus, compensation tests require ethical judgments, the avoidance of which provided the motivation for their development. Furthermore, the nature of the value judgments are not different from those of the old welfare economics since we must still add individuals' utilities to evaluate a change in social state.

The realization that the value-free new welfare economics, no matter how hard its proponents tried, could not be used to identify optimal social states with the notion of Pareto optimality and that the compensation tests implied the old unjustified summation of utilities led to the founding of the next school. The new welfare economics was either incomplete or not new; its practitioners came up against a self-imposed stone wall.

It should be pointed out that the compensation tests laid the foundation for benefit/cost analysis, the most widely used tool for the analysis of public investments. If we call *DE* in Fig. 8-1 $\Delta U_1(\mathbf{x})$ and *FG*, $\Delta U_2(\mathbf{x})$, where $\mathbf{x}$ is a vector of decision variables representing a public project, then the objective function implied by the compensation tests for the two-person problem is

$$\text{maximize} \quad \Delta U_1(\mathbf{x}) + \Delta U_2(\mathbf{x}) \tag{8-11}$$

For a more general problem, this becomes

$$\text{maximize} \quad \sum_{i=1}^{p} \Delta U_i(\mathbf{x}) \tag{8-12}$$

Thus, the objective function is to maximize the total increase in utility; if increases in utility are larger than the decreases, then there is a net gain to society. Notice that interpersonal comparisons of utility are required. In benefit/cost analysis the change in individuals' utilities due to a public project, expressed in (8-12), is estimated by individuals' willingness to pay for project outputs, i.e., the demand curve for the outputs. In this manner, aggregate social welfare is reduced to the monetary unit of net economic efficiency benefits. In applying benefit/cost analysis, then, the analyst is using the market mechanism, which allows willingness to pay to be expressed, to infer a scale for the interpersonal comparison of utilities.

There are problems with this approach. First, willingness to pay does not reflect the distribution of project benefits and costs. This is another way of saying that the market mechanism is not an adequate device to capture individuals' preferences for distribution. Second, markets do not exist for many project impacts such as environmental quality. The estimation of such nonmonetary objectives is a difficult task. Indeed, distributional consequences of projects and nonmonetary objectives such as environmental quality provide the motivation for multiobjective analysis as a substitute for benefit/cost analysis.

8.1.5 Modern Welfare Economics: The Aggregationists

Bergson (1938) is credited with the founding of modern welfare economics. His reformulation of welfare economics represented an attempt at balancing the ethical implicitness of the old and the moral emptiness of the new welfare economics. The major point, also espoused by Samuelson (1965†) and Little (1952), was that ethical judgments are not inherently bad (indeed, the lesson of the prior era was that they are needed if economic theory is to be useful) but that one simply had to be explicit about them, unlike the old welfare economists' implicit assumption of equal weights.

The mathematical contribution of Bergson (1938), relevant to the present discussion, was the development of a "social welfare function." The function is an explicit real-valued statement of how and to what degree the utilities of the individuals in society should be valued. The function is entirely general, the only requirement being that it lead to an unambiguous identification of the optimal social state.

The social welfare function may be thought of as a mathematical relationship in which individual utilities are the arguments. This is entirely analogous to the multiattribute utility functions defined over objectives as in Chapter 7. In general form, the social welfare function would be written as

$$W[U_1(\mathbf{q}_1), U_2(\mathbf{q}_2), \ldots, U_p(\mathbf{q}_p)] \tag{8-13}$$

Recall that one approximation to a multiattribute utility function is a weighted sum of the objectives, where the weights represent constant marginal rates of substitution (see Chapter 7). Thus, the social welfare function may be approximated as the weighted sum of utilities

$$\sum_{i=1}^{p} \lambda_i U_i(\mathbf{q}_i) \tag{8-14}$$

† Originally published in 1947.

This representation is identical to the old welfare economist's objective function except that different relative weights are possible.

Modern welfare economics resulted in the reintroduction of explicit value judgments and interpersonal comparison of utilities into economic science. In doing so, however, a set of new but related questions was generated: How do we find a social welfare function? Who should identify that function? Should individual values count? How should they be counted? Bergson and his colleagues responded to these questions in a way that is most appropriately considered in our next section on counseling techniques. Another set of answers, consistent with the aggregation concept, was provided in 1952 when Arrow's (1963) "Social Choice and Individual Values" was originally published.

Professor Arrow's book presented a challenge to welfare economists since its major conclusion was that an ordering of social states that is consistent with several conditions could not, in general, be developed from an aggregation of individual preferences. While many economists cried "foul" and attempted to refute Arrow's conclusions by attacking his conditions and axioms, others concluded that the analysis was not relevant to welfare economics.

Arrow (1963, p. 3) viewed the central problem of welfare economics as "achieving a social maximum from individual desires." Arrow also spoke in terms of a social welfare function, but while he agreed with the significance and necessity of a social welfare function, he rejected the interpersonal comparisons of utility that are inherent in the Bergson social welfare function and that are made by the economist. His main point was that in a democratic society the social welfare function should represent an aggregation of the preferences of the individuals who are members of that society. The bulk of "Social Choice and Individual Values" is a consideration of the possibilities for such a social welfare function.

In terms of the previous general mathematical form, Arrow's social welfare function is identical with that of Bergson. Thus, the only difference between Arrow's and Bergson's approaches is the origin of the social welfare function. While this difference may not be important mathematically it is essential for one's interpretation and use of welfare economics (and one's philosophy for public sector problem solving, in general).

Arrow's consideration of the existence of a social welfare function as he interpreted it proceeded on the hypothesis of two axioms and five conditions that he felt a social ordering built on the orderings of individuals in a democratic society should satisfy. He then proved his "general possibility theorem," which showed that a transitive social ordering (social welfare function) is not, in general, possible if that ordering is not to be imposed or dictatorial.

A demonstration of Arrow's theorem is provided by the paradox of majority voting, which has been thoroughly discussed in the literature of political theory [see Riker (1961) for a comprehensive survey]. The paradox, which is possible whenever there are more than two voters and alternatives, results in intransitive orderings of social states. Suppose there are three individuals 1, 2, and 3 and three social states A, B, and C. Suppose also that the three people would order the three social states as in Table 8-1. Thus, for example, individual 1 ranks the projects $A > B > C$, i.e., A is preferred to B, which is preferred to C. If the individuals were now to vote on the social states through a series of pairwise comparisons the results would be those shown in Table 8-2. Table 8-2 shows, for example, that in voting on projects A and B, the vote would be two (individuals 1 and 3) to one (individual 2) in favor of project A. With majority rule as the mechanism for decision making, we see from Table 8-2 that our three-person society would have an intransitive ordering: $A > B$ and $B > C$, which implies $A > C$, but this contradicts the third vote in Table 8-2. This is the paradox of majority voting. Transitivity can be obtained only by imposing an ordering on at least one individual or by allowing one individual to dictate what the social ordering should be.

A controversy arose after Arrow originally published his monograph in 1951. The response was rather strenuous because of the serious implications of Arrow's "general possibility theorem" for welfare economics. As Riker (1961, p. 903) has indicated, "great attention was given to Arrow's theorem when it appeared for it seemed to destroy the foundations of welfare economics." Furthermore, "since the mathematical argument ... is conclusive, most economists have attacked his premises." Riker cites several authors who considered each of Arrow's conditions in the hope of discounting or in some way relaxing them.

The impact of Arrow's analysis on welfare economics was profound. It became the point of departure for all of the work on aggregation of individual preferences since its publication. Arrow's work also forced Bergson and his

TABLE 8-1

ORDERINGS OF THREE SOCIAL STATES BY THREE INDIVIDUALS

Individuals		
1	2	3
A	B	C
B	C	A
C	A	B

TABLE 8-2

RESULT OF VOTES ON PAIRWISE COMPARISONS OF PROJECTS, USING THE PROJECT RANKINGS IN TABLE 8-1

	Yes	No
$A > B$?	2	1
$B > C$?	2	1
$A > C$?	1	2

colleagues to define more clearly the role that they saw for the economic analyst in the public decision-making process. This latter impact is taken up in Section 8.2.

8.1.6 Other Thoughts on Aggregation and Collective Choice

Sen (1970) has developed a comprehensive unification of the approaches to the aggregation of individual preferences. Sen shows that a social welfare function is a special type of social ordering; i.e., it is a real-valued function that ranks all alternatives. A more general form of a social ordering is a "social decision function," which may not provide a ranking of all social states, but in all situations an unambiguous choice will be forthcoming. Returning to the majority voting situation displayed in Tables 8-1 and 8-2, we saw that a social welfare function based on majority rule is not generally possible since an unambiguous ranking of social states A, B, and C may not be obtainable. Majority voting is, however, a valid social decision function since in every pairwise comparison of alternatives, a choice of one social state is indicated: when A and B are compared, choose A; when B and C are the options, the choice will be B; and when the alternatives are A and C, choose C.

Sen discussed majority voting and other "collective choice rules" in the context of three interdependent problems: measurability of individual welfare (utility), interpersonal comparability of individual utilities, and the form of the function that specifies collective preference, given assumptions about the first two problems.

All of welfare economics, excluding the Bergsonian school of thought, has assumed that measurability and comparability were possible and that individual utilities could be added to arrive at a social preference. Pareto backed away from measurability and interpersonal comparisons, but that did not provide enough for economic analyses; i.e., social preferences were not unambiguously defined by Pareto's optimality condition. Further, the compensation tests of the new welfare economics did allow the definition of social preferences, but required measurement and interpersonal comparisons of individual utilities. Thus, at least in terms of aggregating individual preferences, welfare economics has not changed a great deal since the days of Bentham.

If we turn away from welfare economics, a number of other rules for aggregating individual preferences can be found, as discussed in Sen (1970, Chapters 8–11). Nash (1953), in looking at bargaining problems, proposed a multiplication of individual utility gains and losses, measured relative to the status quo. Rawls (1958, 1963a,b), a political philosopher, proposed a choice rule that maximizes the minimum individual utility. Such a criterion was used by Brill *et al.* (1976) to identify equitable water pollution abatement plans in the Delaware River. Other aggregation schemes have been proposed by Harsanyi (1955)—the maximization of expected utility—and by Suppes (1966). Sen (1970) provides an excellent discussion of all of these methods. Another widely used technique is the Delphi method (Dalkey and Helmer, 1963; Dalkey, 1969, 1976). The emphasis of the technique is the aggregation of the opinions of a "panel of experts."

Whichever approach an analyst may choose to take, the aggregation of individual preferences will always be a controversial and risky business. The three problems of aggregation mentioned above are problems, indeed: The measurement of individual preferences, the interpersonal comparison of those preferences, and their aggregation into a social preference function are operations that are loaded with the most fundamental of value judgments; judgments that are concerned with the distribution of income, aggregate gains versus the distribution of those gains, and social justice. Of course, these value judgments are unavoidable since collective actions must, by their nature, impact social welfare. At issue, however, is who should make those judgments. The aggregation of individual preferences puts the burden of social decision making on the shoulders of the analysts since the value judgments are imbedded in the analytical procedure. The approaches in the next section attempt to shift this burden to decision makers who are required to articulate the necessary value judgments.

8.2 COUNSELING AN INDIVIDUAL DECISION MAKER: IN SEARCH OF THE PUBLIC INTEREST

We turn now to approaches to multiple-decision-maker problems that are predicated on a different view of the public interest and that prescribe a different role for the analyst. The burden of articulating the public interest is now shifted to a decision maker—the analyst's client; the analyst assumes the role of an information provider. The rationale for this approach is first that society's preferences do not derive from an aggregation of individual preferences; even if aggregation is correct, it is a difficult or an impossible task to accomplish reliably. Second, public decision makers are appointed or elected to fill just that role prescribed for them by the methods in this category, i.e., to articulate the preferences of society or a segment of society.

The discussion will begin with a more detailed consideration of Bergsonian welfare economics that provides further justification for this approach. We shall then consider multiobjective programming techniques in the Bergsonian context.

8.2.1 The Bergsonian View of Welfare Economics

Bergson's (1938) restatement of the problem of welfare economics that allowed explicit value judgments to enter the formulation preceded Arrow's (1963†) work by 14 years. While Bergson and his colleagues disagreed with Arrow's analysis, apparently it was Arrow's work which forced Bergson and Little to define their brand of welfare economics more clearly. Thus, it was in reaction to Arrow's general possibility theorem that this direction of modern welfare economics became fully developed.

A major issue was the appropriate role for the welfare economist. Bergson (1966, Chapter 2‡) fundamentally disagreed with Arrow's attempt to aggregate preferences. Bergson conceived of the welfare economist's role as that of a counselor to citizens in general and public officials in particular. The analyst must first contemplate the ethical values implied by counseling one person or another; the next step would be a determination of whom to counsel. The first step, Bergson (1966, p. 36) contends, is assumed away by Arrow by "his supposition that individuals have some definite values on social states and from his apparent decision not to go behind these values."

On the question of whom to counsel, Bergson (1966, p. 37) reiterated that "the concern of welfare economics is to counsel individual citizens generally." Public officials in a decision-making capacity are to be served in the same manner. Bergson (1966, pp. 37–38) summarized the fundamental differences between his and Arrow's approaches in the following passage:

> According to this view [Arrow's] the problem is to counsel not citizens generally but public officials. Furthermore, the values to be taken as data are not those which might guide the official if he were a private citizen. The official is envisaged instead as more or less ethically neutral. His one aim in life is to implement the values of other citizens as given by some rule of collective decision-making. Arrow's theorem apparently contributes to this sort of welfare economics the negative finding that no consistent social ordering could be found to serve as a criterion of social welfare in the counseling of the official in question.

While Bergson conceded that Arrow's welfare function may be applicable to political decision making, both he and Little (1952) could see no relevance of Arrow's formulation to welfare economics. Little is especially insistent in his article that the only requirement for an analysis in welfare theory is an ordering, not necessarily a "social" ordering, determined by a political process. Thus, concluded Little, it is irrelevant to welfare economics to pursue the direction set out by Arrow.

† Originally published in 1952.

‡ Originally appeared as "On the Concept of Social Welfare," *Q. J. Econom.* **68**, 233 (1954).

Little further defined the differences between the two approaches. Bergson's function—the one believed by Little to be relevant for welfare economics—is based on the value judgments of an individual. The important aspect of the social welfare function for evaluating social states is the ethical content of the function. Arrow's function, on the other hand, is the output from a political decision-making process—the product of a "machine." Since a machine is incapable of value judgments, one cannot give any ethical content to Arrow's welfare function. Therefore such an ordering is irrelevant to welfare theory. Samuelson's labeling of Arrow's function as a "constitutional" welfare function and his support of Little's criticism served to heighten further the distinction between the two approaches.

Bergson recognized an important problem in counseling the public official. Although the welfare economists' efforts should be most welcome by the public decision maker, he may be thwarted by the official's overriding concern with political or other "extra-economic" constraints. Bergson seems ultimately to take a stance very close to Pareto and consistent with the philosophy of generating methods of multiobjective programming. This is evident from his statement that "the welfare economist may prefer simply to leave it to the public official finally to evaluate the new measure in light of his own values on income distribution and his appraisal of political constraints" (Bergson, 1966, Chapter 3, p. 74).

8.2.2 Multiobjective Programming and Multiple-Decision-Maker Problems

The analyst's role prescribed by Bergsonian welfare economics is a comfortable one that we can fill with previously developed methods. The applicable context is an analyst working for a single decision maker who must trade off the impacts of a decision on several objectives and/or several interest groups. In this context we can define objectives, if required, as the benefits or costs (in monetary or physical units) that accrue to various groups of people whose welfare is the concern of our decision maker.

If the analyst is clearly responsible to a single decision maker, then a preference-based method (Chapter 7) may be used to assess the decision maker's preferences, which define a best-compromise solution, or a generating technique (Chapter 6) may be used to estimate the noninferior set, which indicates the range of choice available to the decision maker. Freimer and Yu (1974, 1976) applied compromise programming (Section 7.3.1) to multiple-decision-maker problems. A more general context has multiple objectives or impacts on interest groups *and* multiple decision makers, and the analyst may not be clearly responsible to a single decision maker. An example for this context is a policy analyst in a federal bureaucracy whose analysis is directed at a problem the ultimate consideration of which may come from the legislature. The administrator of the bureau may represent the analyst's client, but only in a limited sense since responsiveness to the legislature's multidimensional conflicting desires is required.

Analytically, a multiple-decision-maker context in which the analyst can identify a client, i.e., the person or authority that pays the analyst's salary, is no more complex than the multiobjective problem. We should keep in mind, however, that this is only one, perhaps simplified, context.

8.3 PREDICTING POLITICAL OUTCOMES

The most cynical and perhaps realistic view of the public decision-making process is that of conflict among interest groups competing to gain as much as possible for their own special interests. There is no point in aggregating individual preferences if the so-called "will of the people" must defer to special interests that control public decision making. Nor is it adequate to counsel a single decision maker or interest group if the bargaining and gaming among the special interests are what define the political process. Individual preferences matter only to the extent that they are represented by an interest group. A single participant's preferences matter only to the extent that the participant's political power can promote them.

The context for the methods in this section, then, is one of conflict among many competing interests, the resolution of which defines public decisions. The analytical goal for these methods is prediction: Given the nature of the choices, who gains and who loses? What is the ability of participants to bargain and the relative political power of the participants? Which decision will be chosen, or a similar question, which alternative is most politically feasible?

Several methods for the prediction of outcomes from a multiple-interest decision problem have been proposed. Paretian analysis, elementary game theory, and vote-trading algorithms are discussed below. However, none of these methods captures all of the aspects of the general multiobjective multiple-decision-maker problem. Before embarking on the specific methods, then, we shall consider the general problem.

8.3.1 The Multiobjective Multiple-Decision-Maker Problem

Before getting into the prediction methods, the multiobjective multiple-decision-maker (MOMDM) problem will be formulated. This will serve as a paradigm of sorts for the methods in the remaining sections.

The form that the MOMDM problem takes depends on the specific decision-making context being studied. We shall consider r political decision makers each with a constituency C_l where

$$C_l = \{i \mid \text{individual } i \text{ is in } l\text{'s constituency}\} \qquad (8\text{-}15)$$

Each decision maker has several objectives, but the objectives of each decision maker may not be the same; i.e., in addition to attaching different weights on the same objective, the decision makers may not even agree on what is important.

It will be assumed that one of the objectives important to each decision maker is the maximization of net economic efficiency benefits accruing to the decision maker's constituents. In terms of the notation in (8-12), decision maker l would like to maximize

$$\sum_{i \in C_l} \Delta U_i(\mathbf{x}) \tag{8-16}$$

where $i \in C_l$ means individual i is in the constituency of decision maker l.

We shall also assume that there is a total of $p - 1$ other objectives, denoted by $Z_2(\mathbf{x}), Z_3(\mathbf{x}), \ldots, Z_p(\mathbf{x})$. These other objectives may include the maximization of net economic efficiency for the nation as a whole, environmental quality objectives, and distributional considerations apart from the constituency benefits in (8-16). We can think of each decision maker, then, as having a p-dimensional vector of objectives:

$$\text{maximize} \quad \left[\sum_{i \in C_l} \Delta U_i(\mathbf{x}), Z_2(\mathbf{x}), Z_3(\mathbf{x}), \ldots, Z_p(\mathbf{x})\right] \tag{8-17}$$

For the purposes of the formulation, it will be assumed that each political decision maker has, at least implicitly, a multiattribute utility function—a social welfare function—defined over the p objectives. If decision maker l does not believe that a particular objective is important, then the social welfare function will reflect this by attaching zero weight to that objective. Given a social welfare function W_l, the lth decision maker's objective function becomes

$$\text{maximize} \quad W_l\left[\sum_{i \in C_l} \Delta U_i(\mathbf{x}), Z_2(\mathbf{x}), \ldots, Z_p(\mathbf{x})\right] \tag{8-18}$$

Considering the social welfare functions of all decision makers, the MOMDM problem is

$$\begin{aligned} \text{maximize} \quad \Bigg\{ & W_1\left[\sum_{i \in C_1} \Delta U_i(\mathbf{x}), Z_2(\mathbf{x}), \ldots, Z_p(\mathbf{x})\right], \\ & W_2\left[\sum_{i \in C_2} \Delta U_i(\mathbf{x}), Z_2(\mathbf{x}), \ldots, Z_p(\mathbf{x})\right], \ldots, \\ & W_r\left[\sum_{i \in C_r} \Delta U_i(\mathbf{x}), Z_2(\mathbf{x}), \ldots, Z_p(\mathbf{x})\right]\Bigg\} \end{aligned} \tag{8-19}$$

$$\text{s.t.} \quad \mathbf{x} \in \mathbf{F}_\mathrm{d} \tag{8-20}$$

where (8-20) requires that the public project be feasible in terms of the constraints imposed on its design.

The solution of (8-19) and (8-20)—the multiobjective multiple-decision-maker problem— is obviously very complex. This complexity was avoided by the generating techniques of Chapter 6 and the preference-based methods of Chapter 7 by taking a "black box" view of the decision-making process. The decision makers performed their own optimization process either after receiving information on the alternatives from the generating techniques or prior to the solution process, as with the preference techniques. In both Chapters 6 and 7 it was tacitly assumed that there was a single decision maker, i.e., $r = 1$ so that (8-19) reduces to the familiar multiobjective problem.

The problem in (8-19) and (8-20) can be viewed as a bilevel vector optimization problem, but one in which the two levels must be solved similtaneously. The outcome of the decision-making process must be noninferior in terms of the objectives (i.e., there cannot be another alternative that yields more benefits accruing to one objective without decreasing the benefits for another objective) and Pareto optimal in terms of the decision makers' social welfare functions (i.e., there cannot be an alternative that will provide more welfare for one constituency or decision maker without decreasing the welfare of another). There are value judgments implied by choosing a noninferior solution and there are other value judgments implied by the selection of a Pareto optimum. Furthermore, the ultimate selection is an outcome of a bargaining process. This is a messy problem that has yet to be considered adequately. The methods discussed below capture parts of the problem.

8.3.2 Paretian Analysis

Paretian analysis was proposed by Dorfman and Jacoby (1970) to identify politically feasible alternatives for environmental control problems. The method has also been applied to power plant siting by Gros (1975). As its name implies, the method concentrates on the set of Pareto-optimal solutions with the intention of predicting the most likely outcome from a political point of view.

Dorfman and Jacoby (1970) made a simplification by reducing the general MOMDM problem of the previous subsection to a single-objective multiple-decision-maker problem. Although there were two objectives important to their water treatment problem—economic efficiency and environmental quality—Dorfman and Jacoby reduced the problem to the maximization of economic efficiency benefits accruing to each participant by assuming that the environmental quality objective was satisfied by each alternative. This simplification led the authors to write "Fortunately we do not have to assign a numerical magnitude to these [environmental quality] benefits since they will be enjoyed under all the decisions within the range of possibility, and do not form any basis for choosing among decisions" (Dorfman and Jacoby, 1970, p. 201). While it is true that all of the alternatives increased water quality (measured as dissolved oxygen concentration—a conventional water quality indicator) and thereby increased environmental quality benefits, the assumption is correct only if the participants are insensitive to the *level* of this objective. That is, it must be assumed that water that is cleaner than a prespecified lower bound on cleanliness provides added utility to nobody.

The single-objective multiple-decision-maker problem was then formulated by Dorfman and Jacoby as a weighted sum of the participants' utilities, which were measured by net economic efficiency benefits accruing to the constituency of each participant. The problem is identical to the weighted problem in Chapter 6, and the following formulation gives Pareto-optimal solutions under the same conditions and arguments that the weighting method gave noninferior solutions:

$$\text{maximize} \quad \sum_{l=1}^{p} w_l U_l(\mathbf{x}) \tag{8-21}$$

$$\text{s.t.} \quad \mathbf{x} \in \mathbf{F}_{\mathrm{d}} \tag{8-22}$$

where w_l is the relative weight on the lth decision maker, and $U_l(\mathbf{x})$ is the economic efficiency benefits generated by a feasible solution $\mathbf{x}$ that accrue to the lth decision maker's constituency. In this case, w_l is a measure of a decision maker's political influence or "clout" in the decision-making body.

Dorfman and Jacoby varied the weights in (8-21) over a range of values that they felt were politically feasible. By examining the results and using their knowledge of the political situation, the authors selected a range of outcomes that could reasonably be expected to result from the decision-making process.

The discussion of the computational characteristics of the weighting method in Chapter 6 applies as well to Paretian analysis. The method is computationally sensitive to the number of decision makers, in the same way that the weighting method is sensitive to the number of objectives.

One of the strengths of this method is that it does not rely on preference information from decision makers, but recall that this is avoided by reducing all considerations to the monetary units of economic efficiency benefits. If this cannot be done, then a multiattribute utility function must be defined for each decision maker. Note the predictive nature of this technique. There is no attempt to identify what should be done, but rather what is most likely to be implemented.

8.3.3 Elementary Game Theory: Two-Person Zero-Sum Games

We turn now to the situation in which the *interaction* among decision makers is explicitly considered. The nature of the interaction is such that the utility gained by a decision maker depends on the actions of all participants in the decision-making process. Such problems are the focus of game theory.

Game theory had its beginnings in the work of von Neumann (1928) and von Neumann and Morganstern (1967†). Since that time, work has been done on extending the theory to progressively more complex and realistic situations, as a glance at the *International Journal of Game Theory* will quickly confirm. At this time, game theory is still primarily a theoretical area of inquiry that has not enjoyed wide application. As theoretical advances continue to be made, however, the possibilities for application will improve.

Some Basic Notions

In this section we shall deal with the most basic problem of game theory: two-person zero-sum games. An excellent discussion of this category of games is provided by Singleton and Tyndall (1974). Readers who desire more detail than that given in this brief discussion are referred to that book.

Two-person zero-sum games have only two players, and the payoff that one player receives from the game is a cost to the other player. For example, the game of "odds and evens" is a zero-sum game. If you and I were playing, we would each put out one or two fingers: If the sum of the fingers were odd (3), then I would win \$1, but if it were even (2 or 4), then you would win \$1. This game is zero sum since whatever you win, I lose and whatever I win, you lose.

In the terminology of games, each player has a set of strategies. If the two players are called X and Y, then player X has, in general, m strategies $X_1, X_2, \ldots, X_m$, and player Y has n strategies $Y_1, Y_2, \ldots, Y_n$, where m need not equal n. In the game of odds and evens, for example, the strategies are

X_1: player X will put out one finger
X_2: player X will put out two fingers
Y_1: player Y will put out one finger
Y_2: player Y will put out two fingers

Each pair of strategies, one for each player, results in a payoff to each player. For example, if player X selects X_1 and Y chooses Y_1, then (X_1, Y_1) results in an even sum (2). If X wins on odds and Y on evens, then the payoff to X of (X_1, Y_1) is $-\$1$ while Y earns \$1. One way to display the payoffs from the game is in a matrix where each cell of the matrix shows the payoff to one

† Originally published in 1944.

TABLE 8-3

PAYOFF TABLE FOR THE GAME OF "ODDS AND EVENS"

		Player Y	
		Y_1	Y_2
Player X	X_1	-1	1
	X_2	1	-1

of the players. Since the game is zero sum, the other player's payoff is just the negative of the displayed value.

Consider Table 8-3, which shows the payoffs to player X from all pairs of strategies for the game of odds and evens. For example, the element in the X_1 row and the Y_1 column is -1, indicating that player X will lose \$1 if X_1 is chosen (one finger) and if player Y chooses strategy Y_1 (one finger). Since this is a zero-sum game, player X's loss of \$1 implies that Y will earn \$1.

All two-person games can be displayed in a matrix such as Table 8-3. Another game is displayed in Table 8-4, where player X has two strategies and player Y has three. This new matrix shows the dependency of each player's payoff on the other players actions. X must choose X_1 or X_2, but Table 8-4 shows us that the payoff from either strategy will depend on the strategy pursued by Y. Which strategy should player X pursue, X_1 or X_2 for the game of Table 8-4? If X_1 is played, then player X will win \$2 if player Y plays Y_1, lose \$3 if Y pursues Y_2, and lose \$2 if Y plays Y_3. If player X chooses X_2, then a loss of \$1 is incurred if Y plays Y_1, X loses \$1 if Y chooses Y_2, but X wins \$6 if player Y plays Y_3. It is assumed that the game is such that neither player can know in advance which strategy the other will choose. Player X's problem, then, is to choose a strategy that somehow is best under all situations.

The notion that has been developed for strategy selection in such games is that of *maximin* strategies: Player X should choose that strategy that will be best in the worst situation; i.e., the maximin strategy will maximize the

TABLE 8-4

A TWO-PERSON 2×3 GAME

		Player Y		
		Y_1	Y_2	Y_3
Player X	X_1	2	-3	-1
	X_2	-2	-1	6

minimum payoff to X. If X chooses, X_1 then in the worst case (Y_2) X will lose \$3. If, on the other hand, X_2 is selected, then player X could lose, at most, \$2 (if Y chooses Y_1). A rational player will choose X_2 since this strategy maximizes the minimum payoff.

If we consider player Y's choice in Table 8-4, a similar criterion can be developed. In this case, however, the numbers in the table indicate the amount that player Y must pay. The larger the number in Table 8-4, the more Y loses. Without prior knowledge of X's strategy, player Y should choose that strategy that minimizes the maximum payoff, since this minimax strategy will be best for Y in the worst situation. Returning to Table 8-4, we see that if Y_1 is selected the worst case is X_1, where X wins \$2 and therefore Y loses \$2. If Y_2 is selected, the worst that Y can do is a gain of \$1 (X chooses X_2). The worst case when Y chooses Y_3 is a loss of \$6 when X selects X_2. Therefore the maximum payoffs (to X) are 2, -1, and 6 for strategies Y_1, Y_2, and Y_3, respectively. A rational player Y will choose that strategy that minimizes this maximum loss, i.e., strategy Y_2.

If one of the players were our client, we could stop here since our recommendation of a strategy has been made. However, game theorists go beyond this level of analysis to see if a stable solution to the competitive situation exists; i.e., is there a solution (a pair of strategies) such that each player cannot improve the payoff by switching to a different strategy? Such a solution is called a *saddle point* solution in the literature of game theory.

For the game in Table 8-4, player X's maximin strategy was X_2 with a minimum payoff of -1; player Y's minimax strategy was Y_2 with a maximum payoff of 1 to Y. If we assume that both players are rational and that they pursue X_2 and Y_2, then neither player can improve the payoff by changing to another strategy. Table 8-4 shows that player Y can improve by changing to Y_1 when X chooses X_2, but player X will immediately switch to X_1 where X wins \$2. Y is no slouch so we can expect X's change to X_1 to precipitate a shift by Y to Y_2, but now X will return to X_2 and both players are back at the original saddle point.

At the saddle point solution for the game in Table 8-4, player X pays player Y \$1; i.e., the payoff of (X_2, Y_2) is -1. The payoff of the saddle point solution is called the *value* of the game (to player X since the numbers were chosen arbitrarily to reflect X's payoff). A game is said to be *fair* if the value of the game is zero. The game in Table 8-4 is unfair.

When do we know that a saddle point solution has been found? Whenever the payoff from the maximin strategy is equal to the payoff from the minimax strategy, the saddle point solution has been found and consists of the maximin and minimax strategies. In fact, all two-person zero-sum games have a saddle point solution, but it may consist of *mixed strategies* rather than *pure strategies*.

We shall return to the game of odds and evens in Table 8-3 to get at the notion of mixed strategies. The maximin strategy of player X in Table 8-3 is X_1 *or* X_2 since both yield a minimum payoff of -1. The minimax strategy for player Y is either Y_1 *or* Y_2 since both yield a maximum payoff of 1. Neither player has received guidance from our rationality rules since the strategies are equally good or bad in both cases. In exasperation, player X flips a fair coin to pick a strategy: "Heads I'll throw out one finger (X_1); tails I'll throw out two (X_2)." Player X has chosen to *randomize* the strategies; i.e., X_1 will be chosen with probability 0.5 and X_2 also with a probability of 0.5. Player X has created a mixed strategy $X_3 = (0.5X_1 + 0.5X_2)$ out of the pure strategies X_1 and X_2. Now, if player Y chooses Y_1, the *expected* payoff to X from X_3 is $0.5(-1) + 0.5(1) = 0$. If Y pursues Y_2, then X receives the same expected payoff.

Player Y can (and should) also flip a coin to pick a strategy. This will create the mixed strategy $Y_3 = 0.5Y_1 + 0.5Y_2$ out of the pure strategies Y_1 and Y_2, and the expected payoff from Y_3 is zero for any of X's strategies. Thus, in terms of the expected payoffs, neither player can do better than the mixed strategies X_3 and Y_3. This is a saddle point solution with a value of the game of zero.

Two-Person $m \times n$ Games

We can generalize the choice of a saddle point solution by considering the generalized game of Table 8-5. In this game player X has m strategies, $X_1, X_2, \ldots, X_m$, while Y has n strategies, $Y_1, Y_2, \ldots, Y_n$. The payoff from the solution (X_i, Y_j) to player X is a_{ij}, where $i = 1, 2, \ldots, m$ and $j = 1, 2, \ldots, n$. If we allowed only pure strategies to be selected, player X should search each row of Table 8-5 for its minimum value. The best pure strategy is the one that maximizes these minima in each row. However, if we allowed mixed strategies, player X would proceed by attaching a probability p_i to each strategy X_i. We would like to choose values for these unknown probabilities, thereby

TABLE 8-5

A TWO-PERSON $m \times n$ GAME

		Player Y			
		Y_1	Y_2	$\cdots$	Y_n
Player X	X_1	a_{11}	a_{12}	$\cdots$	a_{1n}
	X_2	a_{21}	a_{22}	$\cdots$	a_{2n}
	$\vdots$	$\vdots$	$\vdots$		$\vdots$
	X_m	a_{m1}	a_{m2}	$\cdots$	a_{mn}

defining a mixed strategy for player X, in a manner that will maximize the minimum expected payoff to player X. This problem can be structured as a linear program.

Consider what the expected payoff to X will be if player Y chooses Y_1. The expected payoff is just the payoffs from the pure strategies $X_1, X_2, \ldots, X_m$, weighted by the probability of choosing that strategy. Looking at the first column of Table 8-5, we get

$$E\left[\sum_{i=1}^{m} p_i X_i,\ Y_1\right] = p_1 a_{11} + p_2 a_{21} + \cdots + p_m a_{m1} \tag{8-23}$$

where $E[\quad]$ denotes the expected value of the payoff from the indicated strategy pair, a mixed strategy for X and the pure strategy Y_1 for player Y.

We can generalize (8-23) to the jth strategy of player Y:

$$E\left[\sum_{i=1}^{m} p_i X_i,\ Y_j\right] = \sum_{i=1}^{m} p_i a_{ij}, \qquad j = 1, 2, \ldots, n \tag{8-24}$$

Now, without setting the values of p_i, $i = 1, 2, \ldots, m$, we do not know which column of Table 8-5 will lead to the minimum expected payoff to player X. But if we call the minimum expected payoff u, we do know that

$$\sum_{i=1}^{m} p_i a_{ij} \geq u, \qquad j = 1, 2, \ldots, n \tag{8-25}$$

That is, since u is the minimum expected payoff to X, all expected payoffs must be greater than or equal to u. The n inequalities in (8-25) represent constraints on the p_is, the decision variables in our linear program. Other constraints are that the probabilities must be nonnegative and sum to one:

$$p_i \geq 0, \qquad i = 1, 2, \ldots m \tag{8-26}$$

$$\sum_{i=1}^{m} p_i = 1 \tag{8-27}$$

Finally, by letting u be a decision variable, player X's objective function is to

$$\text{maximize} \quad u \tag{8-28}$$

i.e., maximize the minimum expected payoff.

The optimal solution to the linear program in (8-25)–(8-28) defines an optimal strategy for player X that is pure if $p_i = 1$ for some i and mixed if this is not the case. Furthermore, this optimal solution is also a saddle point solution and the value of the game is the value of u at the optimal solution of the linear program. This last result, which comes from duality theory of linear programming (see Chapter 3), shows an intimate relationship between game theory and linear programming.

To further our insight into this last result we shall formulate player Y's minimax problem as a linear program. We shall define q_j as the probability that player Y will choose strategy Y_j and call v the maximum expected payoff (to player X) for the choice by player X of any pure strategy $X_1, X_2, \ldots, X_m$. Then the following constraints must hold:

$$\sum_{j=1}^{n} a_{ij} q_j \leq v, \qquad i = 1, 2, \ldots, m \tag{8-29}$$

Notice the direction of the inequality in (8-29).

Furthermore, the probabilities must be nonnegative and sum to one:

$$q_j \geq 0, \qquad j = 1, 2, \ldots, n \tag{8-30}$$

$$\sum_{j=1}^{n} q_j = 1 \tag{8-31}$$

The objective function of player Y's problem is to minimize the maximum payoff to X. Thus, the objective function is

$$\text{minimize} \quad v \tag{8-32}$$

The minimization of (8-32) subject to (8-29)–(8-31) defines player Y's problem.

The proof that the optimal solution to X's problem, (8-25)–(8-28), is a saddle point solution comes from duality theory [see Section 3.9 and Hillier and Lieberman (1967, Chapter 15)]. First, Y's problem, (8-29)–(8-32), is the dual of X's problem. That is, if you associate q_j, a dual variable, with each constraint in (8-25), and another dual variable v with (8-27), then the formulation of the dual to problem X will lead to (8-29)–(8-32). Keep in mind, when constructing the dual, that u is a decision variable.

Second, since the two problems are their respective duals, the optimal value of X's objective function must equal the optimal value of Y's objective function; i.e., $u^* = v^*$, where the asterisk indicates optimality. Since for a game this equality of solutions occurs only at a saddle point, the optimal solution to either linear program yields a saddle point solution to the game.

8.3.4 Logrolling Models: Multiple-Person and Cooperative Games

Most real-world "games" neither are zero sum nor include only two people; the typical situation is characterized by several (more than two) players and by cooperation (nonzero sum). A rich and extensive literature on more complex game theoretic contexts has evolved. Rapoport (1970) presents a comprehensive treatment of n-person game theory. Cooperative games are discussed in many places; Burger (1963) presents a mathematical treatment of the topic. Applications of cooperative games to public sector problems include Rogers' (1969) analysis of an international river basin development problem and the identification of cost allocation plans for regional wastewater treatment programs by Giglio and Wrightington (1972).

In the remainder of this section we shall consider log-rolling models, a voting context analog of n-person cooperative games. During this discussion the reader should keep in mind the multiobjective multiple-decision-maker (MOMDM) problem of (8-19) and (8-20). It is this problem that we are attempting to solve. Relative to this problem, Harsanyi (1965) mentions two specific problems faced by a player in a game: the "efficiency problem" and the "bargaining problem."

The efficiency problem is the identification of all Pareto-optimal noninferior alternatives. The bargaining problem is the selection of one of the Pareto-optimal alternatives. These two problems represent the two levels of optimization for the MOMDM problem in (8-19) and (8-20).

While n-person cooperative games include both levels of optimization inherent in the MOMDM problem, the efficiency problem is frequently assumed away. That is, game theory usually takes a finite set of pregenerated alternatives as a point of departure. We know from Chapter 6, however, that the generation of all noninferior alternatives is not a trivial exercise. Furthermore, if a few alternatives are to be selected from the (generally) infinite number of noninferior alternatives, value judgments must be made. The emphasis of multiple-decision-maker methods has been the bargaining problem, but the efficiency problem is an important aspect that should not be forgotten.

Another simplification usually made in game theory is to assume that a single measure of value can be ascribed to each decision maker; i.e., the multiobjective part of the MOMDM is assumed away. Haith's (1971) method, discussed below, maintains the multiobjective character of the problem in a log-rolling context. Zeleny (1975) formulated the multiobjective game [see also Blackwell (1956)] but further analytical work has not been pursued.

Haefele (1970), among others, has presented a model for the analysis of the bargaining problem. His general problem is the computation of vote-trading probabilities for the participants in a group decision problem in which majority rule prevails. The key to Haefele's model, and to those presented below, is the concept of logrolling, i.e., trading votes on one issue to secure favorable votes on another. Haefele considers possible coalitions among voters, based on Riker's (1962, pp. 32–33) hypothesis that "participants create coalitions just as large as they believe will ensure winning and no larger." The possible winning coalitions and vote-trading preferences of the decision makers are enumerated to compute vote-trading probabilities that are then used to predict probable outcomes. Haefele's model is limited to problems with three issues and five decision makers.

Haith (1971) extended the various bargaining models such as Haefele's to multiobjective situations. Haith also begins with a given set of alternatives (thereby avoiding the efficiency problem), but he does discuss how decision makers' preferences can be used to generate the set (Haith, 1971, pp. 125–133). The key to Haith's model is the estimation of "preference probabilities" in the case of a single decision maker or "choice probabilities" in the case of many decision makers. These probabilities, which are motivated by the uncertainty of individuals' preferences (or by the uncertainty in quantifying those preferences), are defined as the probability that a decision maker will prefer or choose a given alternative. Thus, for each alternative i there is a corresponding probability p_i^j that each decision maker j will choose alternative i. For large decision units the problem becomes unmanageable, so Haith redefines p_i to be the fraction of the decision unit that prefers or will choose the ith alternative.

The choice probabilities of each decision maker or group are then used to compute a probability for each alternative of it being chosen by the decision unit. Implicit in the probability estimates for each decision maker are the effects of "implicit logrolling," in which the decision makers trade off the objectives of the various interest groups in their constituencies, and "explicit logrolling," the sacrifice of one objective to gain more of another, which occurs among the decision makers themselves. Bargaining is not explicitly modeled by Haith; it is instead captured implicitly by the probabilities. This appears to be a weakness of the model since it puts great emphasis on the estimation of preferences.

Russell *et al.* (1972) present another vote-trading model. Decision makers' preferences are quantified by minimum acceptable levels of the objectives and by priorities on the objectives. The goal of the model is then to vary the minimum acceptable levels so as to maximize the support for a given alternative. A decision maker is assumed to vote "yes" only if an alternative meets the minimum acceptable levels for that decision maker's two highest priority objectives. Loucks (1975) points out that one weakness of the model is the constant value of the priorities during the analysis. He asserts, and Russell *et al.* agree, that one should expect that priorities may change during a real bargaining problem.

An evaluation of the vote-trading algorithms is difficult because they are still largely untested. For real-world problems, one would expect that the methods would be computationally burdensome. Extensive prior generation and analysis of the noninferior set is required before the methods can be applied. The vote-trading models are thus restricted by the number of objectives and decision makers that can be realistically considered. Haith's approach to large decision groups avoids the computational problem, but fails to capture explicitly the logrolling aspects of the bargaining situation.

The assessment of preferences is another obstacle for the vote-trading methods. Haith considers this aspect of the problem in great detail. He suggests an approach that is an extension of a method proposed by Raiffa (1968). The decision maker's preferences are estimated by presenting a sequence of hypothetical situations that force the tradeoffs among the objectives to be considered. The procedure becomes increasingly burdensome for analyst and decision maker alike as the number of objectives and decision makers increases.

CHAPTER 9

Multiobjective Analysis of Water Resource Problems

9.1 INTRODUCTION

Many of the methodological developments and applications of multiobjective programming and planning have been accomplished by water resource systems analysts. This has happened because water resource problems are inherently multiobjective, which has been officially recognized by the United States government, and because water resource planning has traditionally been the proving ground for new methodologies. It is the purpose of this chapter to discuss applications of multiobjective analysis to water resource problems. In so doing, new practical issues of multiobjective planning and programming will be identified.

After water resource problems are briefly characterized, applications of multiobjective analysis to water resources are reviewed and a detailed account of a river basin planning problem in Argentina is presented. The summary of the chapter includes a discussion of some of the lingering issues of multiobjective water resource planning.

9.2 CHARACTERISTICS OF WATER RESOURCE PROBLEMS

The planning of water resource systems must be responsive to the physical water system itself, to the economic system that generates the demands for water, and to the political system that makes planning decisions. All three of these systems are terribly complex so modeling is frequently a difficult task. The physical system is characterized by hydrologic, biological, and chemical complexities while the economic and political systems introduce those complexities that always seem to result when humans are involved. A great deal of effort has been expended on the development of optimization and simulation models for water resource systems that are simultaneously computationally feasible and realistic.

Water resource problems are generally of two types: river basin planning problems, which are related primarily to quantity (too much, not enough, or both at different times) and water quality problems. Although quantity and quality problems are closely related, planning exercises tend to concentrate on one or the other, but not both. These two types of problem are discussed further below.

9.2.1 River Basin Planning

River basin planning is directed at the development of a water body to allow the beneficial use of its water. The primary water uses are municipal water supply, industrial water supply (including cooling), hydroelectric energy production, recreation, commercial fishing, flood control, irrigation, and navigation. The structural alternatives for meeting these demands include dams for storing water, hydroelectric power plants, municipal and industrial water treatment and distribution systems, irrigation distribution and drainage systems, recreational facilities, locks and channels for navigation, and water conveyance channels for transfers not directly related to water uses (e.g., for interbasin transfers). Nonstructural alternatives include various regulatory and management procedures such as restrictions on location in floodplains and peak-load water pricing to alter demand patterns. The emphasis in river basin planning has been on structural alternatives, in part because of the traditional concentration of engineers on constructing facilities.

The typical questions addressed in a river basin planning study are Which structures should be built and to what size? How should the system be operated? When should the various elements of the system be implemented? All of these questions are interrelated, as are the elements of the system.

It is worth considering a small part of the river basin planning problem in order to understand the complexities that are frequently encountered. An important component of many river basin plans is one or more multipurpose reservoirs, i.e., an impoundment of water that is managed so as to satisfy several uses. Unfortunately, most water uses conflict. An extreme but frequently encountered case of competitive water uses is flood control and municipal water supply. The two uses dictate opposite operating policies: One should keep the reservoir as empty as possible for flood control purposes and as full as possible to augment water supplies when natural flows are low. The problem is made complex by the inherent uncertainty of the streamflow: One can never know what the natural flow will be.

Even apparently complementary water uses may conflict. Hydroelectric energy production and irrigation are both conservative uses in that water stored in a reservoir during wet years for use during dry years would benefit both uses. But during a given year, hydropower and irrigation dictate very different temporal patterns for reservoir releases. It is interesting to note that one of the earliest applications of multiobjective analysis was by Thomas and ReVelle (1966), who explored the tradeoffs between hydropower and irrigation for the operation of the High Aswan Dam.

The number of water uses, their inherently competitive nature, the uncertainty of streamflows, and the size of many river basins are manifestations of the physical complexity of the problem. The economic and political nature of river basin planning presents another and perhaps more intricate level of complexity. The use of a river's water almost always has an effect that transcends the local impact of that water use: Upstream water use may alter, reduce, or preclude downstream uses. This fundamental fact of life is at the heart of the multiobjective nature of river basin planning and is a major complicating factor in economic analysis and political decision making for water resources.

The traditional economic approach for the analysis of water resource projects derived from the Flood Control Act of 1936, which said in part that the benefits "to whomsoever they accrue" should exceed the costs. The Act institutionalized benefit/cost analysis as a tool for public investment decision making by requiring that only those projects with the highest benefit/cost ratio [actually the highest *net* benefits are the theoretically correct criterion—see Major (1977, Chapter 2)] be implemented. The benefit/cost ratio is measured by dividing the present value of economic efficiency (or "national income") benefits by the present value of capital and operating costs. The benefits are measured by the willingness of individuals to pay (the area under the demand curve) for project outputs, e.g., water supply or recreation. A great deal of theoretical and practical research has gone into economic efficiency benefit computation [see Prest and Turvey (1965) for a general review and Howe (1971) for a discussion specific to water resources], but there are at least two considerations that make it difficult to put all project impacts into a single monetary dimension as demanded by the benefit/cost criterion.

First, there are questions of distribution of project impacts—the classical upstream–downstream conflict—that the economic efficiency objective cannot address. Most rivers that are attractive for development flow through many political jurisdictions and many regions, some developed and some depressed. Since water is an important resource for initiating and sustaining economic development, river basin plans must be responsive to the differential regional impacts of water resource development. A pure efficiency criterion tends to favor further development in developed regions since infrastructure costs can frequently be avoided. Plans that favor developed regions may not be consistent with federal views of desirable strategies for a nation's growth, and they will certainly be contrary to the developing region's perception of what is best. A multiobjective analysis that trades off efficiency against distribution is necessary for well-informed river basin decision making in such cases. The case study in Section 9.4 demonstrates how this can be done.

Second, the development of river basins, particularly in a developed country, can create environmental impacts that are considered by many to be undesirable. The construction of a dam stills a freely flowing river and may inundate a significant amount of valuable or potentially valuable land. Some of these effects defy monetary quantification. The value of a flowing river or of inundated land may be purely aesthetic in nature, yet for economic efficiency, benefit/cost analysis demands that these aesthetic values be quantified in monetary terms. While problems of this sort are generally difficult to analyze, multiobjective analysis represents a significant improvement over a single-dimensional benefit/cost analysis by allowing environmental impacts to be quantified in natural nonmonetary units. A few examples will illustrate this point.

Major (1974) applied multiobjective analysis to the proposed Big Walnut Reservoir in Indiana. The pool created by the dam would have encroached on a unique ecological area. Major generated the range of tradeoffs between net economic efficiency benefits (from water supply, recreation, and flood control) and environmental quality measured as the acres of the unique area inundated by the reservoir pool. The analysis was instrumental in altering the original design by the U.S. Army Corps of Engineers.

In many river basin design situations, environmental groups object to reservoirs because a freely flowing river will be stilled. The underlying concern is aesthetic in nature and in part based on the belief that any ecological disturbance should be avoided. Cohon *et al.* (1977) handled this kind of general environmental concern with a surrogate objective, which was to minimize total reservoir capacity. The NISE method (Section 6.3) was used to find the tradeoffs between this environmental quality objective and net economic efficiency benefits.

Environmental quality is generally multidimensional so a single environmental objective may be difficult to identify without introducing controversial value judgments into the definition of objectives. Miller and Byers (1973) studied the proposed development of the West Boggs Creek watershed in Indiana. They identified 11 different environmental quality parameters related to the impact of sediments carried by natural runoff and potentially trapped in a series of proposed reservoirs. The authors took a multiobjective approach to avoid the monetary quantification of the impacts of the anticipated sediment load. Net economic efficiency benefits were traded off against an aggregate environmental quality index that attached equal weights to each of the 11 indices. The aggregate index was used to avoid the computational and display complexities associated with the disaggregated 12-objective problem. Each of the 11 quality indices were weighted equally "since there is little guide for measuring the relative social importance of each component of the environmental quality objective" (Miller and Byers, 1973, p. 17). The authors point out that any weighting system could be used in forming the aggregate indicator, but it should be clear that the choice of weights may be very important. Computational convenience was accomplished in this case by making a possibly strong value judgment.

River basin planning is complex for many reasons, one of which is its multiobjective nature. We shall return to multiobjective river basin planning in the case study of Section 9.4.

9.2.2 Water Quality Planning

Unlike the multiplicity of uses characteristic of river basin planning, water quality management concentrates on one water use: the capacity of water bodies to assimilate water-borne municipal, industrial, and agricultural wastes. Other water uses enter into the problem because poor water quality will make some uses more expensive, e.g., municipal and industrial water supply, and may even preclude others, e.g., some recreational activities and fishing. However, our major concern, and the motivation behind U.S. policy, seems to be the ecological threat of overloading water bodies with waste. There is an overriding national concern to keep our waters as pure as possible, ignoring their assimilative capacity.

There are both structural alternatives and management tools that may be used in abating water pollution. A range of general and pollutant-specific waste treatment facilities and processes exist. The most important ones are mechanical, chemical, and biological processes for the removal of organic oxygen-demanding wastes. Nonstructural alternatives include effluent standards, effluent charges, and other economic incentives. The implementation of nonstructural alternatives promotes the use of structural alternatives and encourages changes in waste-producing processes by the individual polluter. Most water quality plans attempt to blend together structural and nonstructural alternatives.

The two basic issues addressed in a water quality planning exercise are desirable quality levels (and therefore the allowable waste discharge) and the level of treatment each discharger should pursue. The first issue—required quality levels—is usually assumed to have been answered *a priori* by legislation, regulation, or current policy. The typical planning exercise addresses the waste load allocation problem only.

Water quality planning objectives are similar to those discussed in the previous section. Efficiency, represented as the minimization of treatment costs, is important as is distribution since upstream–downstream conflicts are still present. The benefits of a plan, which are usually part of an economic efficiency objective, and environmental quality would be obviated by the predetermination of quality levels.

The first large-scale water quality systems analysis in the United States was performed for the Delaware River (Thomann, 1963; U.S. Water Pollution Control Administration, 1966). An early formulation yielded a treatment allocation that minimized total costs, but the plan was politically infeasible due to distributional considerations. The formulation was modified (U.S. Water Pollution Control Administration, 1966; Smith and Morris, 1969) to impose more acceptable relationships among the dischargers' treatment levels. Two formulations were developed: the uniform treatment model (i.e., all dischargers must treat at the same level) and the zoned uniform treatment model (i.e., all dischargers in a certain location with similar production processes or of a certain size must treat at the same level). These two formulations captured a distributional objective, but the range of choice and the richness of the tradeoffs between efficiency and distribution were not generated.

Brill (1972) and Brill *et al.* (1976) reconsidered the Delaware case and proposed metrics for a distribution objective that allowed the explicit consideration of efficiency–equity tradeoffs. Three different metrics for equity were considered: the minimization of deviations from the average treatment level, the minimization of the range of treatment levels, and the minimization of the maximum treatment level. The noninferior sets defined over economic efficiency and equity, using the three alternative metrics, were generated. The analysis was also performed for an effluent charge program.

Dorfman and Jacoby (1970), Monarchi *et al.* (1973), and Haimes *et al.* (1975) considered a hypothetical multiobjective water quality planning situation. The former study used Paretian analysis (see Section 8.3.2) to take into account the many interest groups that influence waste load allocations: local, state, and federal officials, industrial representatives, and environmental groups. Each objective in this hypothetical study measured the monetary impacts of a plan on an interest group. Weights that reflected each group's political power were attached to the objectives in order to predict political outcomes.

Multiobjective analysis can also be useful when quality levels are not set prior to the planning exercise. Models that are predicated on a water quality stream standard, e.g., Thomann (1963) and ReVelle *et al.* (1967, 1968), can be used to generate alternatives that show the tradeoffs among efficiency, distribution, and water quality. An important difficulty, however, is the large number of quality indicators that may be important. The analyst must then confront the tradeoff between computational and display complexity and the prospect of making a controversial value judgment as did Miller and Byers (1973), cited in the previous section.

9.3 THE INSTITUTIONALIZATION OF MULTIOBJECTIVE WATER RESOURCE PLANNING

Water resource planning in the United States, to this author's knowledge, is the only governmental activity in the world that is formally and legally required to be multiobjective. In the early 1960s during the administration of President Kennedy, a cabinet-level body, the Water Resources Council (WRC), was formed to reconsider the methods and procedures of water quantity planning in the United States. After a decade of study and analysis the WRC (U.S. Water Resources Council, 1973) promulgated formal procedures for multiobjective river basin planning.

The so-called "Principles and Standards" developed by the WRC require, at a minimum, that federal river basin planning agencies, i.e., the Army Corps of Engineers, the Bureau of Reclamation in the Department of Interior, and the Soil Conservation Service in the Department of Agriculture analyze two objectives in detail: one that maximizes net economic efficiency benefits and one that is responsive to environmental quality. The impacts of all alternatives on these two objectives and on regional income and the "social well-being" objective (a catchall for those impacts that cannot be put into the other three accounts), where appropriate, must also be measured and displayed. A full multiobjective analysis, in which the full range of choice is identified, is not required, but the WRC's regulations go well beyond traditional benefit/cost analysis. The affected federal agencies have been engaged in an evaluation of procedures for implementing the Principles and Standards [see Buie (1974), Donovan and Jordan (1974), and Porter (1974)].

Formal guidelines for multiobjective water quality planning do not exist, although the National Environmental Policy Act (NEPA) of 1970 (U.S. Code, Volume 42, Section 4321, 1970) subjects to review all federally funded projects that may disrupt natural environments. The motivation for environmental impact statements derives from NEPA. One could claim that an impact statement represents a narrow form of multiobjective planning, narrow in that many alternatives need not be considered. However, the preparation and impact of a statement often results in the identification of new alternatives with a less severe environmental impact.

9.4 A CASE STUDY OF MULTIOBJECTIVE RIVER BASIN PLANNING

In this section the multiobjective analysis of the proposed development of the Rio Colorado in Argentina is discussed. The methodology and results were developed over a 2-yr period from 1970 to 1972 in the Department of Civil Engineering at the Massachusetts Institute of Technology (MIT), under contract to the Republic of Argentina. The project is of particular interest since it represents one of the first attempts at multiobjective programming and planning for a large-scale real-world public investment problem. The reader who wants a more complete account of that project is referred to the book by Major and Lenton (1978). Multiobjective programming and planning aspects are also discussed in Cohon and Marks (1973) and in Cohon *et al.* (1973).

The background and setting for the problem is discussed first, followed by a brief explanation of the modeling methodology that was developed. One model, the screening model, is then discussed in more detail. The multiobjective nature of the problem is then pursued in some detail with a discussion of some of our early and final formulations and results.

9.4.1 The Problem Setting

The Rio Colorado flows from the Andes Mountains in the west to the Atlantic Ocean in the east through the central portion of Argentina as shown in the small map in Fig. 9-1. In the large map of Fig. 9-1 one can see that the Colorado flows through or is on the border of five provinces: Mendoza, La Pampa, Neuquen, Rio Negro, and Buenos Aires. This multiprovincial setting is an important characteristic of the planning problem.

The Colorado is a relatively small river with a mean annual flow of 120 m^3/sec (3200 ft^3/sec), but it represents an important resource nevertheless. It is the only major water resource for La Pampa and the southern tip of Buenos Aires province. In addition, the Rio Colorado is a critical factor for the existing and planned irrigation in Mendoza. The importance of the river's resources is reflected by the number of projects proposed by the provinces: There is not enough water in the river to pursue all of the proposed development. There is a provincial allocation problem here that is exacerbated by

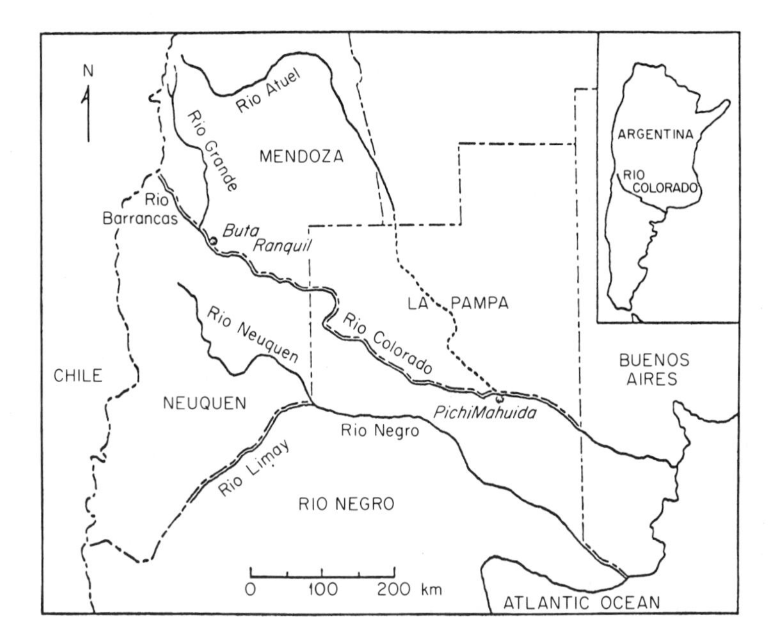

Fig. 9-1. Location map for the Rio Colorado.

current development patterns and by historical conflicts over water. It is important to understand this political and economic background.

Mendoza and Buenos Aires are well-developed provinces while the other three provinces are less developed, particularly La Pampa, which is on a large dry plain that has supported little agricultural or other economic activity. The portion of Buenos Aires province in the basin includes the largest existing irrigation zone on the river. Mendoza is a well-developed and growing province that some call the "California of Argentina." Extensive irrigation has been pursued around the Rio Atuel in Mendoza to support a major wine industry.

The Rio Atuel has played an important role in interprovincial relationships. A careful inspection of the large map in Fig. 9-1 shows that the line representing the Rio Atuel becomes dashed as it enters the province of La Pampa. The river is indicated in this manner because it is now only a river bed; there is no water in the Rio Atuel in La Pampa. There used to be water in the La Pampa reach of the river, but extensive use by Mendoza has eliminated this resource for potential downstream users. This historical fact served to aggravate the usual upstream–downstream conflict among riverine provinces and tended to polarize provincial views on the desirability of projects in the Rio Colorado.

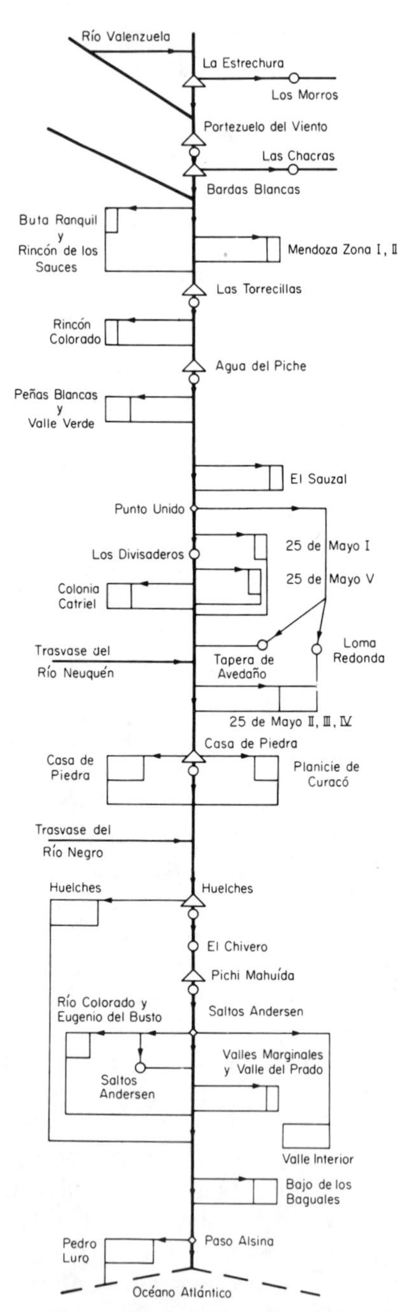

Fig. 9-2. Schematic representation of the Rio Colorado and its development alternatives: △ reservoir; ○ power plant; □ irrigation zone; → transfer.

The proposed projects, shown schematically in Fig. 9-2, included: eight reservoirs for regulating the river and providing "head" (potential energy of stored water measured as the elevation of the water surface) for energy generation; 13 hydroelectric power plants; 17 irrigation zones, which ranged in size from 3500 to 260,000 hectares [one hectare (ha) is 10,000 m^2 or about 2.5 acres]; and four interbasin transfers. The transfer alternatives were particularly controversial since two of the proposed diversions would export up to 80% of the Rio Colorado to the Rio Atuel for irrigation and hydroelectric energy production in Mendoza.

The task of the MIT group was to develop a methodology that could help Argentine decision makers to determine which projects to build, the size of those projects, when to build them, and how to operate them. The Rio Colorado exhibits all of the complexity of the general river basin planning problem. Of most importance for us is the clear multiobjective nature of the problem manifest in the efficiency–distribution tradeoff underlying the interprovincial conflict. We shall discuss the multiobjective aspects of the problem, but first the modeling methodology that was developed will be outlined and one of the models will be discussed in detail.

9.4.2 The Modeling Methodology

The number of questions that must be addressed and the uncertainty of future streamflows preclude the use of a single mathematical model for river basin planning. The MIT methodology included three models: a screening model, a simulation model, and a sequencing model. Each model addresses one or more of the "which, size, operating, and when" questions. By using them together in a sequential matter, good reliable alternatives can be generated as shown in Fig. 9-3.

The screening model is a linear multiobjective programming model (actually, the final version of the model includes some integer variables) that is used to determine which projects to build and their appropriate sizes. The model is static so timing issues are not captured, and deterministic with regard to streamflow so operating policies cannot be considered. The output from the model, which is described in more detail below, is a set of design sizes for each proposed project in the basin. This "configuration" is relatively unreliable due to the optimistic view of the world built into the model by our assumption of hydrologic determinism.

The output from the screening model is used as input to the simulation model described in detail in McBean *et al.* (1972). Using a realistic operating policy and several stochastic representations of streamflow, the input configuration was simulated over several 50-yr sequences to examine its probable behavior. When the input configuration failed to meet preset

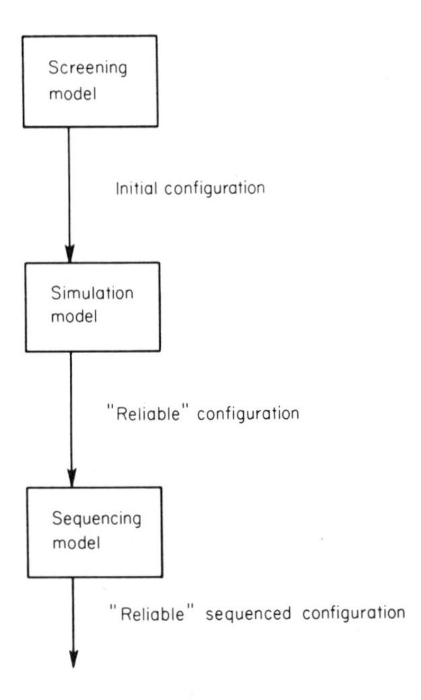

Fig. 9-3. Three models for the analysis of river basin planning problems.

performance criteria it was altered by increasing reservoir sizes or decreasing energy targets or irrigation zone sizes. The redesigned configuration was then tested and new alterations made until satisfactory performance was obtained. Thus, the output from the simulation model was a reliable configuration as shown in Fig. 9-3.

As an aside, it is interesting to note that water resource problems are so complex that the development of realistic models is itself a multiobjective problem. In developing simulation models for groundwater analysis and for linear hydrologic systems, respectively, Neuman (1973) and Neuman and de Marsily (1976) applied multiobjective programming to the estimation of the model parameters. Total estimation errors were traded off against the stability of the model.

The only remaining question—when each project should be built—was addressed by the sequencing model, a linear mixed-integer programming model that is described in Cohon *et al.* (1973). The output configuration from the simulation model was taken as an input set of projects that had to be built by the sequencing model. A single objective of maximizing the present value of new economic efficiency benefits was used. An important constraint was that the number of projects built in any 10-year period could not exceed the number of projects supportable by the expected migration of people into the basin.

The output from the entire methodology, as shown in Fig. 9-3, is a sequenced reliable configuration. One would not expect the methodological design to be as good as one designed by a single model that could address all of the relevant issues, but this suboptimization cannot be avoided when dealing with such complex problems.

The ability to get data for each of the models is obviously important. This will not be discussed in detail here [see Major and Lenton (1978)], although the presentation in the next section will give the reader a feel for the nature of the data requirements for the screening model.

9.4.3 The Screening Model

The multiobjective nature of the Rio Colorado problem was captured in the screening model, the simulation and sequencing models serving the more narrow role of incorporating uncertainty and timing into the noninferior basin plans generated in the screening model. A complete view of the multiobjective programming and planning procedures that were employed requires an understanding of the screening model. In this section the constraint set of the screening model is offered. The discussion is adapted from Marks and Cohon (1975).

Recall that the screening model includes several assumptions, although the strongest of these were relaxed by the further analysis performed with the simulation and sequencing models. The model presented in this section is basically linear; water resource systems, however, are characterized by many nonlinear relationships. To some extent these are accounted for by the use of piecewise linear approximations and some formulation "tricks." However, many of the nonlinearities cannot be included in a linear programming model. River basin systems are also characterized by a high degree of hydrologic stochasticity; i.e., streamflows are uncertain. The present model is deterministic in that mean seasonal streamflows are employed. This represents a fairly strong assumption about the hydrologic processes in the river system. Finally, the model is steady state in that only one year is explicitly modeled and every subsequent year up to the planning horizon (50 yr) is assumed to be hydrologically identical to the first. This assumption precludes the adequate consideration of two important aspects of the planning problem: optimal operating policies for the river basin (i.e., when to store water in reservoirs during wet periods for use in dry periods) and project scheduling (i.e., optimal construction times for projects).

All of the above considerations have led to the use of optimization models for the *preliminary* analysis of river basin planning problems. A model of this type is called a screening model in the water resource literature because it is used figuratively as a screen through which all of the possible combinations of planning alternatives are passed. The use of the model also provides the motivation for a multimodel approach to river basin planning.

The river basin planning problem is basically a resource allocation problem: There are scarce resources (water and usually capital) that must be allocated among water uses (hydroelectric energy production and irrigation in this case) to maximize a set of planning objectives. In addition, there are control alternatives (reservoirs) that allow the resources to be used more effectively. The model consists of an objective function that expresses the set of planning objectives in terms of decision variables representing the release of water from reservoirs, the diversion of water out of the stream for water uses, the realizable production from uses to which water is allocated, and the location and capacities of the structural components of the river system, chosen from among the set of potential projects in Fig. 9-2.

The constraint set consists of continuity constraints, which trace the flow of water through the river system, and constraints on each of the elements or uses of the system: reservoirs, irrigation, hydroelectric energy production, and interbasin imports and exports. Each of these types of constraint is discussed separately below. The objectives are discussed separately in the next section.

Continuity Constraints

Continuity constraints are included in the model to trace the flow of water through the river system by ensuring the conservation of mass at every point in the river at which water is stored, diverted, or imported. A sample site with a reservoir is shown in Fig. 9-4.

The basic continuity relationship can be written, referring to Fig. 9-4, as

$$S_{s,t+1} = S_{st} + Q_{st} + I_{st} - E_{st} - D_{st} \tag{9-1}$$

where the subscripts s and t refer to site and season, respectively. Equation (9-1) says that the storage in the reservoir at the beginning of the next season ($S_{s,t+1}$) must equal the storage at the beginning of the present season (S_{st}) plus any additions during the present season (the inflow Q_{st} and any imports I_{st}) minus any deductions during the present season (the reservoir release D_{st} and any diversions E_{st}).

In Eq. (9-1), all of the variables except Q_{st} represent decisions that are made at site s. On the other hand, the upstream flow Q_{st} depends on natural streamflow and on the decisions made immediately upstream at site $s - 1$. It is necessary to express Q_{st} as a function of these two effects. Figure 9-5 illustrates the following development.

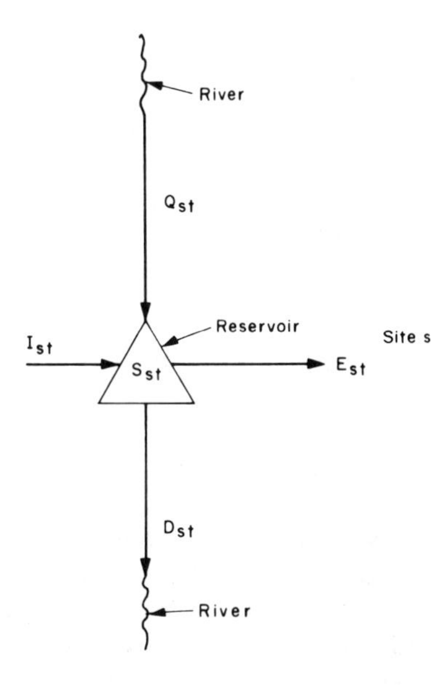

Fig. 9-4. The continuity relationship.

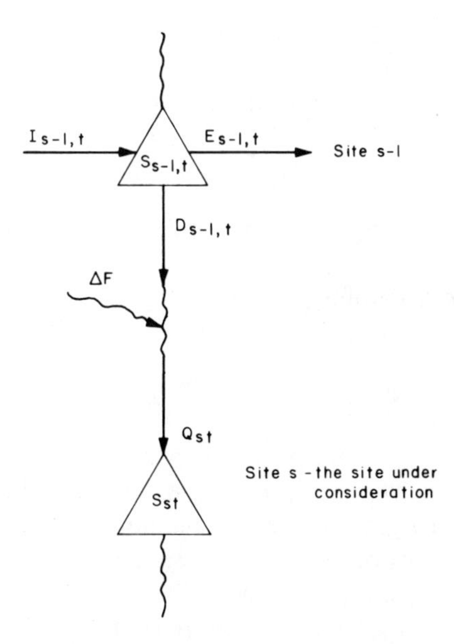

Fig. 9-5. The continuity relationship with an upstream site.

Applying the same conservation of mass principle that was invoked to derive Eq. (9-1), we get from Fig. 9-5

$$Q_{st} = D_{s-1,t} + \Delta F \tag{9-2}$$

where all variables are defined as before and ΔF represents the increment to natural streamflow between sites $s - 1$ and s. Equation (9-2) is substituted into Eq. (9-1) and after rearranging terms so that all decision variables are on the left-hand side and inputs (parameters) are on the right, we get

$$S_{s,t+1} - S_{st} + D_{st} + E_{st} - I_{st} - D_{s-1,t} = \Delta F \tag{9-3}$$

The appropriate value of ΔF to use in Eq. (9-3) represents the major hydrologic question of the model because ΔF is the only representation of natural streamflow in the constraint set.

As mentioned previously, the model presented here is deterministic in that ΔF is based on mean seasonal streamflows that are taken to occur with certainty. The value of ΔF is taken as the difference between the mean seasonal streamflows at sites $s - 1$ and s and represents the flow into the stream from the drainage area between $s - 1$ and s. ΔF is not the total streamflow at site s because $D_{s-1,t}$, which appears in Eq. (9-3), is the flow in the stream at $s - 1$ after development of the basin. Thus ΔF represents the increment (or decrement) to $D_{s-1,t}$ due to natural effects, e.g., runoff from precipitation or snowmelt, tributary flow, infiltration or exfiltration, evaporation, and any existing developmental effects. However, none of these hydrologic processes are treated in the model; rather, ΔF is simply measured as the difference between measured mean seasonal streamflows at sites $s - 1$ and s,

$$\Delta F = F_{st} - F_{s-1,t} \tag{9-4}$$

where the values of F_{st} and $F_{s-1,t}$, the mean seasonal streamflows, are obtained from streamflow records available at gauging stations on the river.

With estimates of streamflows, substituting Eq. (9-4) into Eq. (9-3) results in the stream continuity equation:

$$S_{s,t+1} - S_{st} + D_{st} + E_{st} - I_{st} - D_{s-1,t} = F_{st} - F_{s-1,t} \tag{9-5}$$

The storage terms $S_{s,t+1}$ and S_{st} are expressed in cubic hectometers per season (hm^3/season), where 1 hm^3 = one million cubic meters (m^3). All of the other terms in Eq. (9-5) are average flows expressed as cubic meters per second (m^3/sec). For dimensional consistency the storage terms must be converted to cubic meters per second. This is done by multiplying $S_{s,t+1}$ and S_{st} by (1 hm^3/season)($1/k_t$ season/sec)(10^6 m^3/hm^3) = ($10^6/k_t$)[(m^3/sec)/(hm^3/season)], where k_t is the number of seconds in season t. Multiplying this conversion factor by the storage terms in Eq. (9-5) gives the final form of the stream continuity constraint:

$$\begin{aligned}(10^6/k_t)S_{s,t+1} - (10^6/k_t)S_{st} + D_{st} + E_{st} - I_{st} - D_{s-1,t} \\ = F_{st} - F_{s-1,t} \qquad \forall s, t\end{aligned} \qquad *(9\text{-}6)\dagger$$

Note that Eq. (9-6) is intended to be a general continuity constraint that is written for all sites and seasons. When writing the constraint for a specific site, other terms may appear since I_{st} and E_{st} are surrogate quantities for any inflows or outflows (other than upstream inflow and release). In an actual application these two terms may represent interbasin imports or exports or diversions for irrigation. All of the possible forms that (9-6) might take were omitted here to preserve clarity of exposition.

Reservoir Constraints

There are two purely physical relationships for reservoirs. We require that the storage in a reservoir cannot exceed the storage capacity during any season t or at any site s:

$$S_{st} - V_s \leq 0 \qquad \forall s, t \qquad *(9\text{-}7)$$

in which V_s is the storage capacity of the reservoir in cubic hectometers at site s. Notice that by making V_s a decision variable in the model, the optimal storage capacity of the reservoir can be found.

We also need the storage–head relationship for reasons explained in a later section on constraints for hydroelectric energy production. The constraint says simply that the storage S_{st} is related to the height in meters of water behind the dam A_{st}:

$$S_{st} - \sigma_s(A_{st}) = 0 \qquad \forall s, t \qquad *(9\text{-}8)$$

where $\sigma_s(A_{st})$, a function that relates storage volumes to the water elevation in the reservoir, depends on the shape of the valley at site s. This relationship is generally nonlinear so piecewise linear approximations [see Wagner (1969)] were needed to incorporate it into the model.

Irrigation Constraints

The irrigation process is extremely complex and therefore quite difficult to model by linear programming. This complexity is primarily due to the great number of variables that affect agricultural production. Unlike hydroelectric power production, which depends only on the reservoir head and the

† Throughout this chapter, an asterisk before an equation number indicates the form of the constraint that is included in the model.

flow through the turbines, crop production depends on irrigation water volumes, temporal distribution of irrigation water volumes, water quality (e.g., salinity), solar radiation, precipitation, and a host of soil properties. Furthermore, the significance of each of these variables varies from crop to crop.

What is desired in modeling an irrigation system is a production function, i.e., a function that relates crop yield to quantities of water supplied for irrigation. The agricultural production function, which has many dimensions, one for each of the variables that affect crop yield, has yet to be derived analytically. Indeed, the most widely used approach for estimating the production function has been empirical investigations. By observing crop yields for varying water quantities, an estimate of the production function is found. The basic weakness of this approach is that the other variables that affect the growing process vary from one observation to the next. However, the empirical method is the only workable method that is currently available.

A number of simplifications were made in order to apply the empirical method to this model. These are graphically shown in Fig. 9-6. Not knowing the actual shape of the production function, one point on the function was estimated from empirical data (Fig. 9-6a). Given this point, it was assumed that the annual amount of irrigation water used per hectare τ_s was a constant, giving a constant yield per hectare. Therefore as water used for irrigation is varied the amount of land irrigated varies, not crop yield per hectare. Furthermore, this annual land-to-water relationship is linear with slope τ_s, as shown in Fig. 9-6b. The annual volume of water applied to each hectare must be properly distributed over time to reflect the variation in crop requirements over the year. For example, many crops need most of their annual requirement during the summer and considerably less water during the winter. Figure 9-6c shows an example of how the annual requirement τ_s may be distributed over a year that has been divided into three 4-month seasons. The seasonal proportions are then used to arrive at seasonal irrigation water-use coefficients τ_{st} measured in cubic meters per hectare.

The linear function relating irrigated land L_{st} in hectares to the volume in cubic hectometers of water supplied for irrigation IR_{st} is

$$IR_{st} - (\tau_{st}/10^6)L_{st} = 0 \qquad \forall s, t \qquad *(9\text{-}9)$$

In general, the choice of which crops are to be produced can be a decision variable in the model. It is assumed, however, that a cropping pattern is chosen prior to solution of the model. The choice of crops will dictate the values of τ_{st} to be used in the model.

If benefits from irrigation are to be realized, we must be sure that land irrigated during a season L_{st} also receives its water requirements during other seasons. For example, if 1000 ha are irrigated in season 1, but only

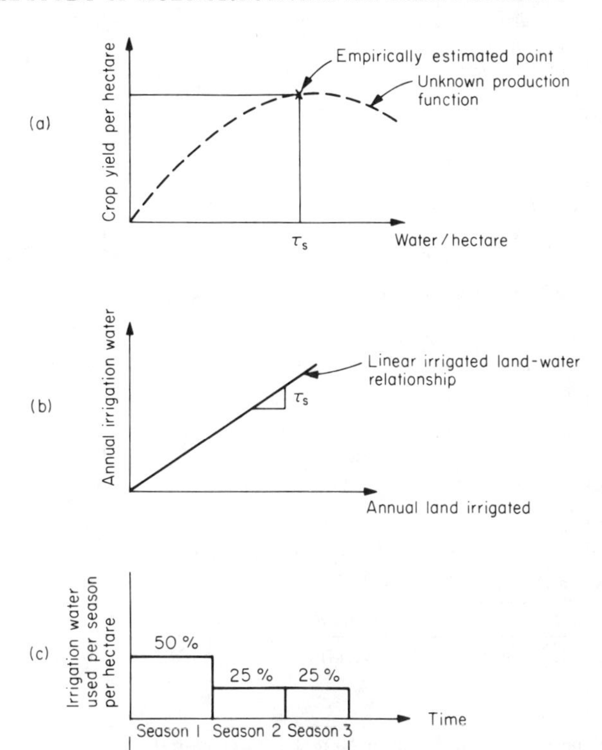

Fig. 9-6. (a) Agricultural production function. (b) Assumed land–water relationship. (c) Temporal distribution of irrigation water requirements.

500 ha receive water during season 2, we cannot expect to reap benefits from the entire 1000 ha. To discourage this type of result in the model, benefits are related to land that receives its irrigation water requirement in every season. This is done by computing the minimum seasonal irrigated land area L_{sm} from the constraints

$$L_{sm} - L_{st} \leq 0 \qquad \forall s, t \qquad *(9\text{-}10)$$

where L_{st} is the land area irrigated during season t. Benefits from agricultural production are related to L_{sm} in the objective function.

The remaining irrigation constraints are derived from a consideration of Fig. 9-7. One constraint relates the volume in cubic hectometers of water that reaches the irrigation site IR_{st} to the flow in cubic meters per second diverted out of the stream E_{st}:

$$IR_{st} - (k_t/10^6)(1 - \varepsilon_{st})E_{st} = 0 \qquad \forall s, t \qquad *(9\text{-}11)$$

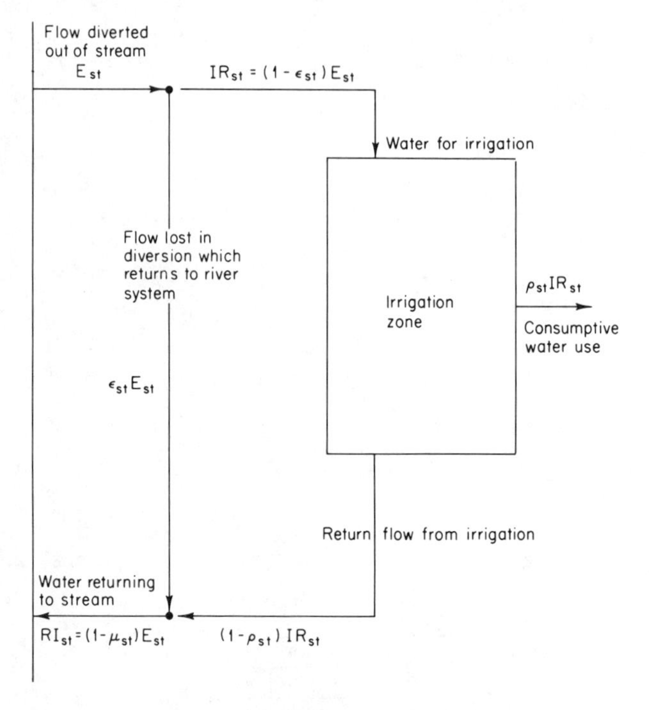

Fig. 9-7. Water losses from irrigation. Note: $RI_{st} = (1 - \mu_{st})E_{st} = \varepsilon_{st}E_{st} + (1 - \rho_{st})(1 - \varepsilon_{st}) E_{st}$; therefore $1 - \mu_{st} = \varepsilon_{st} + 1 - \varepsilon_{st} - \rho_{st} + \rho_{st}\varepsilon_{st}$ and $\mu_{st} = \rho_{st} - \rho_{st}\varepsilon_{st} = \rho_{st}(1 - \varepsilon_{st})$.

where k_t is the number of seconds in season t and converts E_{st} from cubic meters per second to cubic meters, and ε_{st} is a coefficient that represents the water lost in transport from the stream to the irrigation site, which is assumed to return to the stream.

Another constraint relates the flow diverted for irrigation E_{st} to the flow in cubic meters per second that returns to the stream from the irrigation site RI_{st}:

$$RI_{st} - (1 - \mu_{st})E_{st} = 0 \qquad \forall s, t \tag{*9-12}$$

where μ_{st} is the total loss coefficient for irrigation; it represents a combination of the losses due to transport (ε_{st}) and consumptive use requirements (ρ_{st}). As shown in Fig. 9-7, $\mu_{st} = \rho_{st}(1 - \varepsilon_{st})$ when all transport losses are assumed to return to the stream.

Hydroelectric Energy Constraints

The production of hydroelectric energy is a relatively well-defined technical process. There are only three decision variables that affect energy production: the flow through the turbines of the power plant, the head (i.e., potential energy) associated with this flow, and the capacity of the power plant. The relationships of these variables to energy production are the origins of the energy constraints.

The first constraint is the production function for hydroelectric energy,

$$P_{st} - (2.61 \times 10^{-6})ek_t(D_{st}A_{st}) \leq 0 \tag{9-13}$$

where P_{st} is the energy in megawatt-hours (MWh) produced at site s during season t, D_{st} is the average release in cubic meters per second from reservoir s during season t, A_{st} is the head in meters of water in reservoir s at the beginning of season t, e is the power plant efficiency, k_t is the number of seconds in season t, and (2.61×10^{-6}) is a unit conversion factor. Note that the head A_{st} is related to the storage at site s at the beginning of season t, S_{st}, by the storage–head curve previously written as (9-8).

The expression in (9-13) is nonlinear and nonseparable because D_{st} and A_{st}, both decision variables, are multiplied together. The constraint may be made linear by writing it as two constraints, one with an assumed value for the release $\hat{D}_{st}$, and the other with an assumed value for the head $\hat{A}_{st}$. The assumed values of $\hat{D}_{st}$ and $\hat{A}_{st}$ were computed from knowledge of the sites.

$$P_{st} - (2.61 \times 10^{-6})ek_t\hat{D}_{st}(A_{st}) \leq 0 \qquad \forall s, t \tag*{*(9-14)}$$

$$P_{st} - (2.61 \times 10^{-6})ek_t\hat{A}_{st}(D_{st}) \leq 0 \qquad \forall s, t \tag*{*(9-15)}$$

After solution of the model, the assumed values $\hat{D}_{st}$ and $\hat{A}_{st}$ were compared to the computed values. If satisfactory agreement was not found, new assumed values were used and a new solution obtained. This iterative approach was found to converge in all cases in at most two runs. Other linearization techniques, which relied on one constraint rather than two, with either an assumed head or an assumed release, gave physically unrealistic solutions in many cases.

The only other variable to be accounted for is the power plant capacity. The capacity represents an obvious upper bound on energy production,

$$P_{st} - h_tH_s \leq 0 \tag{9-16}$$

where h_t is the number of hours in season t and H_s is the capacity of the power plant in megawatts (MW). Equation (9-16) will be binding only if the plant produces at capacity all of the time. This would be unrealistic and undesirable. Therefore a load factor Y_{st}, which is defined as the ratio of the average daily production to the daily peak production, is introduced into (9-16) to represent the daily variation in production. However, since P_{st} is the *seasonal* energy production, it must be assumed that production does not vary appreciably from day to day. Equation (9-16) becomes

$$P_{st} - Y_{st}h_tH_s \leq 0 \qquad \forall s, t \tag*{*(9-17)}$$

in which Y_{st}, an input parameter, can be between 0 and 1 and is found from assumptions based on loading histories of similar installations and the generating function (base or peak load) that the basin's plants will serve in the national transmission grid.

Interbasin Imports and Exports

Interbasin transfers are modeled simply as diversions of water into or out of the stream. It is required, additionally, that the seasonal transfer does not exceed the channel capacity. For imports, this is written

$$I_{st} - IM_s \leq 0 \qquad \forall s, t \qquad *(9\text{-}18)$$

where I_{st} is the average import at site s during season t and IM_s is the capacity of the import canal at site s, both in cubic meters per second. Similarly, for exports

$$X_{st} - XM_s \leq 0 \qquad \forall s, t \qquad *(9\text{-}19)$$

where X_{st} is the average export from site s during season t and XM_s is the capacity of the export canal at site s, both in cubic meters per second.

9.4.4 Initial Multiobjective Approach

Mathematical Formulation

The identification and quantification of objectives are important steps of a planning exercise that have a profound effect on the nature and usefulness of the results. The Rio Colorado study dealt with a complex decision-making problem that was characterized by many decision makers and conflicting interests.

Decision-making authority for the allocation of the water resources of the Rio Colorado rested with a committee of provincial representatives, one from each of the five riverine provinces. Unanimity was required for a decision. In addition, the federal government could also exercise power through its control of the funding apparatus.

The assumption that each provincial representative would seek to secure an allocation most favorable to his province and the federal government's role in the process led us to an initial set of two objectives. First, the maximization of net discounted economic efficiency benefits was defined to represent the nation's interest in the Rio Colorado; i.e., the water should be allocated so as to maximize the addition to national income. Economic efficiency benefits were associated with irrigation output as a function of L_{sm}, hydroelectric energy production as a function of P_{st}, and exports to Mendoza as a function of X_{st}. Capital, operation, and maintenance costs were associated with reservoir capacity V_s, power plant capacity H_s, distribution systems and infrastructure costs for irrigation as a function of L_{sm}, and import and export channel capcities IM_s and XM_s, respectively.

The mathematical form of the economic efficiency or national income objective is

$$\text{maximize} \quad Z = \sum_s \left[\sum_t (\beta_{st}^{\mathrm{p}} P_{st}) + \beta_s^{\mathrm{i}} L_{sm} + \sum_t (v_{st} X_{st}) \right] - \sum_s [\alpha_s V_s + \delta_s H_s + \phi_s L_{sm} + \gamma_s^{\mathrm{X}} X_{sm} + \gamma_s^{\mathrm{I}} I_{sm}] \tag{9-20}$$

β_s^{p} and β_s^{i} are unit benefits for power and irrigation, respectively; α_s, δ_s, and ϕ_s are unit costs associated with reservoirs, power stations, and irrigation sites, respectively; and, γ_s^{X} and γ_s^{I} are unit costs of exports and imports, respectively. The coefficient v_{st} includes the benefits generated from irrigation and energy production with Rio Colorado water exported to Mendoza. All benefits and costs were actually time streams of future money flows so the coefficients in (9-20) represent present values for these time streams obtained by applying a discount rate of 8%. It should also be noted that while the benefit and cost functions are represented in (9-20) as linear, they are in fact nonlinear. In particular, reservoir and power plant cost functions are concave, exhibiting economies of scale in their construction. This was captured by using piecewise linear approximations and 0–1 integer variables in later formulations [see Cohon *et al.* (1973, Chapter 3)].

The second objective represented an attempt to capture the concern over how water would be distributed among the provinces. Our early results showed that the maximization of net economic efficiency benefits alone would lead to allocations that heavily favored Mendoza and Buenos Aires since their relatively developed status allowed higher net benefits to be generated. Such an allocation was assumed to be unfair to the other three less-developed provinces. Accordingly, a "regional water allocation" objective was established to encourage a more equal distribution of water among the provinces. The objective was to minimize absolute deviations from an equal water allocation: The closer to equality the better.

The mathematical form of the objective employed average water use since if all provinces receive the average provincial use, then an equal allocation is obtained. We also aggregated Neuquen and Rio Negro into one region since the former has few proposed projects of its own. We shall define W_i as the water withdrawn for irrigation in or exported to region i, where $i = 1$ for

Mendoza, 2 for Neuquen–Rio Negro, 3 for La Pampa, and 4 for Buenos Aires. The regional allocation objective is

$$\text{minimize} \quad D = \sum_{i=1}^{4} |W_i - \overline{W}| \tag{9-21}$$

in which D is the total deviation, W_i is the water used by region i, and $\overline{W}$ is the average regional water use, all in cubic meters per second.

It should be reiterated before proceeding with the mathematical development that the choice was to define the objective relative to an equal allocation of water among the four regions. Alternatively, an equitable allocation with unequal regional distributions could have been used. In this case a weighted average would be used in place of the mathematical mean $\overline{W}$. No additional changes in the formulation would be required. Of course, the choice of which allocation to use as a base from which to measure deviations is a controversial value judgment in itself. We are facing here a dilemma similar to the Miller and Byers problem of Section 9.2.1: We have chosen to define a single distributional objective rather than four (or five) regional (or provincial) water-use objectives. A computational and display advantage has been achieved, but only by making a potentially strong judgment as to the definition of equity. A striking aspect of this is that the same set of noninferior alternatives will be generated regardless of which distribution, equal or otherwise, we use as a base in (9-21). Nevertheless, the choice of a base distribution may be, and was in this case, important for the acceptance of the results. The base distribution is immediately labeled as best (since we are minimizing deviations from it), which is an attitude that some provinces would obviously want to discourage.

It should also be pointed out that regional income benefits could be used in place of water withdrawals in (9-21). Net regional income benefits are perhaps a more appropriate measure of the gain that each province realizes from an allocation. Water use was used instead because it was felt that withdrawal was a more meaningful measure for the provincial representatives on the committee.

Proceeding with the mathematical development, we see that the expression in (9-21) is an absolute value that is nonlinear and cannot be included directly in the linear programming model. The transformation that will enable inclusion of (9-21) in the model is

$$\text{minimize} \quad D = \sum_{i=1}^{4} (G_i + T_i) \tag{9-22}$$

subject to the additional constraints

$$W_i - \overline{W} = G_i - T_i, \qquad i = 1, \ldots, 4 \tag{9-23}$$

$$G_i, T_i, W_i, \overline{W} \geq 0, \qquad i = 1, \ldots, 4 \tag{9-24}$$

in which G_i and T_i are the positive and negative deviation of W_i from $\overline{W}$, respectively, and only G_i or T_i, not both, can be nonzero for each of the constraints (9-23). This can be seen from the form of the constraints and the objective function: For a given deviation $G_i - T_i$, the sum $G_i + T_i$ is minimized when G_i or T_i equals 0. Putting (9-23) in the standard form of variables on the left gives

$$W_i - \overline{W} - G_i + T_i = 0, \qquad i = 1, \ldots, 4 \tag{9-25}$$

Two more constraints are required before the formulation is complete. First, the average regional water withdrawal must be related to the individual regional withdrawals:

$$\overline{W} - \frac{1}{4} \sum_{i=1}^{4} W_i = 0 \tag{9-26}$$

Second, each regional withdrawal must be defined in terms of the diversion and export variables used previously in Section 9.4.3. We get

$$W_i - \sum_{s \in R_i} \sum_{t} (E_{st} + X_{st}) = 0, \qquad i = 1, \ldots, 4 \tag{9-27}$$

where R_i is the set of sites in region i and E_{st} and X_{st} are irrigation diversion and export, respectively, defined in the previous section. Note that withdrawals, not consumptive uses, are used in (9-27) and that diversions for energy production and reservoir releases do not enter into the computation of regional water use.

The entire two-objective formulation is summarized in Table 9-1. For convenience we shall call the feasible region in decision space defined by these constraints $\mathbf{F}_d$, and all of the decision variables will be referred to as $\mathbf{x}$. Thus, $\mathbf{x} \in \mathbf{F}_d$ means that the constraints are satisfied by the solution $\mathbf{x}$.

Some of the constraints in the formulation can be eliminated by using the appropriate definitional constraint to substitute for one of the decision variables. For example, W_i, $\overline{W}$, and the constraints (9-26) and (9-27) can be removed by using (9-26) and (9-27) to substitute for W_i and $\overline{W}$ in (9-23). Many of the irrigation constraints can also be removed in this manner. A reduction of the number of constraints saves computational costs, but it also requires that some quantities of interest, such as W_i and $\overline{W}$, be calculated by hand after solving the linear program.

Solution Procedure and Results

The initial multiobjective formulation was applied to a set of potential projects that was smaller than the full configuration shown in Fig. 9-2. The configuration used in the earlier formulation is shown in Fig. 9-8. There are potential diversions for water use in each region: exports at sites 1 and 3 in

TABLE 9-1

SUMMARY OF THE RIO COLORADO SCREENING MODEL WITH THE PRELIMINARY MULTIOBJECTIVE FORMULATION

Maximize [net economic efficiency benefits (9-20)]
Minimize [total deviations from an equal water allocation (9-22)]

Subject to
- continuity constraints (9-6)
- reservoir capacity constraints (9-7)
- storage–head curves (9-8)
- irrigation water requirement constraints (9-9)
- definition of minimum seasonal irrigated land (9-10)
- relation of irrigation water supplied and diversions (9-11)
- definition of irrigation return flow (9-12)
- hydroelectric energy production function (9-14), (9-15)
- power plant capacity constraints (9-17)
- import channel capacity constraints (9-18)
- export channel capacity constraints (9-19)
- definition of deviations from equal allocation (9-25)
- definition of average regional water use (9-26)
- definition of individual regional water use (9-27)
- all variables nonnegative

Mendoza (region 1); irrigation diversions at sites 7, 9, and 12 in Neuquen–Rio Negro (region 2) and in La Pampa (region 3); and an irrigation diversion in Buenos Aires (region 4). Regions 2 and 3 are shown as sharing irrigation sites in Fig. 9-8 because of the peculiarities of our numbering system. It is sufficient only that we can keep track of the water as it is diverted to one region or the other.

The model written for the input configuration of Fig. 9-8, i.e., six reservoirs, five power plants, seven irrigation zones, two exports, and one import, and the assumption of three seasons, had 187 decision variables and 196 constraints. This is a relatively small linear program that cost less than $10 to solve using a commercial simplex algorithm called the Mathematical Programming System (MPS) on an IBM 360/165 computer. Since we had only two objectives, the generation of an approximation of the noninferior set was not computationally burdensome.

Any of the generating techniques in Chapter 6, other than the multiobjective simplex method, which cannot handle a problem of this size, could have been used. We chose the constraint method (see Section 6)—the NISE method had not yet been developed.

The constraint method begins by optimizing each objective individually. If we minimize deviations, i.e., solve minimize (9-22) subject to $\mathbf{x} \in \mathbf{F}_d$, we

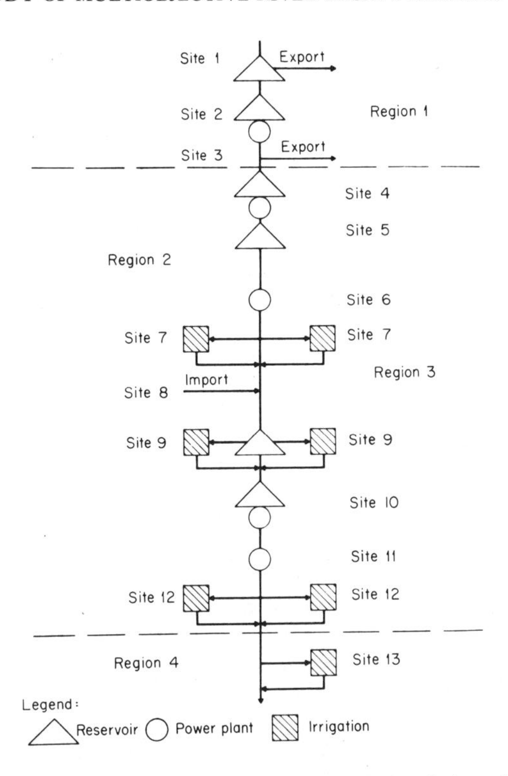

Fig. 9-8. Development alternatives for the initial formulation [adapted from Cohon and Marks (1973)].

get $D = 0$, i.e., an equal water allocation. However, we found that there were alternate optima for $D = 0$; i.e., there are many solutions that yield an equal water allocation. Since some of these solutions may be inferior (see Section 6.2.4), we solved the problem

$$\begin{aligned} \text{maximize} \quad & Z(\mathbf{x}) \\ \text{s.t.} \quad & \mathbf{x} \in \mathbf{F}_d, \qquad D(\mathbf{x}) = 0 \end{aligned} \tag{9-28}$$

That is, we wanted to find that equal water allocation that also maximized economic efficiency benefits. The solution to (9-28) gave $Z = 1.8 \times 10^{12}$ pesos (10^3 pesos $\simeq$ 1 dollar at that time; keep in mind that this is the present value of a 50-year stream of net benefits) and of course, $D = 0$. This is our first noninferior solution and is shown as point A in Fig. 9-9 and listed as such in Table 9-2. The D axis in Fig. 9-9 is decreasing from the origin because we are minimizing D.

Maximizing Z individually gave a unique optimum with $Z = 2.10005 \times 10^{12}$ pesos and $D = 436$ m^3/sec. This solution is labeled J in Fig. 9-9 and

TABLE 9-2

DESIGN CAPACITIES FOR POINTS IN THE NONINFERIOR SET[a,b]

Site No.	Region No.	Alternative[c]	Points on transformation curve									
			A	B	C	D	E	F	G	H	I	J
1	1	RES	0	0	27.5	77	88	100	112	121	121	121
		EXP	0	0	0	0	0	0	0	0	0	0
2	1	RES	7595	3258	3338	3388	3399	3411	3448	3431	3431	3431
		PP	200	200	200	200	200	200	200	200	200	200
3	1	EXP	84	91	98	103	104	105	106	107	107	107
4	2, 3	RES	45	313	225	223	215	207	229	596	805	805
		PP	0	0	0	0	0	0	0	0	0	0
5	2, 3	RES	353	0	0	0	0	0	0	0	0	0
6	2, 3	PP	116	116	123	111	108	105	102	93	92	92
7	2, 3	IRR	70,200	62,843	51,657	46,212	44,981	43,711	42,317	23,158	15,000	3500
8	2, 3	IMP	130	130	130	130	130	130	130	130	130	130
9	2, 3	RES	951	1238	1397	1454	1451	1444	1431	1418	1417	1417
		IRR	75,100	80,277	83,886	85,964	86,694	87,507	88,485	100,000	35,210	35,000
10	2, 3	RES	206	0	0	0	0	0	0	0	0	0
		PP	0	0	0	0	0	0	0	0	0	0
11	2, 3	PP	0	0	0	0	0	0	0	0	0	0
12	2, 3	IRR	66,500	66,500	66,500	66,500	66,500	66,500	66,500	66,500	66,500	66,500
13	4	IRR	133,259	129,920	129,361	126,751	126,496	125,000	122,999	126,127	166,472	173,272

Water use and benefit information										
W_1	251	272	295	309	312	316	319	322	322	322
W_2	251	251	244	241	241	241	241	237	113	99
W_3	251	231	193	173	171	170	169	154	125	114
W_4	251	251	244	241	238	236	233	238	314	327
$\overline{W}$	251	251	244	241	241	241	241	238	218	215
D (m^3/sec)	0	41.0	102.1	136.6	143.6	150.6	157.7	168.0	397.7	436
Z (pesos $\times 10^{12}$)	1.8	1.9	2.0	2.05	2.06	2.07	2.08	2.09	2.10	2.10005

[a] See Fig. 9-9.

[b] Adapted from Cohon and Marks (1973, Table 2).

[c] RES, reservoir in cubic hectometers; PP, power plant in megawatts; IRR, irrigation in hectares; EXP, export in cubic meters per second; IMP, import in cubic meters per second; W_i, water use in region i in cubic meters per second; and $\overline{W}$, average regional water use in cubic meters per second.

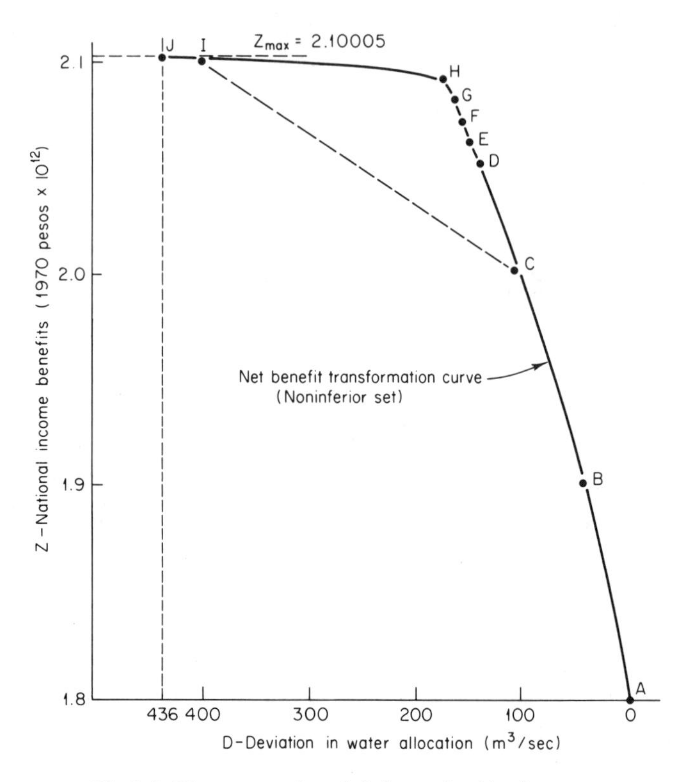

Fig. 9-9. The generated noninferior set in objective space.

Table 9-2. The constraint method proceeds by optimizing one objective while all other objectives are constrained to values that vary through a range of feasible values. We selected D arbitrarily for optimization and constrained economic efficiency benefits:

$$\begin{aligned} &\text{minimize} \quad D(\mathbf{x}) \\ &\qquad \text{s.t.} \quad \mathbf{x} \in \mathbf{F}_{\mathrm{d}}, \qquad Z(\mathbf{x}) \geq B \end{aligned} \tag{9-29}$$

where B (referred to as L_k in Section 6.2) is a preset lower bound on Z. With more than two objectives we would construct a payoff table (see Section 6.2.3) to define a range of values for B. With only two objectives we can simply observe that B must be less than or equal to 2.10005×10^{12} (the maximum of Z) for feasibility and greater than or equal to 1.8×10^{12} (the value of Z at the minimum of D that gave a noninferior solution) to insure noninferiority.

With the range $1.8 \times 10^{12} \leq B \leq 2.10005 \times 10^{12}$ determined, we chose a step size for the variations of B. We began with a step size of 0.1×10^{12}, solving the problem in (9-29) for $B = 1.9, 2.0$, and 2.1×10^{12}. This was done through parametric variation of the right-hand side of the constraint on Z (see Section 3.9.3). The solutions labeled B, C, and I in Fig. 9-9 and Table 9-2 were obtained.

At this stage we had five noninferior solutions: A, B, C, I, and J. It was obvious from an inspection of the dashed curve in Fig. 9-9 that $D(\mathbf{x})$ was changing rapidly between points C and I relative to its rate of change elsewhere. We then applied the constraint method again over the range of rapid variation by solving (9-29) with B varying from 2.05 to 2.09×10^{12} in steps of 0.01×10^{12}. This yielded five more noninferior solutions labeled D–H in Fig. 9-9 and Table 9-2. This approximation of the noninferior set, the solid curve in Fig. 9-9, was considered adequate and the procedure was terminated.

A summary of the solution procedure is shown in Table 9-3. The total cost to find these solutions was about \$70. The use of parametric programming resulted in a savings of about \$20; i.e., it would have cost \$90 to solve the problem ten times at \$9 per solution. The value of the dual variable associated with the constraint on Z is listed for solutions B–H in Table 9-3. The dual variable tells us the tradeoff between D and Z at the solution (see Section 6.2.2). For example, at solution C an increase in economic efficiency benefits (Z) of 1×10^9 pesos (or 0.001×10^{12} pesos) would result in an increase in total deviations (D) of 0.64 m^3/sec. It was also shown in Chapter 6 that the slope of the noninferior set in objective space is equal to the reciprocal of the tradeoff between the two objectives. For point C, Table 9-3 shows that the

TABLE 9-3

COMPUTATIONAL SUMMARY OF THE APPLICATION OF THE CONSTRAINT METHOD TO THE RIO COLORADO SCREENING MODEL

Step	Z (pesos) ($\times 10^{-12}$)	D (m^3/sec)	Dual variable (m^3/sec peso) ($\times 10^9$)	Slope of N_0 (1/dual variable) (peso sec/m^3) ($\times 10^{-9}$)	Solution
Minimize D	1.8	0	—	—	A
Maximize Z	2.10005	436	—	—	J
$1.9 \times 10^{12} \le B \le 2.1 \times 10^{12}$	1.9	41.0	0.45	2.22	B
Step size $= 0.1 \times 10^{12}$	2.0	102.1	0.64	1.56	C
	2.1	397.7	47.87	0.02	I
$2.05 \times 10^{12} \le B \le 2.09 \times 10^{12}$	2.05	136.6	0.69	1.45	D
Step size $= 0.01 \times 10^{12}$	2.06	143.6	0.70	1.43	E
	2.07	150.6	0.70	1.43	F
	2.08	157.7	0.72	1.39	G
	2.09	168.0	1.14	0.88	H

reciprocal of the dual variable is 1.56×10^9 peso sec/m^3. We can compute the slope at C by taking the change in Z and D in moving from point B to point D: $\Delta Z = 0.15 \times 10^{12}$ pesos, $\Delta D = 95.5$ m^3/sec, and $\Delta Z/\Delta D = 1.56 \times 10^9$ peso sec/m^3, which matches the reciprocal of the dual variable. Note that the slope of the curve is positive since D is being minimized and we have inverted the D axis.

Figure 9-9 shows in a concise way the conflict between efficiency and distribution. As we move along the noninferior set from A to J economic efficiency benefits continually increase at the expense of an increasingly unequal distribution of water. There is no point in considering solutions to the left of J since distributions with deviations greater than 436 m^3/sec will not yield higher economic efficiency benefits.

It is important to consider the tradeoffs between the two objectives at the project level. Table 9-2 lists the sizes of each project, i.e., all of the capacity variables from our formulation—V_s, H_s, L_{sm}, IM_s, and XM_s—for each noninferior solution. The water use for each region is also listed. Considering the water uses, as we move from A to G more water is given to Mendoza (region 1) through the export channel at site 3 at the expense of the other regions, particularly La Pampa (region 3). This reallocation of water is reflected in the decrease in the size of the irrigation area at site 7 from 70,200 ha at solution A to 42,317 ha at solution G. As we move further to solution H the distribution becomes even more unequal, but now water use in Buenos Aires (region 4) is also increased at the expense of regions 2 and 3. Moving to I and J results in the further decrease of La Pampa's water use and a very rapid decline in water allocated to Neuquen–Rio Negro (region 2), requiring the decline of irrigation capacity at site 7 from 23,158 ha at H to 3500 ha at J and at site 9 from 100,000 ha at H to 35,000 ha at J. Buenos Aires (region 4) is the beneficiary of the change in allocation: Mendoza's water use peaks at H and remains constant at I and J while the irrigation zone at site 13 increases from 126,127 ha at solution H to 173,272 ha at solution J.

It is interesting that the magnitude of the changes in design capacities is correlated with the distance between points in the noninferior set in Fig. 9-9. Solutions that are close together, such as D–H, give very similar designs. Solutions that are far apart, such as H and I, yield very different designs. Of course, this observation should not surprise us since the noninferior set in objective space is an image of the noninferior set in decision space, with the objectives serving as linear mapping functions.

The solutions in the noninferior set are rather similar in that there is a group of projects that appear in all of the solutions. A large reservoir is built at site 2 for hydroelectric energy production and to provide regulation for the export to Mendoza at site 3. Another large reservoir is selected for site 9 to regulate the imported water that enters the basin at site 8 for the downstream irrigation zones. The nature of the allocation is also similar with two tiers of regional competition for the water. First, Mendoza competes with all of the potential downstream development: The model exports from 84 to 107 m^3/sec each season to Mendoza on the basis of the high economic efficiency benefits that can be realized there. Second, the downstream regions compete among themselves for the remaining water and the imported water. La Pampa (region 2) is generally the least successful in this second level of competition. When very unequal distributions are allowed (points I and J) Buenos Aires (region 4) clearly dominates the other two regions since site 13 could use the water most efficiently due to the existence of a 60,000-ha irrigation zone prior to the study.

While there are similarities among the solutions, the range of choice is a rich one. These are initial results, so definitive conclusions as to the best-compromise solution cannot be made. One can argue rather convincingly, however, that a solution such as H on the elbow of the curve in Fig. 9-9 would be a good candidate. There is a wide range of weights and a large set of indifference curves that would result in H as the best-compromise solution. Table 9-3 substantiates the visual result of Fig. 9-9: The dual variable has a value of 0.72×10^9 m^3/sec peso at point G and a value of 47.87×10^9 m^3/sec peso at point I. These limits represent a range of weights on Z that would lead to H, or a point very close to it, as the best-compromise solution.

We can argue further for point H based on Fig. 9-9. Any point to the left of H would yield little additional net economic efficiency benefits while the distribution would be considerably more unequal. The argument for solutions to the right is not quite so convincing because the slope is not so rapidly changing. However, it does seem plausible that the movements toward equality on the right portion of the curve would be outweighed by the decrease in benefits.

The sharp elbow in the curve of Fig. 9-9 suggests that the noninferior set estimation method (Section 6.3) would have been particularly efficient for this application. A very good approximation could have been obtained with only three solutions: A, H, and J.

9.4.5 Final Multiobjective Approach

The initial approach was modified for a variety of reasons. One of the most compelling criticisms was the difficulty of specifying a base distribution from which to measure deviations. Provincial representatives were not prepared to accept the notion of an equal water distribution as a goal, nor were they willing to specify an alternative base allocation. A new consideration that was incorporated into the final approach and that demanded alterations in the initial approach was the controversy over interbasin transfers—imports and exports—of water. The concern over transfers transcends the usual upstream–downstream conflict because transfers are so final and represent a tampering with the natural way of things. While the initial approach was significantly modified, the importance of the initial results in increasing our understanding and insight into the problem cannot be denied.

Mathematical Formulation

The final approach employed three objectives, although two of them were concerns more than they were objectives in the usual sense. First, the economic efficiency objective as defined and formulated in the previous section [see Eq. (9-20)] remained. There were no changes here.

Second, it was communicated to the project that an import of water into the Rio Colorado was not probable in the near future. Furthermore, the concern over exports to Mendoza was becoming apparent to the analysts. It was decided then that the analysis would explore various combinations of transfer alternatives: no interbasin transfers, export but no import, import but no export, and all transfers allowed. There were a series of conditions that were tested, but a mathematical statement of an objective that could be optimized was not made. Instead, the model was solved with the appropriate variables removed in order to capture each condition. (Actually, appropriate variables were constrained to zero; e.g., for the no-export case the constraints $XM_s = 0$ for all export sites were added.)

Third, while provincial distribution was captured to some extent by the transfer "objective," there surfaced another distributional issue concerning water uses. After plowing through a set of documents regarding the Rio Colorado, we discovered one particularly interesting document that set forth a series of resolutions that the Rio Colorado committee had formed. One of the resolutions required that consumptive uses be given precedence over nonconsumptive uses. For the Rio Colorado, this meant that irrigation should be treated as a high-priority use relative to hydroelectric energy production, a nonconsumptive use. This had implications for provincial water allocation as well since the benefits of exports and imports were dependent on the water uses allowed.

Our third objective became known as the "complementary power objective" since it required that only power that complemented irrigation could be implemented. As with our transfer objective the complementary objective cannot be stated as an optimizable function. It is either required or it is not. When complementary power was not required, modifications in the formulation were not needed. When complementary power was enforced, we proceeded by first constraining all energy production to zero, i.e.,

$$\sum_s \sum_t P_{st} = 0 \tag{9-30}$$

and solving the problem to give us a set of design sizes for the irrigation zones called $\hat{L}_{sm}$. We then solved the model again, removing (9-30) but requiring that

$$L_{sm} \geq \hat{L}_{sm} \qquad \forall s \tag{9-31}$$

The resulting solution gave us a total river basin plan in which energy production did not compete with irrigation.

The final formulation was similar to the initial formulation in Table 9-1 except that the water allocation objective (9-22) and its associated definitional constraints (9-25)–(9-27) are removed. In addition, certain refinements in the continuity and irrigation constraints and in the cost functions were incorporated. The transfer and complementary power objectives are not really part of the mathematical formulation since they cannot be stated as optimizable functions.

Results

The final screening model was formulated for the full input configuration in Fig. 9-2. This model had 629 constraints, 665 continuous decision variables, and eight 0–1 integer variables. The model was solved several times with the Mathematical Programming System Extended (MPSX) [IBM (1971)] on an IBM 370/155 computer. Solving the model just as a linear program cost \$30–\$50 for each run. Integer solutions required an additional \$25–\$50 of computer time.

There were 28 final runs of the screening model, which included sensitivity analyses of various economic and physical parameters. The runs of major interest here were one in which net economic efficiency benefits were maximized and the other two objectives were ignored, six runs in which economic efficiency benefits were maximized under various transfer requirements that included imposed levels of imports and exports, and three sets (two runs in each set) of complementary power runs that included three import and export combinations.

The results for the most important runs are shown in Table 9-4. The runs are labeled 1–5 for convenience. Run 1, which represents a base result in that exports, imports, and power were unconstrained, yielded the maximum net benefits of 254×10^9 pesos. (These benefits are on the order of 10% of the benefits generated by the early formulations because of an updating of economic data.) Run 1 includes an export to Mendoza of 43 m^3/sec each

TABLE 9-4

NET ECONOMIC EFFICIENCY BENEFITS FOR VARIOUS INTERBASIN TRANSFER ALTERNATIVES AND WITH COMPETITIVE AND COMPLEMENTARY POWER

	Run	Transfer combination (m^3/sec) export	Transfer combination (m^3/sec) import	Net economic efficiency benefits (pesos) ($\times 10^{-9}$)
Competitive power	1	43	100	254
	2	43	0	227
	3	0	0	242
Complementary power	4	43	0	215
	5	0	0	196

season and an import from the Rio Negro basin of 100 m^3/sec each season.

Comparing runs 2 and 3 with 1 demonstrates the intricate provincial competition within the basin. Disallowing imports and requiring an export of 43 m^3/sec results in a loss of 27×10^9 pesos in net benefits in run 2. The decrease in benefits is a result of less irrigation in the lower basin (see Fig. 9-2) due to the lower water availability caused by the removal of the imported water. In run 2 all of the lower basin irrigation sites are smaller than they are in run 1.

The results of run 3 show that disallowing exports as well as imports *increases* net benefits by 15×10^9 pesos relative to 2, in which 43 m^3/sec were exported. It is more beneficial not to export when there is no import (runs 2 and 3 show this), but it is beneficial to export when there is an import (runs 1–3 show this). The competitive relationship is a delicate one: When there is an import, the lower basin gets all the water it needs so that Mendoza competes only with the middle basin from Bardas Blancas to Trasvase del Rio Negro (see Fig. 9-2); in this case Mendoza can compete successfully. When there is no import, Mendoza must compete with all of the potential downstream development; in this case exports are not beneficial.

The complementary power runs 4 and 5 invert the competitive relationship. When competitive power is disallowed benefits are lower (compare run 2 with 4 and run 3 with 5), but it is beneficial to export in this case even when there is no import (compare runs 4 and 5). This happened because power was allowed in Mendoza even though it was disallowed in the Rio Colorado basin; the resolution, alluded to above, covered power only in the Rio Colorado. Thus, when competitive power is allowed in Mendoza and only complementary power is pursued in the lower basin, Mendoza can compete successfully with the downstream provinces.

9.4.6 Outcome of the Project

There were 28 different configurations generated with the final screening model formulation. Three of these (runs 1, 4, and 5 in Table 9-3) were selected for further analyses in the simulation and sequencing models (see Fig. 9-3) because they were considered to capture best the range of interest for the Argentine decision makers. All of the results were communicated to the decision makers, and the Argentine professionals who had worked on the project at MIT returned with the models to Argentina for a year of discussions with the provincial representatives and their staffs and for further analysis with the models.

At the end of the additional year, a configuration that was a modification or run 4 in Table 9-4 was selected for implementation. The ultimate choice of the no-import complementary power result confirmed our observations as to the political difficulty of transferring water out of the Rio Negro and the desirability of irrigation. It was decided to pursue a no-import alternative with the possibility of implementing an import at some time in the future.

Notice that the selection of an alternative similar to run 4 requires that more than 15% of the maximum possible net benefits (obtained in run 1) be sacrificed. It is interesting that the decision showed a willingness to trade economic efficiency for other considerations. Those other considerations included: a desire to emphasize irrigation, which probably does more to promote the settlement and development of undeveloped areas than does hydroelectric energy production for the national grid; a decision to avoid some of the political ramifications of major interbasin transfers of water; and a requirement for an acceptable provincial distribution of the Rio Colorado's waters—the selected configuration did provide a distribution that is closer to equal than the pure economic efficiency result.

9.5 LINGERING ISSUES

Water resource development and management will continue to lead to classic situations of conflict: distribution of benefits and costs among users or polluters and development versus environmental quality. These situations lend themselves to multiobjective analyses, but the use of multiobjective programming methods may not always be a straightforward process. There are many issues that the analyst must confront in the analysis of real-world multiobjective problems.

First, objectives must be identified and quantified in a meaningful way. Identification may not be an easy step because of the inaccessibility of decision makers, their unwillingness or inability to articulate objectives, or simply because the identity of all of the decision makers is unknown. The Rio Colorado case study showed some of the procedures that the analyst may use and their pitfalls. We proposed the objective of minimizing deviations from an equal water allocation based on our understanding of the problem, i.e., that distribution was an important consideration. We did not realize, however, that the use of our equal allocation as the base distribution represented a controversial value judgment or that the decision makers would be unwilling to articulate an alternative unequal base allocation.

The import–export and complementary power "objectives" were identified as the project progressed and the decision makers had the opportunity to respond to our results. This points out an important characteristic of many planning exercises: The analyst should allow for new objectives to be identified or for old objectives to be modified. Planning is an educational process in which decision makers and analysts gain an understanding of the system and its problems. Therefore it should not be surprising that new objectives will surface after a period of time when decision makers can use the new knowledge that the analysis has provided to them. In the case of the complementary power objective, it was not until the second year of the project that the document that led us to that criterion came to light.

Second, the set of objectives has implications for computational burden and the nature of the results. In our early formulation we had a two-objective problem to which the constraint method was applied. The results were presented concisely and dramatically on a two-dimensional plot of the approximated noninferior set in objective space. In the later formulation, however, our objectives did not allow this kind of programming analysis or a graphical display. The objectives were discrete all-or-nothing considerations.

Another computational consideration is the cost of dealing with a large number of objectives. One of our formulations, which was not discussed in this chapter, had six objectives: economic efficiency and the maximization of net provincial income benefits for each of the five provinces. The use of the constraint or weighting methods to approximate the noninferior set for this formulation was considered too expensive, so only three noninferior solutions were generated. Notice that a single objective of minimizing the deviations from a base distribution of provincial incomes in place of the five provincial income objectives would reduce this to a two-objective problem. Of course, this brings us back to the dilemma of specifying a base distribution.

In spite of these problems, multiobjective programming and planning will continue to be an important tool for the analysis of water resource problems. A careful analysis that emphasizes the identification of objectives and the range of choice open to decision makers provides results that can be of enormous value to sound decisions about the use of our water resources.

CHAPTER 10

Multiobjective Analysis of Facility Location Problems

In this chapter two applications of multiobjective facility location models are presented. First, the analysis of fire station location alternatives in Baltimore, Maryland is discussed. This provides a very practical case study in which the emphasis will be on the problem and the role multiobjective analysis was able to play in its solution. The second application deals with a multiobjective facility location model for power plant siting. This is a problem of great practical importance, but at the time of this writing the model has only been formulated. Thus, this study is of a hypothetical nature, but is useful in further illustrating the role of multiobjective programming in the analysis of a wide range of problems.

Facility location is a natural framework for multiobjective analysis, as the two applications in this chapter are intended to demonstrate. Before we discuss the two applications, a brief review of location modeling will be presented.

10.1 FACILITY LOCATION ANALYSIS

The location of facilities, where a facility can be any structure erected to manufacture, store, or distribute a product or provide a service, is a common decision-making problem that is confronted by managers of a wide range of activities. Facility location is both a private and public sector concern, although the two application areas may present very different types of problems. A typical private sector problem is to locate a factory, warehouse, distribution facility (e.g., a truck terminal), or service center so as to minimize the firm's cost of supplying a given market. Mathematical programming models such as the plant location formulation have been presented to deal with this problem.

Our concern is with public sector location problems, in which a frequently encountered difficulty is the absence of a monetary measure, such as cost, for the assessment of a facility's performance. We can think of three types of public sector facility: ordinary, emergency (or extraordinary), and noxious facilities. Ordinary public facilities, such as libraries or health clinics, are established so as to provide a service to as many people as possible. Facilities that provide an emergency service, such as fire stations, must be located so as to provide as much coverage as possible; i.e., as many emergencies as possible must be covered. The location of noxious facilities, such as solid-waste disposal facilities, should be sensitive to their impact on neighbors.

Optimization models for public sector facilities of the three types mentioned above have attracted increasing attention from operations researchers and regional scientists in the past 15–20 years. ReVelle and Church (1978) provide a complete review of the important theoretical developments of this period.

While most of the optimization models developed to date have used a single objective, it is clear that the location of each of the three facility types is an inherently multiobjective problem. In each case—ordinary, emergency, and noxious type—the facility's performance, which is measured by a non-monetary criterion, must be compared to cost since all agencies—even critical ones such as fire and police departments—operate with fixed finite budgets. Of course, if benefits are associated with the services provided by the facility, then performance and cost are commensurable since they would share a common monetary unit. There are, however, critical problems with estimating the benefits (or disbenefits) of some facilities. In many cases there are *multiple* performance measures so commensuration among the measures as well as with cost is required. Even if a single criterion is relevant, the benefit estimation procedure may require value judgments that an analyst will not care to make: In the case of fire station location, what is the value of a life?

The two problems discussed in this chapter demonstrate the role that multiobjective analysis has to play in public facility location decisions. Emergency facilities (fire stations) and noxious facilities (power plants) are considered, but the basic constructs of the multiobjective programming models that are presented and the manner in which they are used to generate noninferior location alternatives should be applicable to the wide range of location problems that exists.

10.2 FIRE STATION LOCATION IN BALTIMORE, MARYLAND

Analysts at The Johns Hopkins University were contracted by the City of Baltimore to study the location of new fire stations and the relocation of existing fire companies. The study lasted for 15 months, terminating in July 1976 with a set of alternative locations and relocations that were presented to the Fire Chief and the Mayor of Baltimore. In this section the methodology used in the new location phase of the project and selected results are discussed.

The problem is discussed in the next subsection, followed by a presentation of the location model. The objectives of the location analysis are then developed, followed by a discussion of computational considerations and results.

10.2.1 The Problem

Baltimore is in the rather comfortable position of having a fire department that is one of the most highly rated among large urban areas in the United States and is recognized for its innovative approaches to management and prefire planning. The fire protection system exists, however, like most systems that serve people, in a changing environment where decisions made in the past may now be no longer optimal or even acceptable. The changing character of Baltimore's residential and commercial patterns has created a situation in which there is room to improve the fire protection system.

At present (1977) fire-fighting companies and the stations that house them are concentrated in the central part of Baltimore, providing excellent (and in some locations excessive) coverage to the center of the city, but leaving some parts of the periphery uncovered or inadequately covered. There are good historical reasons for this configuration of facilities. Many of the fire stations in Baltimore were built prior to 1920 at a time when the city's boundaries fell well within its current boundaries. As land was annexed from the surrounding county, the city grew out from its center so that previously located stations are now distant from some portions of the annexed areas. Another historical motivation for the concentration of stations in the central city was the great conflagration of 1904, which virtually leveled the portion of Baltimore that is today the high-value district (central business district). It was this catastrophe that promoted the intensive period of fire station construction in the central city prior to 1920.

There is yet another incentive for concentrating fire-fighting capability in the central city that is still important today and is not unique to Baltimore. The rating of fire departments and the determination of fire insurance premiums is controlled by the Insurance Services Office (ISO). This organization exists primarily to assess a fire department's ability to protect *property*. Since property value tends to be concentrated in the center of the city (hence the name "high-value district"), the fire department is under some pressure to focus its effort on that area. The nature of ISO's influence on location decisions of the fire department is formal and specific since it is communicated by the ISO standards for the coverage of structures. The standards played an important role in our analysis and will be discussed further in subsequent subsections.

It was perceived by the Fire Chief of the Baltimore Fire Department that two related problems existed in the current configuration of fire stations. First, some peripheral areas of the city were inadequately covered; second, there were several old fire stations with outmoded facilities in locations that were less than optimal. It was the charge of the analysts to identify alternative locations for new stations and determine which old stations could be closed. The problems are interrelated to the extent that companies in closed stations can be relocated to a new station. For the purposes of analysis, however, the two problems were treated separately. The station-closing problem, although it presents an interesting analytical challenge, is not discussed here. In the next section a model for determining new fire station locations is presented.

10.2.2 A Fire Station Location Model

A variety of problem-specific models for the analysis of facility locations exists. ReVelle and Church (1978) present a complete review and mathematical development of many of these models. Within the usual categorization of location models, we are interested in a *prescriptive* model for the analysis of *public* facilities that provide *extraordinary* (emergency) services over an area described as a *network*. Each of the italicized terms in the previous sentence represents a characteristic of the fire station problem that sets it apart from other facility location problem areas.

First, we are interested in a prescriptive or normative model that will identify (prescribe) optimal (actually, noninferior in the multiobjective case) location alternatives. We are not interested in the descriptive type of model that regional scientists and geographers use to explain demographic or industrial location patterns. Our interest in a prescriptive model implies the use of optimization (vector optimization in the multiobjective case).

Second, fire stations are public facilities in that they are owned and operated by a governmental agency. Third, the service provided is of an extraordinary nature in that users of the facility do not employ its services on a regular basis. Both of these characteristics affect the nature of the model's objective function. In the private facility case, the objective is usually straightforward: Minimize cost while providing a given service, or maximize profit. The public facility problem is more complex in this respect in that the notion of the payoff or benefits of a facility is dependent on the welfare gains from the facility. In the case of ordinary public facilities, e.g., libraries, a useful surrogate for welfare is to minimize average distance of people from the facility, thereby maximizing accessibility to the services provided. A more useful notion for extraordinary facilities is the maximum distance or time of a potential user from a facility such as a fire station. An objective for a fire station location problem is to maximize coverage, for which a point is covered if it is within the maximum allowable distance or time of a facility.

Fourth, location models are specific to the manner in which the area of analysis is represented. If location is allowed anywhere in a continuous land area, then a planar model is appropriate. In the Baltimore case study, the city was represented by a network with approximately 600 points, each of which was located at the centroid of an area roughly $\frac{1}{9}$ square mile in size. All adjacent points were connected by arcs that measured the corresponding interpoint distances. See Fig. 10-1 for an example of this network representation. (Note that time of travel could be used in place of distance. Distance was used, however, since time of travel was found to relate almost linearly with distance and this relationship did not vary from arc to arc.) This grid of points and arcs form a network so a network model, which restricts the analysis of potential facility locations to the grid points, was used.

Church and ReVelle (1974) formulated a network optimization model called the "maximal covering location problem" (MCLP) for public facilities that provide an extraordinary service. Simply stated, the MCLP maximizes coverage for a given number of facilities to be located. Coverage is defined in the following way: point i is covered if there is a facility located within S miles of i, where S is the "maximum service distance," which for the case of fire stations has been specified by ISO standards.

In the general statement of the MCLP there are two sets of points: I, the set of all "demand" points, and J, the set of all potential facility locations. The two sets may overlap partially or completely; i.e., points in the network may be both demand and potential facility points. We define x_j as a 0–1 integer variable that equals 1 if a facility is placed at point j and 0 otherwise. One constraint in the model is that exactly α facilities must be sited, where α is a prespecified number,

$$\sum_{j \in J} x_j = \alpha \tag{10-1}$$

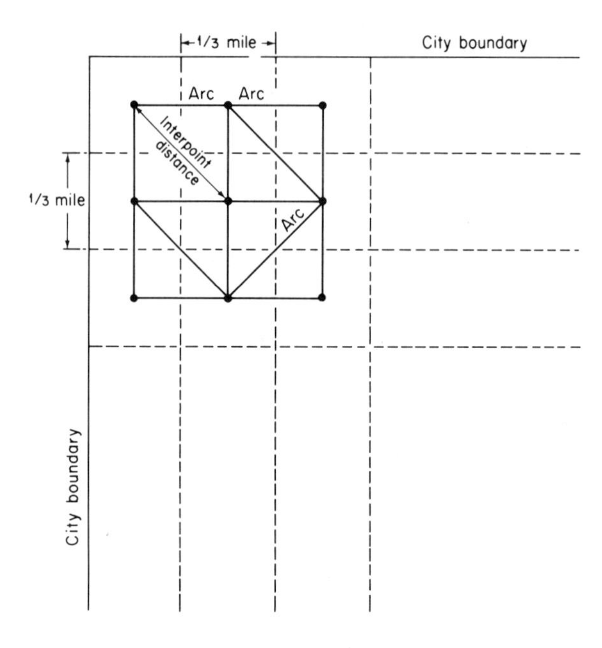

Fig. 10-1. Network representation of an area.

The notion of coverage can be incorporated by defining y_i, a 0–1 variable that equals 1 if there is a facility within S miles of point i and 0 otherwise, and N_i, the set of all potential facility sites within S miles of point i. The coverage constraints are

$$y_i \leq \sum_{j \in N_i} x_j \qquad \forall i \in I \tag{10-2}$$

or

$$y_i - \sum_{j \in N_i} x_j \leq 0 \qquad \forall i \in I \tag{10-3}$$

The constraint requires at least one $x_j, j \in N_i$, to equal 1 if y_i is to equal 1; i.e., point i can be considered covered only if at least one facility is located within S miles of it.

The objective function of the MCLP is to maximize coverage. In general, there may be some attribute of a demand point that should serve as a weight on the importance of covering it. For example, property value is an important attribute in a fire station problem and we should be more inclined to provide coverage to a point with high property value than to a low-value point.

Calling a_i the amount of the attribute, e.g., property value, at point i, the objective function is

$$\text{maximize} \quad Z = \sum_{i \in I} a_i y_i \tag{10-4}$$

The entire MCLP is the maximization of (10-4) subject to (10-1), (10-3), and integrality restrictions on y_i $\forall i \in I$ and x_j $\forall j \in J$. It is of importance for computational reasons that the MCLP frequently terminates all-integer [upward of 80% of the time—see Church and ReVelle (1974)] when the integrality requirements are ignored, i.e., when the MCLP is solved as a linear program. Integrality is, of course, important since a half of a fire station is of no practical significance: We want to build a whole facility at a site ($x_j = 1$) or we do not ($x_j = 0$), but we are not interested in fractional solutions (e.g., $x_j = 0.5$). The techniques that exist for resolving fractional solutions, e.g., the branch and bound algorithm, can be expensive for the large problems encountered in practice. The MCLP is appealing, then, since relatively inexpensive linear programming routines may be used in lieu of expensive integer programming routines.

The MCLP incorporates a single attribute into its objective function. As we shall see in the next subsection, however, fire station location problems are multiobjective in nature. The Baltimore study gave rise to the multiobjective facility location (MOFLO) problem presented in Schilling (1976). MOFLO employs the same constraint set as the MCLP, but it allows for more than one coverage objective. Defining a_{ik} as the amount of the kth attribute at point i, the objective function of MOFLO is

$$\text{maximize} \quad Z = \left[\left(\sum_{i \in I} a_{i1} y_i \right), \left(\sum_{i \in I} a_{i2} y_i \right), \ldots, \left(\sum_{i \in I} a_{ip} y_i \right) \right] \tag{10-5}$$

where, in general, there are p objectives.

The MOFLO problem was used in the analysis of new fire station location alternatives in Baltimore. The objectives used in the analysis are discussed in the next subsection.

10.2.3 Planning Objectives

The point of departure for the analysis of fire station location alternatives in Baltimore was an assumption that location policy was influenced to a large extent by the property coverage standards of ISO. While ISO does exert influence over departmental decisions, as it must, several meetings with the Fire Chief and Deputy Chiefs indicated that many objectives were important. By the end of the study, six coverage objectives were identified and used in the analysis.

The primary ISO activity is the establishment of standards for fire fighting that will encourage at least adequate protection for property. As a mathematical function for MOFLO, the ISO objective is to maximize property value covered by new fire stations

$$Z_1 = \sum_{i \in I} a_{i1} y_i \tag{10-6}$$

where a_{i1} is the property value of the $\frac{1}{9}$-square-mile area represented by point i. Values of the a_{i1} were obtained from Baltimore's Real Property File, which includes property values for all assessed structures in the city. Note that in our case there are some areas of high property value that are currently covered that we do not wish to cover again. To prevent further redundant coverage, the set I is defined to include only currently uncovered points.

Siting a limited number of facilities on the basis of the ISO objective alone may lead to a substantial lack of coverage in densely populated low-property-value areas. In fact, most American cities have such low-value dense residential areas so the exclusive dependence on ISO standards for location decisions may be generally ill advised. Of course, the ISO cares about people, but it is their role to insure protection of structures, which may not necessarily include the best protection for populations. When there is a sufficient budget to allow all points to be protected, the problem is moot: All people and all property are covered. When budgets are constrained, as they are, protection directed at property only may not be the best policy.

The Fire Chief articulated a population coverage objective at an early stage in the analysis. The form of the objective to maximize population coverage is identical to (10-6), only it will be called Z_2, and the population at point i is a_{i2}. Values for the a_{i2} were obtained from census data disaggregated from tract levels.

The MOFLO formulation has a form that promotes the rapid quantification of objectives. All coverage objectives are of the same form, as in (10-6), so that one need only develop attributes of interest at the grid points to identify new objectives. In this spirit, a third objective of maximizing area covered was formulated to interject a sensitivity to future changes in housing and industrial location patterns. Some areas with few people or structures would receive very low weights in the first two objectives. The area coverage objective assigned equal weight to all points, since all points represented equal areas, so that currently undeveloped areas were treated as having the same importance as developed areas to reflect their potential future development. The area coverage objective is Z_3 with $a_{i3} = 1 \; \forall_i$.

The first three objectives did not consider the importance of relative fire frequency for fire station location: Stations should be placed where fires are most likely to occur. Three additional objectives were developed with the Fire

Chief and Deputy Chiefs in order to incorporate this consideration. A fourth objective was to maximize expected fires covered. Calling this objective Z_4, the a_{i4} are the expected number of fires at point i. Values for the a_{i4} were estimated from a sample of historical fire data maintained by the Fire Department.

The fire coverage objective Z_4 did not capture the variation of population and property value from point to point, and the property and population coverage objectives Z_1 and Z_2 did not include the variation of fire frequency from point to point. Measures that approached overall indices were developed by literally combining fire frequency with property value and population. The fifth objective Z_5 was to maximize *property hazard* covered, where a_{i5} is the product of fire frequency with property value at point i, and the sixth objective Z_6 was the maximization of *population hazard* covered, where a_{i6} is the product of fire frequency with population at point i. The hazard objectives are weighted versions (by fire frequency) of objectives Z_1 and Z_2.

Some of the six objectives may be collinear (see Section 4.5.2) in the general situation, especially objectives Z_2, Z_4, and Z_6. One would expect this if densely populated areas have high fire frequency, which they often do. If this is the case, then two of the population, fire, and population hazard objectives may be redundant. Even with the possibility of redundancy, it was decided to use the full set of objectives since there was interest in all of them and because the extent of the collinearity was unknown *a priori*.

10.2.4 Computational Considerations

The MOFLO problem is a multiobjective 0–1-integer programming problem that has a special constraint structure that accelerates its solution time. These two characteristics of the problem—the integrality requirements and the special structure of the constraints—have important implications for the multiobjective solution technique that may be employed. In fact, the implications result in a dilemma—the analyst's own multiobjective problem.

Of the many solution techniques available (see Chapters 6–8) our emphasis was on generating methods (see Chapter 6). Recall that generating techniques find an approximation of the noninferior set; they focus on the range of choice available to the decision maker. This emphasis on alternatives, without a recommendation of an optimal alternative from the analysts, appealed to the Fire Chief. Indeed, it is fair to claim that it was our proposed use of multiobjective generating methods that sparked the Fire Chief's initial interest in our study. He liked not being told what was "best."

As discussed in Chapter 6, however, there are several generating methods and it is here that the dilemma arises. Methods that are based on weights on the objectives—either the weighting method or the noninferior set estimation

(NISE) method—have the advantage of not disturbing the constraint set of MOFLO, but such methods may not be capable of finding certain integer-valued noninferior points. The constraint method avoids this last difficulty, but it disturbs the constraint set.

The problem of skipping some noninferior solutions can be understood with Fig. 10-2. In an integer programming problem the noninferior set may be nonconvex. This means that noninferior integer solutions may lie below the line segments connecting other adjacent noninferior solutions. Figure 10-2 shows three solutions in objective space. If this were a continuous (non-integer) linear programming problem, we could conclude that point C must be inferior since there are points on the line segment connecting A and B (convex combinations of A and B) that dominate C. For the discrete (integer) case, however, if these convex combinations of A and B are not all-integer solutions (by assertion for this example), C is not dominated and is non-inferior.

If the weighting method were applied to the problem depicted in Fig. 10-2, we should never discover point C, even though it belongs to the noninferior set. Suppose the weight on Z_2 is set to one (we only care about relative weights); then if $w_1 \leq \beta$, point A will optimize the weighted problem and

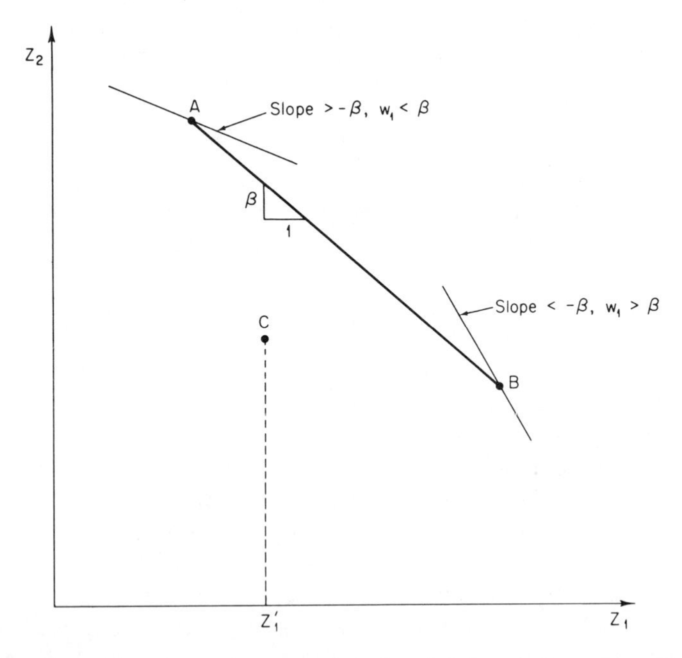

Fig. 10-2. The weighting method may not find all noninferior solutions of a multiobjective integer programming problem.

if $w_1 \geq \beta$, point B will optimize the weighted problem. There is no value for w_1 such that C is the optimal solution of the weighted problem. This phenomenon, which results whenever the noninferior set is nonconvex—integer problems being one case where nonconvexity arises—is often referred to as a "duality gap."

The constraint method can be used to find such nonconvex noninferior points as C in Fig. 10-2 by constraining Z_1 to be greater than or equal to Z_1' and maximizing Z_2. This method works because the constrained problem represents a new constraint set that is, in effect, forced to be convex. The duality gap is filled by lopping off part of the feasible region.

Our dilemma with the MOFLO model, then, was whether the ability to find points such as C in Fig. 10-2 was worth the price of potentially higher solution costs. It is not well understood, but the MCLP constraint set is such that a linear programming solution will be all-integer, and therefore optimal, quite frequently. The frequency of integer linear programming solutions decreases and the need for expensive integer programming routines increases when the constraint set is disturbed. Our own strategy was to ignore the possibility of skipping some points since our interest was in generating an approximation, not an exact representation, of the noninferior set. We used the weighting method, but the problems associated with multiobjective integer programming continue to be of some research interest.

The weighted objective function for the MOFLO model was

$$\text{maximize} \quad Z(\mathbf{w}) = \sum_{k=1}^{6} \sum_{i \in I} (w_k a_{ik} y_i) \tag{10-7}$$

where w_k is the weight on objective k and index k indicates the property value ($k = 1$), population ($k = 2$), area ($k = 3$), expected fires ($k = 4$), property hazard ($k = 5$), and population hazard ($k = 6$) objectives. This basic formulation was used to generate the results discussed below.

10.2.5 Results

Two sets of results are discussed here. The first set is for a hypothetical problem with two coverage objectives and thirty grid points presented in Schilling (1976). This simplified problem was solved for the noninferior locations of three facilities. Five different noninferior alternatives were generated with the weighting method and displayed as in Fig. 10-3. Notice that the noninferior set is represented as a collection of discrete points since MOFLO is an integer programming problem.

Although the results are for a hypothetical problem, there are two aspects of the results that have practical implications. First, if Z_1 is property value coverage and Z_2 is population coverage, the value of a multiobjective

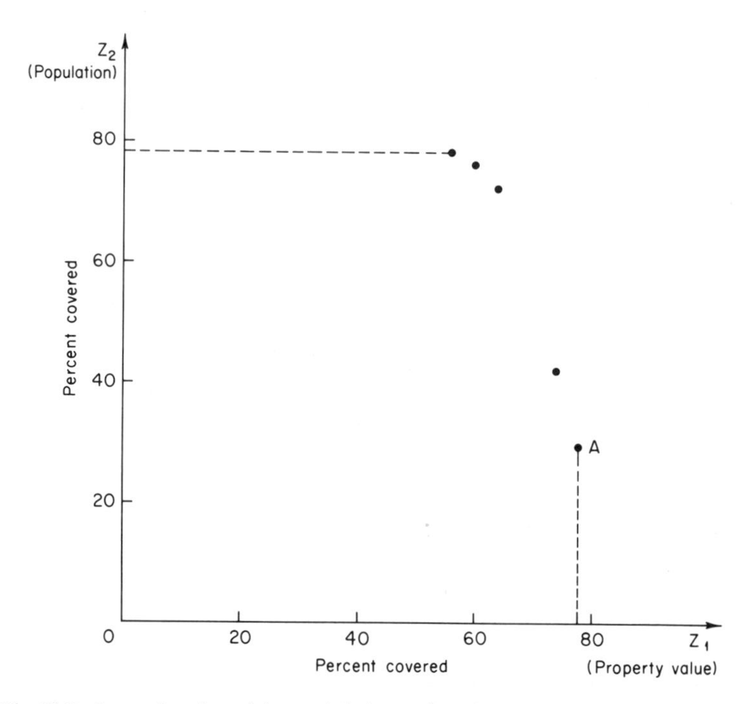

Fig. 10-3. Approximation of the noninferior set for a hypothetical 30-node problem.

approach to facility location can be seen immediately from Fig. 10-3. Point A, which is the property value coverage maximizing solution, covers only 31% of population coverage. The solution at A is the optimal solution under ISO guidelines. Figure 10-3 shows, however, that substantial gains in population coverage can be achieved with a relatively small sacrifice in property value coverage. Real situations, as shown below, may not generally demonstrate such a wide range in the tradeoff between property and population, but a multiobjective approach is necessary to avoid an unbalanced solution such as point A. This point takes on even more importance if one considers the possibility that point A in Fig. 10-3 could have been an alternative optimum; thus, a solution that provides the same amount of property coverage but *less* population coverage could have been selected if a single-objective approach were used.

Second, the noninferior set in Fig. 10-3 was generated with the number of facilities α, set at 3. A different value for α would lead to a different noninferior set: Higher values of α would result in a set farther out, i.e., to the northeast, since more facilities would allow more points and therefore more property *and* more population to be covered; lower values of α would displace the noninferior set toward the origin. If the number of facilities could be taken as a surrogate for cost—it is a particularly good surrogate if all potential facilities have approximately equal costs—then a cost objective can be considered by varying α. Church and ReVelle (1974) demonstrated this use of α for cases with a single coverage objective.

The second set of results was generated for the real location problem in Baltimore. The MOFLO problem consisted of 1362 decision variables, 688 constraints, and the six objectives discussed in the previous subsection. Results were obtained with the weighting method, implemented on the Mathematical Programming System (MPS) on an IBM 370/145 computer operated by the City of Baltimore. A typical run required 3 min of central processing unit (cpu) time. On a typical academic computer facility, which charges about $500 per cpu hour, a model run for a given set of weights would have cost about $25.

Displaying alternatives presented a problem since the six coverage objectives obviously precluded the use of a conventional graphical presentation. The value path technique described in Section 6.5.3. was developed for this problem. Value paths for a typical set of alternatives is shown in Fig. 10-4. A value path shows the impact of a single locational alternative, generated with MOFLO, on each of the six objectives. Each objective was measured in

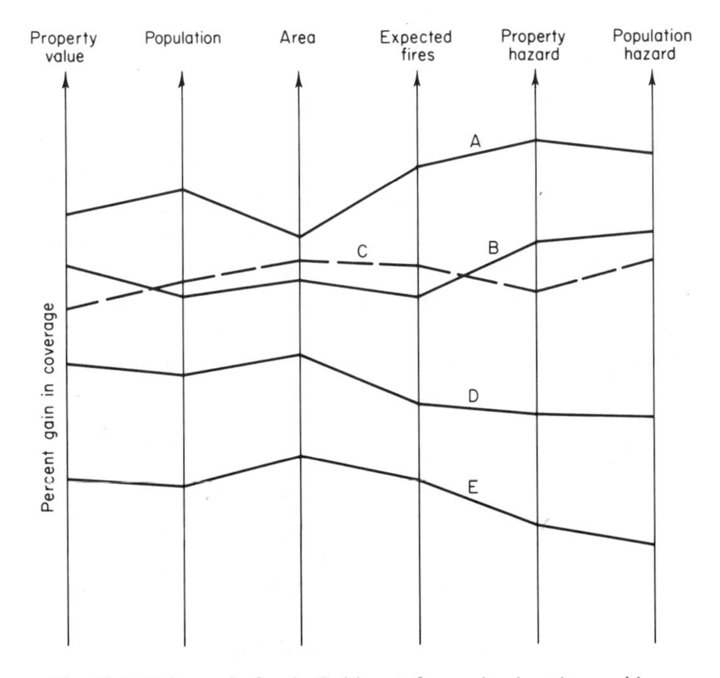

Fig. 10-4. Value paths for the Baltimore fire station location problem.

terms of the improvement in coverage relative to current coverage. This allowed us to measure each objective in the same units of percentage of improvement.

Value paths for noninferior alternatives must cross at least once since noninferior solutions are noncomparable: A noninferior alternative yields more of one or more objectives and less of one or more other objectives when compared with any other noninferior alternative. In Fig. 10-4, the value path labeled A lies above all of the other value paths, i.e., alternative A dominates the other alternatives shown in the figure. The implication of this is that there is a single noninferior and therefore optimal solution. That is, the coverage objectives do not conflict: A single solution maximizes all of the coverage objectives simultaneously. This occurred in this case study because the currently *uncovered* areas of the city exhibit collinearity among the objectives: The most highly populated uncovered areas also have the highest property values and fire frequencies among uncovered areas. This will not be true in a general case.

Several alternatives are shown in Fig. 10-4; if the location of a single station had been the problem, then alternative A would have been enough. In our problem, the number of new stations was uncertain so several other alternatives were generated. Solutions B–E give progressively lower coverage in all objectives. Notice that the value paths for B and C cross, signifying their noncomparability.

10.2.6 Summary and Conclusions

The Baltimore fire problem was more complex than the location of new fire stations and equipment. The problem of optimal station closing and company relocation was a large portion of the overall analysis. Other considerations such as company workloads and backup coverage, in addition to the six coverage objectives, were introduced into the analysis. Thus, the model and results discussed in this section tell only a part, although a major one, of the story.

At the time of this writing (1977) the Fire Chief is considering the results of this analysis so that a capital improvements program for the Fire Department can be developed. It may be premature to call the study a success in that a final decision has yet to be made. We were successful, however, since the results generated by the study were perceived as useful and relevant, and we are confident that the analysis had a beneficial impact on the decision-making processes of the Fire Department.

The Baltimore study represents the first practical application of multiobjective location analysis. This is a reflection of the youth of the analytical techniques rather than the rarity of such problems. The power plant siting case in the next section demonstrates another important problem for which multiobjective analysis may prove to be very useful.

10.3 A REGIONAL ENERGY FACILITY LOCATION MODEL

In this section a multiobjective model for the analysis of regional energy facility locations is presented. The model emphasizes electrical energy generating facilities, i.e., power plants, but the structure of the model may be applicable to other types of facilities related to energy, such as nuclear processing centers, facilities for liquified natural gas, and oil refineries.

The model presented here was developed for Brookhaven National Laboratory as an exploratory study of the way in which multiobjective analysis and location modeling could be applied to regional energy problems. Church and Cohon (1976) present the results of this work. The model must be considered a theoretical effort at this time, given the exploratory nature of the work. A current effort, at the Applied Physics Laboratory of Johns Hopkins, supported by the Electric Power Research Institute, is directed at the implementation of a similar model for the analysis of energy facility location in a six-state region of the eastern United States.

The regional energy facility location problem and the major elements of a planning methodology are discussed in the next subsection. Several subsections on the specific constraints of the model are then presented, followed by a discussion of the computational and data requirements of the model.

10.3.1 The Problem and an Overview of the Model

Energy—its supply, conversion, and demand—is a major national and international issue. The inadequacy of the United States' domestic reserves of oil and natural gas have been blamed for the severe economic dislocations of the winter of 1976–1977 and for the erosion of the United States' global economic position. The importance of energy and the conflicts and complexity of solving the "energy crisis" were reflected in the creation of the Department of Energy, a new cabinet-level department, in 1977 and in the Carter Administration's National Energy Plan, which literally consumed the first session (1977) of the 95th Congress and was still pending near the end of the second session (summer 1978).

The supply of energy requires a wide range of facilities for extraction and fuel preparation, e.g., coal and uranium mines and nuclear fuel enrichment and fabrication plants; fuel transportation; conversion, e.g., power plants and coal gasification plants; transmission, e.g., electric transmission lines and pipelines for oil and natural gas; and waste disposal facilities, e.g., nuclear waste repositories. The appropriate mix of these facilities and their size and location are major energy planning issues at the national and regional levels. Our focus in this section is on regional electrical energy facility planning, a relatively new problem that is not only of obvious importance to regions but is also required to assess environmental impacts in support of national energy planning.

The regional energy facility siting problem is a very complex one for several reasons. First, there are many objectives that one would like to optimize simultaneously in selecting plant locations, making it conceptually difficult from the perspective of the analyst as well as the decision maker. The second complexity is the size of the problem; i.e., the land area included in the region and the additional generating capacity that must be located are both large. The region's size and the level of detail at which it is analyzed are important determinants of model size and computational burden.

Third, the dependence of system costs (one of the planning objectives) on transmission distance, as well as on plant size and location, results in analytical complexity. The transmission line–generating plant relationship has the effect of creating an enormous number of location alternatives. After locating a generating facility, which is itself a difficult problem in the present context, one then must continue to search for the best transmission route.

A fourth complexity arises from the economic forces at work in the problem. Our basic consideration in the analysis of additional generating capacity is the development of sources that can supply electrical energy sufficient for regional needs. While this objective is undoubtedly perceived by all, quasi-public utilities must also be concerned with the maintenance of a rate of return that is high enough to attract required capital. The rate of return on investments in generating capacity is a function of revenue from energy sales and the costs of new facilities, which are functions of consumption, price, plant location, and transmission line routing. Furthermore, consumption is a function of price so in the end all of these quantities that are affected by siting decisions interact. These interactions should be taken into account in regional energy planning.

A regional energy facility location model (RELM) that merges multiobjective and location analyses is presented in this section. The model captures a good deal of the regional energy planning problem defined above, but there are certain aspects that are not included. The model selects gross locations (not specific sites) and sizes and types of power plant while taking into account capacity requirements, population safety, environmental quality and natural resource requirements, the equity of capacity distribution among political jurisdictions or subregions, and plant, transmission, infrastructure, and water transfer costs. The economic consequences of demand for energy are not included, and the transmission problem is not captured in its entirety. Therefore it is suggested that RELM be viewed as only a part of a larger planning methodology.

One possible planning methodology that would accommodate RELM is presented in Church and Cohon (1976, p. 38). The methodology attempts to compensate for the assumption in RELM that generating facilities can be preassigned to load (demand) centers and for the absence in RELM of an explicit consideration of utilities' rate of return requirements. The model, however, is relatively flexible and the importance of assumptions in RELM varies from one problem setting to another. The appropriate methodology will depend on the specific problem setting and those assumptions in RELM that are considered most limiting.

Of course, any methodology that employs RELM will be directed at the development of interesting and useful alternatives for regional energy planning. To this end, it seems advisable to concentrate on the generation of several noninferior alternatives and extensive sensitivity analyses on various parameters of RELM, such as demand for new energy generating capacities and fuel mix. The use of a generating method in conjunction with sensitivity analyses would be expected to provide a great deal of insight into reasonable planning possibilities within a region, the degree to which objectives conflict, and the sensitivity of potential solutions to the validity of assumptions about prevailing regional conditions. Keep in mind, however, that the regional scale of the analysis does not allow RELM to identify specific facility locations. Rather, gross locations such as at the county or multicounty level are found. The identification of specific locations would require further detailed analysis.

A multiobjective location model, RELM has the following form:

minimize facility costs
minimize water transfers
maximize equity of plant distribution
minimize population safety impact

Subject to

minimum capacity and concentration constraints
fuel mix constraint
water quantity requirements
water requirements for heat dissipation
air pollution constraints
constraints required for formulation of the above objectives

In the subsections that follow, each of these components of the model is discussed in detail, after which some of the issues relative to implementation of the model are presented.

The version of RELM presented in this section assumes a fuel mix; fuel availability is not modeled. Another version for coal-fired energy facilities, RELMC (RELM for coal), is presented in Church and Cohon (1976, Appendix D). This model takes into account coal availability and the transportation requirements for moving the coal to conversion facilities (power and gasification plants).

10.3.2 Minimum Capacity and Concentration Requirements

There are two types of generating capacity constraints in RELM: minimum new generating capacity required to meet demands and maximum allowable concentration at any site or at a group of sites. Both of these constraints are discussed in this section.

The theoretically correct approach to determining capacity additions to an existing supply system is grounded in neoclassical economics (assuming that economic efficiency is the only relevant criterion). The optimal capacity addition should take into account consumers' surplus and revenue from increased consumption of electricity and the costs of constructing and operating new capacity. These effects are not included in the model, however, due to the complexities related to estimation of consumption and prices.

The approach taken here is to specify new capacity for the region *a priori*. This capacity requirement is denoted as D and represents the total capacity addition measured in megawatts (MW) that is required to meet electricity demands in the region. The capacity requirement is

$$\sum_j \sum_k S_{jk} = D \tag{10-8}$$

where S_{jk} represents the new capacity of type k located at site j. It is assumed throughout RELM that the region has been discretized into a set of points in a grid system, as in the fire station example (see Fig. 10-1) so that site j refers to the area represented by point j. The type index k may refer to fuel type, e.g., nuclear or fossil; cooling technology, e.g., once through or cooling tower; and to any other plant characteristic that is considered important. It should be pointed out that many different assumptions or scenarios for demand may be tested by parametrically varying D.

Another set of constraints relates a 0–1-integer variable F_j for each site to the capacity variables in a way that controls the degree of concentration at a site:

$$S_{jk} - Q_{jk} F_j \leq 0 \qquad \forall jk \tag{10-9}$$

where F_j equals 1 if a plant is located at j and 0 otherwise, and Q_{jk} is the maximum number of units of type k that may be located at j. Constraint (10-9) could be written more simply as an upper bound on S_{jk}. The variable F_j is included here, however, in order to establish the relationship between S_{jk}, the capacity of type k at site j, and F_j, a variable that indicates whether there is any capacity at all at site j. Notice that if $S_{jk} > 0$ for any k, then (10-9) requires that $F_j = 1$ (since F_j can only equal 0 or 1). If $F_j = 0$, then $S_{jk} = 0$ for all k.

It should also be pointed out that Q_{jk} may be the same at all sites; e.g.,

$Q_k = Q_{jk}\ \forall j$ would represent a policy of no installations at any site of a given type larger than a prespecified limit. Alternatively, Q_{jk} may vary from site to site to reflect geographical or design considerations.

A small modification in (10-9) would allow the capacity of a given type, say nuclear facilities, located within a subregion, e.g., a state, to be constrained. If we call this subregion R, where $R = \{j \mid j \text{ is in the subregion}\}$ and $\tau_N = \{k \mid k$ is a plant type that is nuclear$\}$, then the constraint would be

$$\sum_{j \in R} \sum_{k \in \tau_N} S_{jk} \leq \hat{Q}_N \tag{10-10}$$

where $\hat{Q}_N$ is the maximum number of nuclear-powered units allowable in R.

It is also interesting to note that capacity concentration could be an objective. Parametric variation of Q_{jk} in (10-9) would trace out the tradeoff between capacity concentration and objectives such as the minimization of water transfers. The recent interest in energy centers or "power parks" suggests that such an analysis would be relevant in many energy planning situations.

10.3.3 Fuel Mix Constraints

For various reasons, one may want to require a prespecified mix among the types of capacity. In particular, one may want to constrain the solution so that a certain fraction of the new capacity uses a certain fuel. Such constraints may be motivated by fuel availability or the desire to be independent from foreign sources. Such constraints would tend to be region specific to reflect the relative scarcity of some fuels in a region.

Call τ_N the set of all types of new capacity k that are powered by nuclear energy and α_N the maximum fraction of all new capacity that may be nuclear. Then we can require that no more than α_N of new capacity in the region may be nuclear by the constraint

$$\sum_{j} \sum_{k \in \tau_N} S_{jk} \leq \alpha_N \sum_{j} \sum_{k} S_{jk} \tag{10-11}$$

Recognizing that the sum on the right-hand side of (10-11) is, from (10-8), the total new capacity requirement D, the constraint may be rewritten as

$$\sum_{j} \sum_{k \in \tau_N} S_{jk} \leq \alpha_N D \tag{10-12}$$

In some situations one may wish to site exactly $\alpha_N D$ MW of nuclear capacity. In such cases (10-12) should be written as an equality constraint.

Similarly, there may be a minimum fossil fuel capacity requirement. This can be written as

$$\sum_{j} \sum_{k \in \tau_F} S_{jk} \geq \alpha_F \sum_{j} \sum_{k} S_{jk} \tag{10-13}$$

or

$$\sum_{j} \sum_{k \in \tau_F} S_{jk} \geq \alpha_F D \tag{10-14}$$

where τ_F is the set of all fossil-fueled plant types and α_F is the minimum fraction of new plants that must be fossil fueled in the region.

A further distinction may be made among the capacity types. One may wish to require that at least α_c of all new fossil-fueled plants be coal fired. Defining τ_c as the set of all types that are coal fired, the following constraint would be included:

$$\sum_{j} \sum_{k \in \tau_c} S_{jk} \geq \alpha_c \sum_{j} \sum_{k \in \tau_F} S_{jk} \tag{10-15}$$

or

$$\sum_{j} \sum_{k \in \tau_c} S_{jk} - \alpha_c \sum_{j} \sum_{k \in \tau_F} S_{jk} \geq 0 \tag{10-16}$$

Other fuel mix constraints of this form can be included by defining appropriate new parameters analogous to the αs and τs above.

10.3.4 Environmental Quality and Natural Resource Constraints

Power plants impact the environment in many different ways. In this section we shall concentrate on four environmental quality and resource-use aspects of the problem: land impacts, water use, heat discharges into water, and air quality. It should be kept in mind that although these considerations can be captured in the model, their representation is relatively simplistic; more detailed analyses within the planning methodology may be required.

Land Impacts

A power plant (not including transmission lines) impacts land and its use by disrupting or precluding other potential land uses, altering the landscape, and altering soil composition due to plant construction and emissions of various types. None of these impacts are taken into account explicitly in RELM. They are considered, however, in the definition of the set of feasible facility locations.

All of the land impacts mentioned above tend to be localized in that they occur within the scale of our smallest planning unit of a county-sized cell. Thus, land impacts are best taken into account early in the analysis when potential plant locations are assembled. Those cells in which the known land impacts would be unacceptable (because of political opposition, special soil characteristics, seismicity, zoning, local topography, or the lack of sufficient space for a plant) should be excluded from the feasible set. In effect, land impact is treated here as a strict constraint resulting in the infeasibility of a location if any of the various land impacts are unacceptable.

Environmental Quality—Air Constraints

There are essentially two types of gaseous emissions from power plants. The first type is associated with the production of power, e.g., combustion products, fly ash, or radioactive emissions. However, it is assumed here that radioactive emissions are to be controlled and do not present any real locational problems other than the objectives of maintaining a free zone and minimizing the population within a certain distance of the plant. Radioactive emissions from nuclear plants are taken up again in the context of population impact. The second type is associated with the type of cooling process, e.g., water vapor and warmer air.

Both types of emissions can be considered undesirable if the magnitude of emissions is relatively large. For example, it could happen that due to local meteorological conditions any increase in water vapor could bring on troublesome fogging. This would necessitate a certain type of cooling process; stated in another way, this dictates that a certain other process could not be used. A simple method for ensuring that a particular technology is not used would be to use the following type of constraint:

$$S_{jk} = 0 \tag{10-17}$$

for those sites j where technology k would be incompatible with meteorological or other conditions.

A more realistic approach would be to limit the capacities of units established on the basis of meteorological conditions. For example,

$$S_{jk} \leq \bar{C}_{jk} \tag{10-18}$$

where $\bar{C}_{jk}$ is the maximum allowable number of units of type k at site j due to meteorological or other prevailing conditions.

It is also possible to incorporate a more substantial approach within RELM to maintain air quality standards. For example, the elements of Teller's (1968) fuel substitution model can be included in RELM to model interactive meteorological effects. This is done in the following way:

$$\sum_{k} \sum_{j} \Gamma_{jn}^{m} E_{mk} S_{jk} \leq \varepsilon_{mn} \qquad \forall m, n \tag{10-19}$$

where ε_{mn} is the standard for air pollutant m at monitoring station n, E_{mk} is the emission level of pollutant m per unit of capacity of technology k, Γ_{jn}^{m} is the net transfer of pollutant m from site j to monitoring station n, and S_{jk} is the capacity established at site j using technology k. This latter approach could perhaps be classified at a level of detail greater than that actually called for in RELM. Whether to include this representation or the more simple approaches depends on the analytical context.

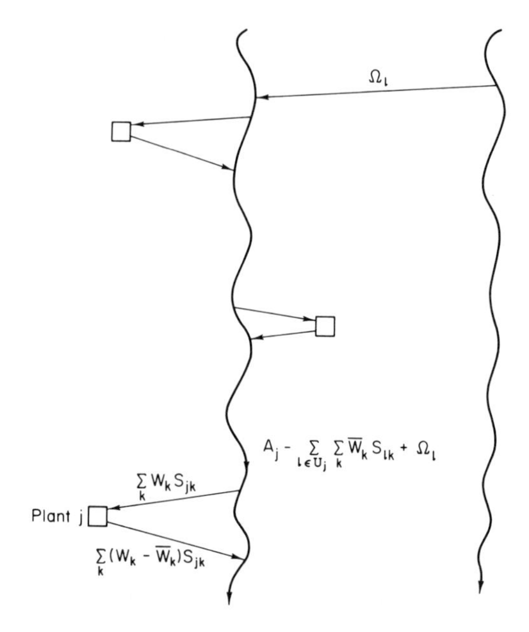

Fig. 10-5. Representation of water use in the regional energy facility location model (RELM).

Water Availability

Power plants need water for cooling, which is taken into account in the model by the water availability constraints. A representation of two river basins is shown in Fig. 10-5. There are three power plants shown (although this is an arbitrary number) each of which has water diverted to it from the stream. The quantity of this diversion at site j is

$$\text{diversion at site } j = \sum_k W_k S_{jk} \tag{10-20}$$

where W_k is the cooling water flow required per period (say a year) by a unit of type k (fuel type, cooling option, etc.) and S_{jk} is the capacity of type k to be installed at site j. Note that if water requirements are site specific, then the water-use coefficients should be further subscripted, i.e., as W_{jk}.

The consumptive use at a plant, i.e., the difference between water diverted and water returned to the stream, is denoted by $\overline{W}_k$, which is again for a unit of type k. Assuming linearity, the total consumptive use at site j is

$$\text{consumptive use at site } j = \sum_k \overline{W}_k S_{jk} \tag{10-21}$$

and total water return from site j is

$$\text{return flow from } j = \sum_k W_k S_{jk} - \sum_k \overline{W}_k S_{jk} = \sum_k (W_k - \overline{W}_k) S_{jk} \tag{10-22}$$

as shown in Fig. 10-5.

The purpose of the water requirement constraints is to ensure sufficient water for power plant operations. This can be accomplished by imposing continuity at each site from which water is diverted from the stream for cooling purposes. For site j we would write

$$\text{water diverted} \leq \text{water available}$$

which can be represented mathematically as

$$\sum_k W_k S_{jk} \leq A_j - \text{water lost upstream} + \text{water added upstream} \tag{10-23}$$

where U_j is the set of all power plant sites, UI_j the set of all import sites, beyond what currently exists.

The inequality in (10-23) can be further developed by considering Fig. 10-5. The sources of upstream water losses are all of the power plants and the potential interbasin transfers (exports) Ω_l upstream of point j. The source for a water addition (import) is the interbasin transfer Ω_l in the opposite direction. Incorporating these flows into (10-23) gives

$$\sum_k W_k S_{jk} \leq A_j - \sum_{l \in U_j} \sum_k \overline{W}_k S_{lk} + \sum_{l \in UI_j} \Omega_l - \sum_{l \in UE_j} \Omega_l \tag{10-24}$$

where U_j is the set of all power plant sites, UI_j the set of all import sites, and UE_j the set of all export sites upstream of site j. Putting decision variables on the left-hand side and knowns on the right yields

$$\sum_k W_k S_{jk} + \sum_{l \in U_j} \sum_k \overline{W}_k S_{lk} - \sum_{l \in UI_j} \Omega_l + \sum_{l \in UE_j} \Omega_l \leq A_j \qquad \forall j \tag{10-25}$$

As mentioned before, A_j represents the water available, assuming no development in the stream other than what currently exists. A_j may be measured as the α%-safe yield or that flow in the stream that, based on past records, would be equaled or exceeded α% of the time, where α is prespecified. The choice of α for the determination of A_j would depend on the analytical point of view.

Another consideration relative to water use is total consumptive use in a river basin in addition to consumption at each site. It may be desirable to constrain total consumptive use in each river basin. This would be represented by

$$\sum_{j \in U_{j^*}} \sum_{k} W_k S_{jk} \leq \sigma_{j^*} \qquad \forall j^* \tag{10-26}$$

where j^* is the point in a basin farthest downstream, U_{j^*} is the set of all sites in the basin upstream of j^*, and σ_{j^*} is the total amount of allowable consumptive use in the basin. σ_{j^*} may be a fraction of a low flow of interest, such as the seven-consecutive-day low flow occurring once in 10 years. One may choose to vary σ_{j^*} to consider the impact of reservoir development on energy facility locations.

Thermal Discharges

Power plants generate waste heat, which is released in discharges of heated cooling water. Since the discharged water is warmer than the receiving water body, a temperature rise of the latter is experienced. The purpose of the thermal discharge constraints is to ensure that temperature-rise standards of receiving water bodies can be met.

The maximum permissible temperature rise in degrees Fahrenheit at site j (allowing for variations in standards from state to state) will be denoted as $T_{j,\max}$. The maximum allowable heat discharge $H_{j,\max}$ from plant j in British thermal units per hour is related to $T_{j,\max}$ by

$$H_{j,\max} = 3600 T_{j,\max} \rho C_p \bar{A}_j \tag{10-27}$$

where ρ is the density of water in pounds per cubic foot, C_p is the specific heat of water in British thermal units per pound per degree Fahrenheit, 3600 is the number of seconds in an hour, and $\bar{A}_j$ is the flow in cubic feet per second at site j. Equation (10-27) can be rewritten as

$$H_{j,\max} = \Psi_j \bar{A}_j \tag{10-28}$$

where

$$\Psi_j = 3600 T_{j,\max} \rho C_p \tag{10-29}$$

The maximum allowable heat input to the stream represents an upper bound on the heat discharge from power plants on controlled water bodies. Taking into account all of the power plants that discharge at j or at points upstream of j leads to the constraint

$$\sum_{k} H_{jk} S_{jk} + \sum_{l \in U_j} \theta_{lj} \sum_{k} H_{lk} S_{lk} \leq H_{j,\max} \tag{10-30}$$

where H_{jk} is the average heat load discharged in British thermal units per hour per unit plant size of type k (including cooling option) at site j, θ_{lj} is an attenuation coefficient that represents the decay of thermal impact of heat discharged from an upstream point l to point j, and all other symbols are defined as before. In effect (10-30) requires that the heat load discharged at j and the heat load remaining from upstream discharges not exceed the maximum permissible heat load in the water body at point j.

The constraint in (10-30) can be rewritten further. Substituting for $H_{j,\max}$ from (10-28), the constraint becomes

$$\sum_k H_{jk} S_{jk} + \sum_{l \in U_j} \theta_{lj} \sum_k H_{lk} S_{lk} \leq \Psi_j \bar{A}_j \tag{10-31}$$

The flow $\bar{A}_j$ can be expressed in terms of the decision variables and parameters used in the last subsection as

$$\bar{A}_j = A_j - \sum_{l \in U_j} \sum_k \overline{W}_k S_{lk} - \sum_k \overline{W}_k S_{jk} + \sum_{l \in UI_j} \Omega_l - \sum_{l \in UE_j} \Omega_l \tag{10-32}$$

The equation defines the flow immediately downstream of site j as the current flow at j, A_j, minus consumptive loss at power plants at site j and upstream of j, plus upstream imports, minus upstream exports.

Substituting (10-32) into (10-31) and rearranging gives the final form of the constraint:

$$\sum_k (H_{jk} + \Psi_j \overline{W}_k) S_{jk} + \sum_{l \in U_j} \sum_k (\theta_{lj} H_{lk} + \Psi_j \overline{W}_k) S_{lk}$$

$$- \Psi_j \sum_{l \in UI_j} \Omega_l + \Psi_j \sum_{l \in UE_j} \Omega_l \leq \Psi_j A_j \qquad \forall j \tag{10-33}$$

It should be noted that the water requirement and thermal impact constraints may take on different forms depending on the physical configuration under consideration. For example, more than one generating site may withdraw or discharge at the same point in the stream. Particular variations such as this can easily be considered by altering the appropriate terms in the constraints. It should also be pointed out that the heat constraints could easily be made seasonal if the temperature rise standards vary from one season to the next. Finally, note that the safe yield A_j is used here. This seems reasonable since one is concerned with, in this case, events of relatively common frequency. Any appropriate value of A_j could, of course, be included.

10.3.5 Facility and Infrastructure Costs

The costs for new capacity arise from capital expenditures and operating and maintenance costs for all of the necessary equipment and structures related to generation, transmission, and cooling and from infrastructure

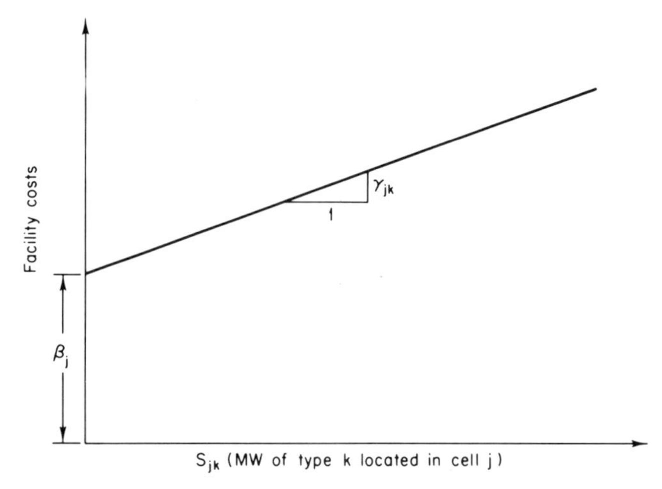

Fig. 10-6. Representation of power plant facility costs.

costs. These costs vary with plant type and may be nonlinear for a given plant. A possible representation of plant costs is shown in Fig. 10-6. It is assumed that costs are composed of fixed costs β_j for site preparation and linear variable costs γ_{jk}, which are a function of plant type. The fact that fixed costs may themselves be a function of plant size and type is not captured by this function. Total plant costs at site j, C_j, are thus

$$C_j = \beta_j F_j + \sum_k \gamma_{jk} S_{jk} \qquad \forall j \tag{10-34}$$

where the fixed costs, which are incurred if a plant of any type or size is built at j, are associated with F_j. Total system costs C are

$$C = \sum_j C_j = \sum_j \left(\beta_j F_j + \sum_k \gamma_{jk} S_{jk} \right) \tag{10-35}$$

Water-use facilities, such as transfers and reservoirs, should also enter into the cost function. These elements should be included as the application of the model requires.

The fixed cost portion of the cost function β_j reflects the economies of scale and the importance of infrastructure in energy facilities. If fixed costs are not significant, then the term $\beta_j F_j$ may be omitted.

Transmission of energy results in another cost factor. A reliable transmission cost estimate requires a load flow analysis. It is possible, however, to include a good representation of the transmission problem within RELM itself. The necessary additions to the model are presented in Church and Cohon (1976, Appendix C). A simpler and less realistic approach is presented here. It is based on the assumption that energy facilities will be tied into an existing transmission grid at the nearest load center to the facility. This assumption is unrealistic when energy centers or concentrated areas of capacity are contemplated. However, the simple approach does promote a wide distribution of facilities close to demand areas, which is the desirable effect of transmission costs.

Define δ_j as the transmission cost or distance from facility site j to its closest load center. This parameter is incorporated into (10-35) to give

$$C = \sum_j \left[(\beta_j + \delta_j)F_j + \sum_k \gamma_{jk} S_{jk} \right] \tag{10-36}$$

which is the total cost function to be minimized. Notice that transmission costs do not vary with capacity. It is also worth reiterating that a RELM solution will generally underestimate transmission costs. For this reason, a detailed load flow analysis may be an important part of the overall methodology.

10.3.6 Water Transfers as an Objective

The interbasin transfer of water is a controversial alternative, but an important one in arid regions, such as the southwest United States. The minimization of interbasin transfers to provide cooling water may be a relevant objective because of the environmental consequences of transfers. The statement of this objective is quite straightforward:

$$\text{minimize} \quad \sum_l \Omega_l \tag{10-37}$$

where Ω_l is as defined previously.

Interesting tradeoffs among transfers and the degree of capacity concentration can be generated. The impact of in-basin water resource development, i.e., reservoirs, on the necessity for interbasin transfers could also be considered by varying A_j in the water supply constraints (10-25) and in the thermal discharge constraints (10-33), and by varying σ_{j*} in the water-use constraints (10-26). It should also be pointed out that it is quite appropriate to include a separate objective for water transfers when these alternatives also enter into the cost objective: This is not double-counting!

10.3.7 Population Safety Impact

Questions have been raised with respect to the safety of nuclear power plants. Unfortunately, there is no easy way to identify the risks of long-term low-level radiation exposure. In addition, although the impact of a significant uncontrolled breakdown is known, the probability of such an occurrence is hard to quantify. There are other safety issues related to nuclear-powered generating facilities, but their severity is not primarily a function of facility location. Therefore these risks are not considered in the model. Long-term low-level exposure is the main consideration.

One approach to minimizing the impact of nuclear reactors, based on the individual dose guidelines of the U.S. Nuclear Regulatory Commission (*Code of Federal Regulations*, Title 10, Part 100, 1966) is to provide an exclusion area immediately around a facility and minimize the number of people within a certain radius from the plant. The exclusion area aspect is dealt with in RELM during the site selection process while other criteria related to the larger impact area are explicitly considered.

Two alternative statements of a population safety objective based on the risk associated with living within a given distance of a nuclear facility are presented below. It should be noted that neither formulation takes into account the spatial distribution within the potential impact distance. Meteorological conditions, e.g., prevailing wind directions, are also not considered.

The first formulation assumes that the more facilities established within the potential impact distance, the larger the risk, and the second assumes that the major risk or impact is associated with the mere presence of one or more plants within the distance. The first approach is represented by the following objective:

$$\text{minimize} \quad \sum_{i} \sum_{j \in IA_i} P_i F_j \tag{10-38}$$

where P_i is the population within the area represented by point i, F_j is a 0–1 variable defined previously, and IA_i is the set of potential facility locations that may impact point i, i.e., that are within the potential impact distance of cell i.

This objective has units of *plant-people*. A plant-person is one person living within a given distance of a nuclear power plant. If that person lives within a given distance of two plants, then the impact is counted twice and equal to two plant-people. The objective is to minimize the total impact in plant-people as a surrogate approach to maximizing safety.

The above approach does not capture the potential dependence of risk to populations on the capacity of plants located within the impact area, nor does it distinguish between fossil-fueled and nuclear capacity. However, these restrictions can be relaxed by using

$$\text{minimize} \quad \sum_{i} \sum_{k \in \tau_N} \sum_{j \in IA_i} S_{jk} P_i \tag{10-39}$$

This objective is measured in units of capacity-people (e.g., 1000-MW-people), where a capacity-person is one person living within a prespecified distance of an established unit of capacity. The objective also excludes non-nuclear plants.

For the second formulation, in which a person is impacted by the mere presence of one or more facilities but not as a function of the number of facilities, one can consider the following approach:

$$\text{minimize} \quad \sum_i P_i C_i \tag{10-40}$$

subject to the additional constraint

$$-F_j + C_i \geq 0 \qquad \forall i \quad \text{and} \quad \forall j \in IA_i \tag{10-41}$$

The variable C_i is equal to one only if one or more facilities are placed within the potential impact distance of area i. Thus, the objective function measures the total population within the potential impact distance of one or more power plants. This formulation also does not distinguish between nuclear and nonnuclear technologies. This can be incorporated, however, by altering the constraint (10-9). Additional 0–1 integer variables for the siting of a nuclear facility at j would be defined and used in (10-9).

10.3.8 Distribution of New Capacity

Without some type of constraint limiting the use of each particular geographical or political area, it is possible that a large fraction of the generating capacity could be assigned to one particular area in the region. This would occur if there were an unusually large water supply close to but with no significant impact on load centers, no appreciable problem with heat dissipation, and no other major environmental difficulty. It could also occur if any one area should prove to be generally better (i.e., have fewer impacts than other areas). Such a configuration of facilities, if it indeed does occur, would most likely generate strong objections from the targeted area and possibly lead to strong criticisms of the overall approach for locating generating capacity.

There are several approaches that can be taken to promote a more equitable or, at least, a more politically acceptable distribution of facilities. The approaches discussed here are predicated on the quantification of a distribution objective, which may take one of several forms. One approach is to require that each geographical area house at least a fraction of the capacity needed for that area; this minimum fraction would then be traded off against the other objectives. This approach is accomplished mathematically by defining G_g as the set of all points i or potential facility locations j that are in the area g, and f_g as the minimum acceptable fraction of new capacity requirements in area g that must be located within area g. These definitions are used in the constraint

$$\sum_{j \in G_g} \sum_k S_{jk} \geq \left(\sum_{i \in G_g} P_i \Big/ \sum_i P_i \right) Df_g \qquad \forall g \tag{10-42}$$

This relationship requires that established capacity in area g be at least as large as that fraction of total demand generated by the area (measured as the fraction of total regional population located in area g) multiplied by the minimum supply fraction f_g.

Even if each area is required to serve a certain fraction of its needs, it could also happen that a particular area could be called upon to serve considerably more than its needs. This could be mitigated by incorporating the following constraint:

$$\sum_{j \in G_g} \sum_k S_{jk} \leq \left(\sum_{i \in G_g} P_i \Big/ \sum_i P_i \right) Df'_g \qquad \forall g \tag{10-43}$$

where f'_g is the maximum allowable fraction of needed capacity established in the immediate area g.

Note that the above representation appears as constraints in the model. Tradeoffs between the capacity distribution objective and other objectives could be generated by varying the minimum or maximum fractions, f_g or f'_g.

An alternative approach to equitably distributing plant capacity is to use an explicit objective function that minimizes the range of the distribution directly:

$$\text{minimize} \quad (y_{\max} - y_{\min}) \tag{10-44}$$

subject to (10-42), (10-43), and the additional constraints

$$f'_g - y_{\max} \leq 0 \qquad \forall g \tag{10-45}$$

$$f_g - y_{\min} \geq 0 \qquad \forall g \tag{10-46}$$

where $y_{\max}$, a decision variable, is the largest fractional generating excess for any area; $y_{\min}$, a decision variable, is the largest fractional generating deficit for any area; and f_g and f'_g are now decision variables.

Another possibility would be to minimize the sum of the deviations of each area's new capacity from its needs. This can be done with the objective

$$\text{minimize} \quad \sum_g (y_g + z_g) \tag{10-47}$$

subject to the additional constraint

$$\sum_{j \in G_g} \sum_k S_{jk} + y_g - z_g = \left(\sum_{i \in G_g} P_i \Big/ \sum_i P_i \right) D \qquad \forall g \tag{10-48}$$

where y_g and z_g are the negative and positive deviations, respectively, of the capacity located in g from the new capacity requirements in g. It is important to note that this objective is presented in terms of capacity excess or deficit whereas the previous approaches are couched in terms of *fractional* capacity excess or deficit. This approach is mathematically similar to the minimum-deviations approach to the water allocation problem in Argentina discussed in Chapter 9 (see Section 9.4.4).

Other approaches could be taken to consider the equity of new capacity distribution, but the alternatives discussed above indicate the range of formulations that are possible. In addition to other mathematical forms for this objective, it may be desirable to redefine the objective in terms of a given capacity type, e.g., nuclear. To do this one need only sum over those k that are of the specified type in (10-42), (10-43), and (10-48). A nuclear plant distribution objective may be included in addition to, rather than in place of, the total capacity distribution objective.

10.3.9 Implementing and Using the Model

The RELM is summarized in Table 10-1. The model is a large mixed-integer programming model with four objectives. At the present stage the model is theoretical since it has not yet been implemented, although two on-going projects are directed at its use for regional energy planning. Successful implementation depends on the extent of data and preanalysis requirements and the computational burden of the model. These aspects of the model are discussed below.

Site and Analysis Area Selection

The RELM is based on a discrete representation of a region. "Sites" or "cells" in the model are actually points in space that represent any land area, which may range in size from a few square miles to several counties. The selection of the area that a point is to represent should be based on computational and analytical considerations. A useful rule would be to select points to represent the largest contiguous area possible that will still capture all of the important aspects of the problem; i.e., the area should be approximately homogeneous with respect to population density, land use, environmental attributes, and existing infrastructure. One implication of this rule is that the area represented may vary from point to point, e.g., small for urbanized areas and relatively large in rural areas.

Site selection is also an initial screening process. Only those areas that are feasible facility locations should be included. Considerations such as land

TABLE 10-1

SUMMARY OF THE REGIONAL ENERGY FACILITY LOCATION MODEL

	Equation No.
Objectives	
(1) Minimize facility and transmission costs	(10-36)
(2) Minimize water transfers	(10-37)
(3) Population safety	
(a) Minimize plant-people within the potential impact distance of facilities, *or*	(10-38)
(b) minimize MW-people within the potential impact distance of facilities, *or*	(10-39)
(c) minimize people within the potential impact distance of facilities	(10-40)
(4) Distribution of new capacity	
(a) Treated as constraints (see constraint 11), *or*	
(b) minimize range of the fraction of needed new capacity located in each subarea of interest, *or*	(10-44)
(c) minimize deviations of located new capacity from required new capacity in each subarea	(10-47)
Constraints	
(5) New capacity and concentration requirements	(10-8) and (10-9) [See also (10-10)–(10-16)]
(6) Air quality constraint	(10-18) or (10-19)
(7) Water supply for cooling constraint	(10-25)
(8) Limit on consumptive use	(10-26)
(9) Limit on thermal impact	(10-33)
(10) Constraint for objective 3c	(10-41)
(11) Constraints for objective 4a	(10-42) and/or (10-43)
(12) Constraints for objective 4b	(10-42), (10-43), (10-45), and (10-46)
(13) Constraints for objective 4c	(10-48)

use, safety exclusion zones, and seismicity will disqualify some areas. Note, however, that spatial attributes such as population must be adequately captured so that location impacts are realistically reflected by the model. This requirement may result in the inclusion of infeasible locations as "demand points" in the model at which facilities may not be placed.

The region to which RELM may be realistically applied is related to the site selection process. Computational requirements of RELM are a function of the number of discrete points that are included in the model. Thus, as the area of analysis increases, the cell size must also increase to maintain computational efficiency. A reasonable scale for the application of the model would be at the level of an energy pool—an integrated group of power utilities that cooperate for reliability purposes.

Data Requirements

The data requirements for RELM are extensive and each piece of the data may imply a significant amount of analysis for its derivation. This difficulty is not necessarily a characteristic of RELM, but is generic to complex problems.

The first major requirement is the development of a discrete representation of the analysis area. This aspect and the site-screening process are discussed above.

Data on capital and operating costs of generating facilities and infrastructure costs for all sites and technological alternatives are required. In addition, water transfer and reservoir costs (if low-flow augmentation is considered) must be derived. These data would probably be the easiest to obtain.

Distances for impact areas of nuclear facilities must be supplied in order to state the safety objectives. There is some uncertainty regarding the appropriate value for this distance, so sensitivity analysis will be particularly important here.

Upper bounds on capacity concentration and fuel mix requirements must be derived. These are essentially policy issues that, again, may require a significant amount of prior analysis and experimentation.

Water requirements, consumptive use, and heat generation by fuel and cooling option must be derived. These data may require some effort, but one would expect little analytical difficulty. A more nebulous task, however, is the estimation of heat and air pollution attenuation coefficients, maximum allowable heat loads, and air quality standards. Sensitivity analyses should be emphasized here in an attempt to deal with the uncertainty surrounding these numbers.

There are undoubtedly other data requirements that would become apparent only when implementation is attempted. The data required are extensive, but regional energy planning demands this complexity.

Computational Requirements

The regional energy facility location model has been formulated as a linear integer programming problem of considerable size. The solution costs of RELM are a function of the number of constraints and integer variables included in the formulation. A problem with, say, 100 discrete points [on the order of a county-level analysis of the Pennsylvania–New Jersey–Maryland (PJM) pool] and five total technological options (fuel and cooling combinations), would have approximately 800–1300 constraints and 100–200 integer 0–1 variables in addition to several hundred continuous variables, depending on the specific form of the safety and distribution objectives employed. This is a large problem, but certainly not beyond the capability of existing linear and integer programming packages such as IBM's Mathematical Programming System Extended (MPSX).

A substantial saving could be realized if the fixed cost portion of facility costs were ignored and the safety objectives that require 0–1 integer variables were not used. In this case the formulation would be a linear programming problem with close to 800 constraints. This is definitely not an inordinately large problem. It would probably cost on the order of $50 for a single solution on a computer with rates consistent with academic computing facilities. Of course, the sacrifice of realism associated with the elimination of integer variables must be carefully examined.

Current projects at Brookhaven National Laboratory and at The Johns Hopkins Applied Physics Laboratory are employing the linear programming (noninteger) form of the model. Brookhaven is using the model for a coal utilization assessment in the northeast United States. The Hopkins project is exploring energy facility location in the Federal Energy Administration's Region III (Pennsylvania, Maryland, Delaware, Virginia, and West Virginia) plus New Jersey to meet the demands of that region in the year 2000. Both projects report reasonable computational costs.

10.4 SUMMARY AND CONCLUSIONS

The goal of this chapter was to demonstrate the applicability of multiobjective analysis to facility location decision making. Fire stations and power plants were considered, representing two different kinds of facilities—emergency and noxious, respectively. The notion of a multiobjective approach was intuitively appealing to at least one decision maker—the Fire Chief of Baltimore, Maryland. He appreciated the role that the analyst filled: Alternatives were generated for his consideration; an optimal alternative was not recommended. It is reasonable to expect that this generating philosophy would also be appropriate for regional energy decision making, which is characterized by a multiplicity of interests in the private sector, in the quasi-public sector (i.e., utilities), and in the public sector at all levels of government.

There is a wide range of location problems, many of which exhibit the multiobjective characteristics shown in this chapter. The analysis of many of these problems would benefit from the use of multiobjective methods, and one can expect multiobjective applications to facility location to increase with time.

CHAPTER 11

Summary and Prospects for Future Development

The focus of this book has been on the range of methods available for the solution of multiobjective problems and the role of those techniques in the public decision-making process. The discussion was pragmatic in that the application of the methods was emphasized. Without repeating the discussions in Chapters 1, 2, and 5, the key points that were made regarding the place for multiobjective programming in the planning process will now be reviewed, and the prospects for future theoretical and practical work will be discussed. The chapter concludes with a "pep talk" for analysts and practitioners.

11.1 THE VALUE OF MULTIOBJECTIVE ANALYSIS

In considering the roles of the analyst and decision maker in the public decision-making process, it was claimed that multiobjective approaches are superior to conventional single-objective methods. The key point here is that analysts should analyze and decision makers should decide. Single-objective models proceed to the identification of a so-called "optimal" solution—the feasible solution that is best in terms of a single measure of value. Decision makers are given the choice of accepting or rejecting this single solution without learning anything about how the solution compares with other feasible solutions. Since in a public decision-making context, a single objective can be defined only by making important and perhaps controversial value judgments, the analyst is forced by single-objective approaches to usurp a large part of the decision makers' responsibilities.

Multiobjective analysis, by contrast, emphasizes the range of choice associated with a decision problem. The important judgments regarding the relative values of objectives are not made by the analyst; the burden of making these value judgments rests, instead, squarely on the shoulders of the decision makers. Furthermore the articulation of the value judgments requires a consideration of the range of noninferior solutions. Thus, in addition to maintaining suitable roles for the analyst and decision maker, multiobjective analyses will generally develop better understanding of and insight into the problem.

If one accepts the premises that decision makers like to decide and that it is their responsibility to decide, then the value of multiobjective methods is difficult to challenge in light of the above discussion. We can extend this claim for the virtue of multiobjective analysis by considering its implications for systems analysis.

Systems analysis has not fulfilled the expectations that many had for its role in public decision making. The mid-1960s were the peak of enthusiasm for systems analysts since they saw their tools as the answer to the complex and multifaceted problems of the public sector. As more and more models for public systems were relegated to bookshelves that rapidly became dusty, systems analysts had to stop and reconsider. This period of reevaluation has been well documented in the journal *Interfaces* (a joint publication of the Operations Research Society of America and the Institute for Management Science) since the early 1970s; indeed, this reassessment is being conducted at the present.

It is the claim and essence of this book that effective public systems analysis requires the use of multiobjective analysis. One of the reasons that systems analysis has not yet realized its potential for the solution of public sector problems is the preponderance of single-objective models that generate an optimal solution. Decision makers wish to decide; they like to have choices and therefore a better understanding and control of the decision situation.

One could claim that single-objective models that have been used to generate a single optimal solution have been misused. That would be a good claim to make, for that is precisely the point of multiobjective analysis. Good systems analyses stress sensitivity analysis for the exploration of the range of choice. Multiobjective analysis is nothing more than systematic sensitivity analysis of the most important value judgments. The emphasis in a multiobjective analysis, however, is on the explicit identification of those value judgments.

Several methods were reviewed in detail in Chapters 6–8. Among the methods there were many differences, but of most importance was the role for the analyst that the use of each method implied. This consideration motivated our categorization of the methods into three classes: generating techniques, preference-oriented methods, and multiple-decision-maker methods. Each of these three categories is briefly reviewed below.

11.1.1 Generating Techniques

The generating techniques reviewed in Chapter 6 included the weighting and constraint methods, the noninferior set estimation (NISE) method, and the multiobjective simplex method. The techniques are most useful when the analyst needs an approximation or exact representation of the noninferior set. The planning procedure proceeds with the decision makers' consideration of this range of choice. The selection of a best-compromise solution from among the generated noninferior solutions provides an implicit articulation of the relative values of the objectives.

The applicability of generating methods is relatively insensitive to the specific nature of the decision-making process—the number, identifiability, and accessibility of decision makers. Generating methods emphasize the development of information, but in doing so usually incur relatively high computational costs. The costs for applying a generating technique increase exponentially with the number of objectives.

11.1.2 Techniques That Incorporate Preferences

The preference-oriented methods reviewed in Chapter 7 included several techniques that require the explicit articulation of preferences either prior to solution or in an iterative manner. Multiattribute utility functions, the prior assessment of weights, geometrical notions of best, the surrogate worth tradeoff method, and iterative techniques were discussed. The analyst's role is generally similar to that prescribed by generating methods. Preference-oriented methods require, however, that the analyst go beyond the passive information-providing role and counsel the decision maker in an attempt to articulate preferences.

The applicability of techniques that incorporate preferences is sensitive to the nature of the decision-making process. These methods require explicit articulation of value judgments so unidentified or inaccessible decision makers will cause problems. Multiple-decision-maker settings also raise the issue of whose preferences should be used. The inclusion of preferences—when it is possible—does reduce the computational costs of analysis.

11.1.3 Multiple-Decision-Maker Problems

This very speculative area of inquiry is directed at situations in which there are several decision makers. It was shown in Chapter 8, however, that some of these situations could be treated as multiobjective problems. In our discussion we considered welfare economics, Paretian analysis, and game theory. It was convenient for the discussion to subdivide the multiple-decision-maker methods further into three classes: techniques for the aggregation of individual preferences, techniques for counseling one of many decision makers, and methods for the prediction of political outcomes.

It is difficult to generalize about the applicability and costs of these methods. Some of them have never been applied and others have been applied to small hypothetical problems.

11.2 PROSPECTS FOR FUTURE THEORETICAL DEVELOPMENT AND APPLICATION

In spite of the impressive array of existing multiobjective methods, new and better methods will undoubtedly be developed in the future. In the generating category, current work is directed at increasing the efficiency of the generating process both for approximating and identifying exactly the noninferior set. Most mathematical programmers have been concentrating on multiobjective simplex algorithms. Approximation of the noninferior set has been generally ignored by multiobjective researchers, with the exception of this author and his colleagues. This author would encourage researchers to direct more attention at approximation, which for most practical problems is the only feasible and desirable thing to do.

Most of the recent work within the preference-based category has been directed at iterative techniques. This is an exciting area of analysis because iterative methods attempt to involve the decision maker in the solution process. There has been a recent emphasis on the development of interactive computer codes that permits decision makers to solve their own problems during a session or two at a computer terminal.

Within multiple-decision-maker methods, the greatest area of activity is in game theory. A great deal of recent and on-going research has gone into games with more than two participants and with complicating characteristics such as cooperation and bargaining. In a related area, voting algorithms are also of interest.

It is in the area of application, however, that the future of multiobjective programming appears most promising and exciting. The discussions of Chapters 1 and 2 indicated a wide range of public sector problems to which multiobjective programming has been or can be applied. Water resources, transportation, land-use and energy planning, school desegregation, and pollution control are just some of the problem areas that lend themselves to multiobjective analysis.

Hypothetical applications are not the subject of interest; rather, real-world problem solving of the sort in Chapters 9 and 10 is the direction in which we must move. It is only here—in the real decision-making arena—that the true value of multiobjective analysis can be tested. The importance of this cannot be overstressed, not only because the methods are applicable to real-world problems, but because it is the real decision-making problem that leads to the most interesting new theoretical developments. Most of the techniques of operations research that we use today were developed in response to a real-world problem. That is precisely the distinction between operations research and mathematics, and it is a compelling reason for the emphasis on applications of multiobjective methods.

11.3 A PEP TALK FOR ANALYSTS AND PRACTITIONERS

It is undoubtedly obvious that the author of this book is fascinated with multiobjective analysis and its role in the solution of public decision-making problems. It is probably also obvious that there are many other authors and researchers—the developers of the methods discussed in this book—who are also fascinated with this topic. Why are we drawn so to multiobjective analysis? It is certainly a new and useful methodology for decision problems, but that is not the only reason. It seems that multiobjective analysis is so interesting because it is a conceptually simple extension of well-known single-objective techniques and, at the same time, an area very rich in substance and mathematical complexity. It also gets to the heart of the psychology of decision making in its emphasis on tradeoffs and conflict. For this last reason, multiobjective analysis not only has fascinated analysts but also has captured decision makers.

Multiobjective analysis includes a new way of thinking about problems. It frees the analyst and decision maker from the search for that unique unambiguous measure of what is best. Instead we can deal with problems on their own terms without assigning value where dollars or some other measure makes no sense. This is a great step forward for systems analysis and decision making.

References

Andrews, W., Burge, R., Capener, H., Warner, W., and Wilkinson, K. (1973). "The Social Well-Being and Quality of Life Dimension in Water Resources Planning and Development," 213 pp. (*Proc. Univ. Council on Water Resources Conf., Utah State Univ., Logan, Utah*).

Arrow, K. (1963). "Social Choice and Individual Values," 2nd ed., 124 pp. Cowles Foundation Monograph No. 12, Yale Univ. Press, New Haven, Connecticut.

Aumann, R. (1964). Subjective Programming. *In* "Human Judgments and Optimality" (M. Shelby and G. Bryan, eds.), pp. 217–242. Wiley, New York.

Bammi, De., Bammi, Da., and Paton, R. (1976). Urban Planning to Minimize Environmental Impact. *Environment and Planning* **8**, 245.

Barber, G. (1976). Land-Use Plan Design via Interactive Multi-Objective Programming. *Environment and Planning* **8**, 625.

Beeson, R. (1971). Optimization with Respect to Multiple Criteria, 140 pp. Ph.D. Thesis, Univ. of Southern California, Los Angeles.

Belenson, S., and Kapur, K. (1973). "An Algorithm for Solving Multicriterion Linear Programming Problems with Examples." *Operational Res. Quart.* **24**, 65.

Benayoun, R., Roy, B., and Sussman, B. (1966). ELECTRE: Une Méthode pour Guider le Choix en Présence de Points de Vue Multiples. Working Paper No. 49, SEMA (Metra International), Direction Scientifique, Paris.

Benayoun, R., deMontgolfier, J., Tergny, J., and Laritchev, O. (1971). Linear Programming with Multiple Objective Functions: Step Method. *Math. Programming* **1**, 366.

Bentham, J. (1948). "An Introduction to the Principles of Morals and Legislation," 378 pp. Hafner, New York.

Bergson, A. (1938). A Reformulation of Certain Aspects of Welfare Economics. *Quart. J. Econom.* **52**, 310.

Bergson, A. (1966). "Essays in Normative Economics," 246 pp. Harvard Univ. Press, Cambridge, Massachussetts.

Blackwell, D. (1956). An Analog to the Minimax Theorem for Vector Payoffs. *Pacific J. Math.* **6**, 1.

Boulding, K. (1952). Welfare Economics. *In* "A Survey of Contemporary Economics" (B. Haley, ed.), pp. 1–34. Richard D. Irwin, Inc., Homewood, Illinois.

Brill, E. (1972). Economic Efficiency and Equity in Water Quality Management, 139 pp. Ph.D. Thesis, Dept. of Geog. and Environ. Engrg., Johns Hopkins Univ., Baltimore, Maryland.

Brill, E., Liebman, J., and ReVelle, C. (1976). Equity Measures for Exploring Water Quality Management Alternatives. *Water Resources Res.* **12**, 845.

Briskin, L. (1966). A Method of Unifying Multiple-Objective Functions. *Management Sci.* **12**, B406.

Buie, E. (1974). Implementation of Multiple Objective Planning by the Soil Conservation Service. *In* "Multiple Objectives Planning Water Resources" (E. Michalson, E. Englebert, and W. Andrews, eds.), Vol. I, pp. 20–23. Idaho Research Foundation, Moscow, Idaho.

Burger, E. (1963). "Introduction to the Theory of Games," 202 pp. Prentice-Hall, Englewood Cliffs, New Jersey.

Charnes, A., and Cooper, W. (1961). "Management Models and Industrial Applications of Linear Programming," Vol. I, 467 pp. Wiley, New York.

Charnes, A., Cooper, W., DeVoe, J., Learner, D., and Reinecke, W. (1968). A Goal Programming Model for Media Planning. *Management Sci.* **14**, B423.

Charnes, A., Cooper, W., Niehaus, R., and Stedry, A. (1969). Static and Dynamic Assignment Models with Multiple Objectives and Some Remarks on Organization Design. *Management Sci.* **15**, B365.

Church, R., and Cohon, J. (1976). Multiobjective Location Analysis of Regional Energy Facility Siting Problems, 90 pp. Rep. No. BNL 50567, Policy Analysis Division, Brookhaven National Lab., Upton, New York.

Church, R., and ReVelle C. (1974). The Maximal Covering Location Problem. *Papers Regional Sci. Assoc.* **32**, 101.

Cochrane, J., and Zeleny, M. (1973). "Multiple Criteria Decision Making," 816 pp. Univ. of South Carolina Press, Columbia.

Cohon, J., and Marks, D. (1973). Multiobjective Screening Models and Water Resources Investment. *Water Resources Res.* **9**, 826.

Cohon, J., and Marks, D. (1975). A Review and Evaluation of Multiobjective Programming Techniques. *Water Resources Res.* **11**, 208.

Cohon, J., Facet, T., Haan, A., and Marks, D. (1973). Mathematical Programming Models and Methodological Approaches for River Basin Planning. 308 pp. Rep. No. R75-37, Dept. of Civil Engrg, MIT, Cambridge, Massachusetts.

Cohon, J., Church, R., and Sheer, D. (1979). Generating Multiobjective Tradeoffs: An Algorithm for Bicriterion Problems. *Water Resources Res.* (to be published).

Dalkey, N. (1969). The Delphi Method: An Experimental Study of Group Opinion. 88 pp. Memo. No. RM-5888-DR, Rand Corp., Santa Monica, California.

Dalkey, N. (1976). Group Decision Analysis. *In* "Multiple Criteria Decision Making, Kyoto 1975" (M. Zeleny, ed.), pp. 45–74. Springer-Verlag, Berlin and New York.

Dalkey, N., and Helmer, O. (1963). An Experimental Application of the Delphi Method to the Use of Experts. *Management Sci.* **9**, 458.

Dalkey, N., Rourke, D., Lewis, R., and Snyder, D. (1972). "Studies in the Quality of Life: Delphi and Decision Making," 174 pp. Heath, Lexington, Massachusetts.

David, L., and Duckstein, L. (1976). Multi-Criterion Ranking of Alternative Long-Range Water Resource Systems. *Water Resources Bull.* **12**, 731.

deNeufville, R., and Keeney, R. (1973). Multiattribute Preference Analysis for Transportation Systems Evaluation. *Transportation Res.* **7**, p. 63.

deNeufville, R., and Marks, D. (1974). "Systems, Planning and Design," 438 pp. Prentice-Hall, Englewood Cliffs, New Jersey.

deNeufville, R., and Stafford, J. (1971). "Systems Analysis for Engineers and Managers," 353 pp. McGraw-Hill, New York.

Donovan, W., and Jordan, J. (1974). Emerging Corps of Engineers Guidelines for Implementing the Planning Requirements of Water Resources Council's Principles and Standards and Related Policies. *In* "Multiple Objectives Planning Water Resources" (E. Michalson, E. Engelbert, and W. Andrews, eds.), Vol. I, pp. 9–20. Idaho Research Foundation, Moscow, Idaho.

Dorfman, R., and Jacoby, H. (1970). A Model of Public Decision Illustrated by a Water Pollution Policy Problem. *In* "Public Expenditures and Policy Analysis" (R. Haveman and J. Margolis, eds.), pp. 173–231. Markham, Chicago.

Dyer, J. (1972). Interactive Goal Programming. *Management Sci.* **19**, 62.

Dyer, J. (1973). An Empirical Investigation of a Man–Machine Interactive Approach to the Solution of the Multiple Criteria Problem. *In* "Multiple Criteria Decision Making" (J. Cochrane and M. Zeleny, eds.), pp. 202–216. Univ. of South Carolina Press, Columbia.

Easton, A. (1973). "Complex Managerial Decisions Involving Multiple Objectives," 421 pp. Wiley, New York.

Ecker, J., and Kouada, I. (1975). Finding Efficient Points for Linear Multiple Objective Programs. *Math. Programming* **8**, 375.

Evans, J., and Steuer, R. (1973). A Revised Simplex Method for Linear Multiple Objective Programs. *Math. Programming* **5**, 54.

Farquhar, P. (1975). A Fractional Hypercube Decomposition Theorem for Multiattribute Utility Functions. *OR* **23**, 941.

Farquhar, P. (1976). Pyramid and Semicube Decompositions of Multiattribute Utility Functions. *OR* **24**, 256.

Farquhar, P. (1977). A Survey of Multiattribute Utility Theory. *In* "Multiple Criteria Decision Making" (M. Starr and M. Zeleny, eds.), pp. 59–90. North-Holland, Amsterdam.

Fishburn, P. (1970). "Utility Theory for Decision Making," 234 pp. Wiley, New York.

Fishburn, P. (1973). Bernoullian Utilities for Multiple Factor Situations. *In* "Multiple Criteria Decision Making" (J. Cochrane and M. Zeleny, eds.), pp. 47–61. Univ. of South Carolina Press, Columbia.

Fishburn, P. (1974). Von Neumann–Morgenstern Utility on Two Attributes. *OR* **22**, 35.

Fishburn, P., and Keeney, R. (1974). Seven Independence Concepts and Continuous Multiattribute Utility Functions. *J. Math. Psych.* **11**, 294.

Fishburn, P., and Keeney, R. (1975). Generalized Utility Independence and Some Implications. *OR* **23**, 928.

Freimer, M., and Yu, P. (1974). The Application of Compromise Solutions to Reporting Games. *In* "Game Theory as a Theory of Conflict Resolution" (A. Rapoport, ed.), pp. 235–260. Reidel, Dordrecht, Holland.

Freimer, M., and Yu, P. (1976). Some New Results on Compromise Solutions for Group Decision Problems. *Management Sci.* **22**, 688.

Gass, S., and Saaty, T. (1955). The Computational Algorithm for the Parametric Objective Function. *Naval Res. Logist. Quart.* **2**, 39.

Geoffrion, A. (1967). Solving Bicriterion Mathematical Programs. *OR* **15**, 39.

Geoffrion, A., Dyer, J., and Feinberg, A. (1972). An Interactive Approach for Multi-Criterion Optimization with an Application to the Operation of an Academic Department. *Management Sci.* **19**, 357.

Giglio, R., and Wrightington, R. (1972). Methods for Apportioning Costs among Participants in Regional Systems. *Water Resources Res.* **8**, 1133.

Graaff, J. deV. (1971). "Theoretical Welfare Economics," 178 pp. Cambridge Univ. Press, London and New York.

Greenberger, M., Crenson, M., and Crissey, B. (1976). "Models in the Policy Process," 355 pp. Russell Sage Foundation, New York.

Gros, J. (1975). Power Plant Siting: A Paretian Environmental Approach. *Nuclear Engineering and Design* **34**, 281.

Hadley, G. (1962). "Linear Programming," 520 pp. Addison-Wesley, Reading, Massachusetts.

Haefele, E. (1970). Coalitions, Minority Representation, and Vote-Trading Probabilities. *Public Choice* **8**, 75.

Haimes, Y. (1973). Integrated System Identification and Optimization. *In* "Control and Dynamic Systems: Advances in Theory and Applications," (C. Leondes, ed.), Vol. 9, pp. 435–518. Academic Press, New York.

Haimes, Y. (1977). "Hierarchical Analyses of Water Resources Systems," 478 pp. McGraw-Hill, New York.

Haimes, Y., and Hall, W. (1974). Multiobjectives in Water Resources Systems Analysis: The Surrogate Worth Trade-Off Method, *Water Resources Res.* **10**, 615.

Haimes, Y., Hall, W., and Freedman, H. (1975). "Multiobjective Optimization in Water Resources Systems: The Surrogate Worth Trade-Off Method," 200 pp. Elsevier, Amsterdam.

Haith, D. (1971). The Political Evaluation of Alternative Metropolitan Water-Resource Plans, 188 pp. Tech. Rep. No. 31, Water Resources and Marine Sciences Center, Cornell Univ., Ithaca, New York.

Harsanyi, J. (1955). Cardinal Welfare, Individualistic Ethics, and Interpersonal Comparisons of Utility. *J. Political Economy* **63**, 309.

Harsanyi, J. (1965). Bargaining and Conflict Situations in the Light of a New Approach to Game Theory. *Amer. Econom. Rev.* **55**, 447.

Hicks, J. (1940). The Valuation of Social Income. *Economica* **7**, 105.

Hill, M. (1973). Planning for Multiple Objectives—An Approach to the Evaluation of Transportation Plans. 273 pp. Monograph No. 5, Regional Science Research Institute, Philadelphia, Pennsylvania.

Hillier, Y., and Lieberman, G. (1967). "Introduction to Operations Research," 639 pp. Holden-Day, San Francisco, California.

Holl, S. (1973). Efficient Solutions to a Multicriteria Linear Program, with Application to an Institution of Higher Education, 194 pp. Ph.D. Thesis, Dept. of Math. Sciences, Johns Hopkins University, Baltimore, Maryland.

Howard, N. (1971). "Paradoxes of Rationality: Theory of Metagames and Political Behavior," 248 pp. MIT Press, Cambridge, Massachusetts.

Howe, C. (1971). Benefit–Cost Analysis for Water Systems Planning. 144 pp. Monograph No. 2, American Geophysical Union, Washington, D.C.

Ignizio, J. (1976). "Goal Programming and Extensions," 261 pp. Heath, Lexington, Massachusetts.

International Business Machines, Inc. (IBM) (1968). Mathematical Programming System/360, (360A-CO-14X) Linear and Separable Programming User's Manual, H20-0476-1. International Business Machines, Inc., New York.

International Business Machines, Inc. (IBM) (1971). Mathematical Programming System Extended (MPSX), Linear and Separable Programming Program Description, SH20-0968-0. International Business Machines, Inc., New York.

Ijiri, Y. (1965). "Management Goals and Accounting for Control," 191 pp. North-Holland Publ., Amsterdam.

Johnsen, E. (1968). "Studies in Multiobjective Decision Models," 628 pp. Student-litteratur, Lund, Sweden.

Kaldor, N. (1939). Welfare Propositions and Interpersonal Comparisons of Utility. *Econom. J.* **49**, 549.

Keeney, R. (1969). Multidimensional Utility Functions: Theory, Assessment and Applications, 180 pp. Tech. Rep. No. 43, Operations Research Center, MIT, Cambridge, Massachusetts.

Keeney, R. (1971). Utility Independence for Multiattributed Consequences. *OR* **19**, 875.

Keeney, R. (1972a). Utility Functions for Multiattribute Consequences. *Management Sci.* **18**, 276.

Keeney, R. (1972b). An Illustrated Procedure for Assessing Multiattributed Utility Functions. *Sloan Management Rev.* **14**, 37.

Keeney, R. (1973a). Decomposition of Multiattribute Utility Functions, 33 pp. Tech. Rep. No. 80, Operations Research Center, MIT. Cambridge, Massachusetts.

Keeney, R. (1973b). A Decision Analysis with Multiple Objectives: The Mexico City Airport. *Bell J. Econom. Management Sci.* **4**, 101.

Keeney, R. (1973c). Utility Function for Response Times of Engines and Ladders to Fires. *Urban Analysis* **1**, 209.

Keeney, R. (1974). Multiplicative Utility Functions. *OR* **22**, 22.

Keeney, R., and Raiffa, H. (1976). "Decisions with Multiple Objectives: Preferences and Value Tradeoffs," 569 pp. Wiley, New York.

Kirkwood, C. (1976). Parametrically Dependent Preferences for Multiattributed Consequences. *OR* **24**, 92.

Koopmans, T. C. (1951). "Activity Analysis of Production and Allocation," 404 pp. Wiley, New York.

Kuhn, H., and Tucker, A. (1951). Nonlinear Programming. Proc. Berkeley Symp. Math. Statist. Probability, *2nd*, (J. Neyman, ed.), pp. 481–492. Univ. of California Press, Berkeley.

Lee, S. (1972). "Goal Programming for Decision Analysis," 387 pp. Auerback, Philadelphia, Pennsylvania.

Lee, S., and Clayton, E. (1972). A Goal Programming Model for Academic Resource Allocation. *Management Sci.* **18**, B395.

Lee, S., and Moore, L. (1977). Multi-Criteria School Busing Models. *Management Sci.* **23**, 703.

Little, I. (1952). Social Choice and Individual Values. *J. Political Economy* **60**, 422.

Loucks, D. (1975). Planning for Multiple Goals. *In* "Economy-Wide Models and Development Planning" (C. Blitzer, P. Clark, and L. Taylor, eds.), pp. 213–233. Oxford Univ. Press, London.

Loucks, D. (1977). An Application of Interactive Multiobjective Water Resources Planning. *Interfaces* **8**, 70.

MacKrimmon, K. (1973). An Overview of Multiple Objective Decision Making. *In* "Multiple Criteria Decision Making" (J. Cochrane and M. Zeleny, eds.), pp. 18–46. Univ. of South Carolina Press, Columbia.

Major, D. (1969). Benefit–Cost Ratios for Projects in Multiple Objective Investment Programs. *Water Resources Res.* **5**, 1174.

Major, D. (1974). Multiobjective Redesign of the Big Walnut Project. *In* "Systems Planning and Design" (R. deNeufville and D. Marks, eds.), pp. 322–337. Prentice-Hall, Englewood Cliffs, New Jersey.

Major, D. (1977). "Multiobjective Water Resource Planning." Water Resources Monograph No. 4, 81 pp. American Geophysical Union, Washington, D.C.

Major, D., and Lenton, R. (1978). "Multiobjective, Multi-model River Basin Planning: The MIT-Argentina Project." Prentice-Hall, Englewood Cliffs, New Jersey.

Major, D., Cohon, J., and Frydl, E. (1974). Project Evaluation in Water Resources: Budget Constraints. 285 pp. Tech. Rep. No. 188, Parsons Lab. for Water Resources and Hydrodynamics, MIT, Cambridge, Massachusetts.

Manheim, M. (1974). Reaching Decisions about Technological Projects with Social Consequences: A Normative Model. *In* "Systems, Planning and Design" (R. deNeufville and D. Marks, eds.), pp. 381–397. Prentice-Hall, Englewood Cliffs, New Jersey.

Marglin, S. (1962). Objectives of Water Resource Development: A General Statement. *In* "Design of Water-Resource Systems" (A. Maass, M. Hufschmidt, R. Dorfman, H. Thomas, Jr., S. Marglin, and G. Fair, eds.), pp. 17–87. Harvard Univ. Press, Cambridge, Massachusetts.

Marglin, S. (1967). "Public Investment Criteria," 103 pp. MIT Press, Cambridge, Massachusetts.

Marks, D., and Cohon, J. (1975). An Application of Linear Programming to the Preliminary Analysis of River Basin Planning Alternatives. *In* "Studies in Linear Programming" (H. Salkin and J. Saha, eds.), pp. 251–273. North-Holland Publ., Amsterdam.

McBean, E., Lenton, R., Vicens, G., and Schaake, J. (1972). "A General Purpose Simulation Model for the Analysis of Surface Water Allocation Using Large Time Steps," 245 pp. Tech. Rep. No. 160, Parsons Lab. for Water Resources and Hydrodynamics, MIT, Cambridge, Mass.

Meisel, W. (1973). Tradeoff Decisions in Multiple Criteria Decision Making. *In* "Multiple Criteria Decision Making" (J. Cochrane and M. Zeleny, eds.), pp. 461–476. Univ. of South Carolina Press, Columbia.

Miller, W., and Byers, D. (1973). Development and Display of Multiple-Objective Project Impacts. *Water Resources Res.* **9**, 11.

Monarchi, D., Kisiel, C., and Duckstein, L. (1973). Interactive Multiobjective Programming in Water Resources: A Case Study. *Water Resources Res.* **9**, 11.

Nash, J. (1953). Two-Person Cooperative Games. *Econometrica* **21**, 128.

Neuman, S. (1973). Calibration of Distributed Parameter Groundwater Flow Models Viewed as a Multiple-Objective Decision Process under Uncertainty. *Water Resources Res.* **9**, 1006.

Neuman, S., and de Marsily, G. (1976). Identification of Linear Systems Response by Parametric Programming. *Water Resources Res.* **12**, 253.

Nijkamp, P., and Vos, J. (1977). A Multicriteria Analysis for Water Resource and Land Use Development. *Water Resources Res.* **13**, 513.

Orne, D., and Wallace, W. (1974). Alternative Mathematical Programming Approaches to New Town Planning. *In* "Urban Simulation: Models for Public Policy Analysis" (M. Whithed and R. Sarly, eds.), pp. 71–85. Sijthoff Publ., Leiden, Netherlands.

Pareto, V. (1971). "Manual of Political Economy," 504 pp. A. M. Kelley, New York.

Philip, J. (1972). Algorithms for the Vector Maximization Problem. *Math. Programming* **2**, 207.

Pigou, A. (1920). "Economics of Welfare," 976 pp. Macmillan, London.

Pollak, R. (1967). "Additive von Neumann–Morgenstern Utility Functions," *Econometrica* **35**, 485.

Porter, D. (1974). Philosophy of Multiple Objective Planning in the Bureau of Reclamation. *In* "Multiple Objectives Planning Water Resources" (E. Michalson, E. Englebert, and W. Andrews, eds.), Vol. I, pp. 23–28. Idaho Research Foundation, Moscow, Idaho.

Pratt, J., Raiffa, H., and Schlaifer, R. (1965). "Introduction to Statistical Decision Theory," 931 pp. McGraw-Hill, New York.

Prest, A., and Turvey, R. (1965). Cost–Benefit Analysis: A Survey. *Econom. J.* **75**, 683.

Price, W., and Piskor, W. (1972). The Application of Goal Programming to Manpower Planning, *INFOR—Canad. J. Operational Res. and Information Processing* **10**, 221.

Raiffa, H. (1968). "Decision Analysis," 309 pp. Addison-Wesley, Reading, Massachusetts.

Raiffa, H. (1969). Preferences for Multi-attributed Alternatives, 118 pp. Memo. No. RM-5868-DOT/RC, Rand Corp., Santa Monica, California.

Rapoport, A. (1970). "*N*-Person Game Theory," 331 pp. Univ. of Michigan Press, Ann Arbor.

Rawls, J. (1958). Justice as Fairness. *Philosoph. Rev.* **67**, 164.

Rawls, J. (1963a). The Sense of Justice. *Philosoph. Rev.* **72**, 281.

Rawls, J. (1963b). Constitutional Liberty and the Concept of Justice. *In* "Justice" (C. Friedrich and J. Chapman, eds.), pp. 98–125. Atherton Press, New York.

ReVelle, C., and Church, R. (1979). "Design of Locational Systems." Pion Ltd., London (to be published).

ReVelle, C., Loucks, D., and Lynn, W. (1967). A Management Model for Water Quality Control. *J. Water Pollution Control Federation* **39**, 1164.

ReVelle, C., Loucks, D., and Lynn, W. (1968). Linear Programming Applied to Water Quality Management. *Water Resources Res.* **4**, 1.

Riker, W. (1961). Voting and The Summation of Preferences: An Interpretative Bibliographical Review of Selected Developments during the Last Decade. *Amer. Political Sci. Rev* **55**, 900.

Riker, W. (1962). "Theory of Political Coalitions," 300 pp. Yale Univ. Press, New Haven, Connecticut.

Robbins, L. (1935). "An Esasy on the Nature and Significance of Economic Science," 160 pp. MacMillan, London.

Rogers, P. (1969). A Game Theory Approach to the Problems of International River Basins. *Water Resources Res.* **5**, 749.

Roy, B. (1971). Problems and Methods with Multiple Objective Functions. *Math. Programming* **1**, 239.

Russell, C., Spofford, W., and Haefele, E. (1972). "Environmental Quality Management in Metropolitan Areas," 188 pp. Resources for the Future, Washington, D.C.

Salkin, H., and Saha, J. (1975). "Studies in Linear Programming," 322 pp. North-Holland–American Elsevier, Amsterdam and New York.

Samuelson, P. (1965). "Foundations of Economic Analysis," 447 pp. Atheneum, New York.

Schilling, D. (1976). Multiobjective and Temporal Considerations in Public Facility Location. 118 pp. Ph.D. Thesis, Dep. of Geog. and Environ. Engrg., Johns Hopkins University, Baltimore, Maryland.

Schinnar, A. (1976). A Multidimensional Accounting Model for Demographic and Economic Planning Interaction. *Environment and Planning* **8**, 455.

Sen, A. (1970). "Collective Choice and Social Welfare," 225 pp. Holden–Day, San Francisco, California.

Silverstein, A. (1978). Multiobjective School Desegregation Models. M.A. Thesis, Johns Hopkins University, Baltimore, Maryland.

Singleton, R., and Tyndall, W. (1974). "Games and Programs," 304 pp. Freeman, San Francisco, California.

Smith, E., and Morris, A. (1969). Systems Analysis for Optimal Water Quality Management. *J. Water Pollution Control Federation* **41**, 1635.

Spivey, W., and Tamura, H. (1970). Goal Programming in Econometrics. *Naval Res. Logist. Quart.* **17**, 183.

Starr, M., and Zeleny, M. (1977). "Multiple Criteria Decision Making," 326 pp. North-Holland Publ., Amsterdam.

Steiner, P. (1969). The Public Sector and the Public Interest. *In* "The Analysis and Evaluation of Public Expenditures: The PPB System" (U.S. Congress, Subcommittee on Economy in Government), Vol. I, pp. 13–45. U.S. Govt. Printing Office, Washington, D.C.

Stuart, D. (1970). Urban Improvement Programming Models. *Socio-Econom. Planning Sci.* **4**, 217.

Suppes, P. (1966). Some Formal Models of Grading Principles. *Synthese* **6**, 284.

Teller, A. (1968). The Use of Linear Programming to Estimate the Cost of Some Alternative Air Pollution Abatement Policies. *Proc. IBM Sci. Computing Symp. Water Air Resource Management, IBM, White Plains*, 1968, pp. 345–353.

Thomann, R. (1963). Mathematical Model for Dissolved Oxygen. *J. Sanitary Engineering Div., Amer. Soc. Civil Engineers* **89**, 1.

Thomas, H., and ReVelle, R. (1966). On the Efficient Use of High Aswan Dam for Hydropower and Irrigation. *Management Sci.* **12**, B296.

Torgerson, W. (1958). "Theory and Methods of Scaling," 460 pp. Wiley, New York.

United Nations Industrial Development Organization (UNIDO) (1972). "Guidelines for Project Evaluation," 383 pp. United Nations, New York.

U.S. Environmental Protection Agency (1973). "The Quality of Life Concept," 397 pp. U.S. Govt. Printing Office, Washington, D.C.

U.S. Water Pollution Control Administration (1966). "Delaware Estuary Comprehensive Study," 113 pp. U.S. Dept. of the Interior, Philadelphia, Pennsylvania.

U.S. Water Resources Council (1973). Water and Related Land Resources, Establishment of Principles and Standards for Planning. *Federal Register* **38**, 24778.

Vedder, J. (1970). Planning Problems with Multidimensional Consequences. *J. Amer. Inst. Planners* **36**, 112.

von Neumann, J. (1928). Zur Theorie der Gesellschaftsspiele. *Math. Ann.* **100**, 295.

von Neumann, J., and Morganstern, O. (1967). "Theory of Games and Economic Behavior," 641 pp. Wiley, New York.

Wagner, H. (1969). "Principles of Operations Research," 937 pp. Prentice-Hall, Englewood Cliffs, New Jersey.

Wallenius, J. (1975). Comparative Evaluation of Some Interactive Approaches to Multicriterion Optimization. *Management Sci.* **21**, 1387.

Werczberger, E. (1976). A Goal Programming Model for Industrial Location Involving Environmental Considerations. *Environment and Planning* **8**, 173.

Yu, P. (1973). A Class of Decisions for Group Decision Problems. *Management Sci.* **19**, 936.

Yu, P., and Zeleny, M. (1975). The Set of All Nondominated Solutions in Linear Cases and a Multicriteria Simplex Method. *J. Math. Anal. Appl.* **49**, 430.

Zadeh, L. (1963). Optimality and Non-Scalar-Valued Performance Criteria. *IEEE Trans. Automatic Control* **AC-8**, 59.

Zangwill, W. (1969). "Nonlinear Programming," 356 pp. Prentice-Hall, Englewood Cliffs, New Jersey.

Zeleny, M. (1974a). "Linear Multiobjective Programming," 220 pp. Springer-Verlag, Berlin and New York.

Zeleny, M. (1974b). A Concept of Compromise Solutions and the Method of the Displaced Ideal. *Comput. Operations Res.* **1**, 479.

Zeleny, M. (1975). Games with Multiple Payoffs. *Internat. J. Game Theory* **4**, 179.

Zeleny, M. (1976). "Multiple Criteria Decision Making, Kyoto 1975," 345 pp. Springer-Verlag, Berlin and New York.

Zionts, S., and Wallenius, J. (1976). An Interactive Programming Method for Solving the Multiple Criteria Problem. *Management Sci.* **22**, 652.

Index

A 8
B 9
C 0
D 1
E 2
F 3
G 4
H 5
I 6
J 7